Troubleshooting Electronic Devices

Troubleshooting Electronic Devices

Joel Goldberg, Ph.D.

NOTICE TO THE READER

Publisher does not warrant or guarantee any of the products described herein or perform any independent analysis in connection with any of the product information contained herein. Publisher does not assume, and expressly disclaims, any obligation to obtain and include information other than that provided to it by the manufacturer.

The reader is expressly warned to consider and adopt all safety precautions that might be indicated by the activities described herein and to avoid all potential hazards. By following the instructions contained herein, the reader willingly assumes all risks in connection with such instructions.

The publisher makes no representations or warranties of any kind, including but not limited to, the warranties of fitness for particular purpose or merchantability, nor are any such representations implied with respect to the material set forth herein, and the publisher takes no responsibility with respect to such material. The publisher shall not be liable for any special, consequential or exemplary damages resulting, in whole or in part, from the readers' use of, or reliance upon, this material.

Cover photo by Bruce Parker
Cover design by Lisa L. Pauly
Photo layout courtesy of Albany branch office of Honeywell's Home and Building Control Division. Honeywell is a global controls company that provides products, systems and services for home, buildings, industry, and aviation and space.

Delmar staff:
Administrative Editor: Wendy J. Welch
Project Editor: Melissa A. Conan
Production Coordinator: James Zayicek
Art Coordinator: Megan K. DeSantis
Design Coordinator: Lisa L. Pauly

For information, address

Delmar Publishers Inc.
3 Columbia Circle, Box 15015,
Albany, NY 12203-5015

Printed in the United States of America
Published simultaneously in Canada
by Nelson Canada,
a division of The Thomson Corporation

1 2 3 4 5 6 7 8 9 10 XXX 00 99 98 97 96 95 94

Library of Congress Cataloging-in-Publication Data
Goldberg, Joel, 1931–
Troubleshooting electronic devices/by Joel Goldberg.
p. cm.
Includes index.
ISBN 0-8273-4889-4
1. Electronic apparatus and appliances—Maintenance and repair.
I. Title.
TK7870.2.G63 1992
621.381'028'8—dc20 92-40592
CIP

DEDICATION

This book is dedicated to my wife, Alice. Her support, her patience, and her compassion during its writing and production cannot be overstated. My hope is that all of our efforts, each in our own way, will be recognized by those who read and understand what has been written on the following pages.

Joel Goldberg

TABLE OF CONTENTS

FOREWORD 

Planning a vacation involves work. You should take certain steps to ensure a successful trip. You have to make appropriate arrangements. If you are traveling to a foreign country you may want to learn a little of its language. You should get acquainted with the particular history, sights of interest, and special events in that country. Planning a trip requires thorough preparation. The more complete your preparation and planning, the more successful your experience can be.

During the trip or vacation a positive attitude and realistic expectations are important. Accepting difficulty and maintaining a balance between what you can control and change and what you cannot must be acknowledged.

Enrolling in a class, or trying to learn a new subject on your own, is like taking a vacation. It involves work. Preparation is of the utmost importance. A positive attitude is a bonus. The more effort you are willing to contribute to learning the material, the more you will benefit.

Electronic repair work is a skill that involves problem-solving techniques. It is goal-oriented. It has rules that must be followed. It is like using a map while traveling. You use the map to identify a method of moving from one point to another point. The second point may be the end of the journey. Using the map provides the information required to reach your goal. Information learned about electronics provides needed information for those who repair electronic equipment.

You have had to learn new vocabulary for this work. You will now have to learn some new techniques or, at the least, learn how to apply techniques you already know more successfully. Once you become acquainted with these techniques it will be similar to returning to a vacation spot and being able to enjoy it with little effort. Once you learn the technique or the skill it will be yours forever.

How successful you are will depend upon your attitude. The more positive your attitude, the more enjoyable your experience will be. Realistic expectations of what you hope to achieve and a realistic appraisal of the time commitment needed are both important for a successful experience. When a solid foundation is built there is little or no need to rebuild it. If you do not rely on shortcuts while learning the techniques necessary for success in this occupation, you will be much further ahead when involved in the actual service project.

Along with the actual knowledge you will have from learning this material and techniques, you can also give yourself credit for accomplishing a goal. You can feel good about that!

Jacqueline Grekin, Ph.D.
Michigan State University

PREFACE 

Successful electronic service technicians must be able to apply the basic laws to any and all circuits. Fundamentally, these laws have not changed since their introduction. The only thing that has changed over the years is their application. The ability to apply these basic laws in a changing environment leads to the identification of the successful service technician.

The material presented in this book will aid you in the development of a set of skills that may be applied to almost any diagnostic process. While the material is primarily devoted to electronic service-related processes, the techniques may be applied to almost any device. Learning how to use them is the challenge to anyone wishing to become a successful service technician. The applications of these rules remain constant in a changing technological world. Your ability to apply them will determine your success.

This book starts with the premise that the reader or student has a background in the fundamental rules proposed by Ohm, Kirchhoff, and Watt. These rules are basic to your ability to successfully diagnose and repair electronic equipment. There is constant reference to these rules as well as the rules for analysis of series and parallel circuits throughout the text. These are all presented in an effort to aid the student or reader to understand how and why to approach a service problem and how to be successful in the diagnosis of the fault in the device.

The material presented in each chapter builds on the information in the previous chapters. Constant review of the basic rules reinforces the knowledge presented in the book. The ultimate purpose is to aid the reader or student in the development of a set of skills that can be used in the field of electronic service. Once learned, these skills can be applied to almost all electronic circuits. Circuits and systems currently under development, and even futuristic ideas, will most likely use these same basic electronic theories. Again, once you learn them, you should be able to transfer them to newer products and circuits.

Each chapter features:

- Clear and easily understood explanations of how to approach troubleshooting problems.
- Explanations for diagnosis using the concepts presented by Ohm, Watt, and Kirchhoff.
- Review of the topical material covered in the chapter.
- End questions related to material in the chapter.
- Troubleshooting problems to enhance the information in the chapter.
- How to conduct and evaluate specific tests of electronic circuits and components.

The following reprint of the "Fox Trot" comic strip may exemplify the feelings of many instructors of electronics. It is the author's hope that, upon completion of the material in this book, the student/reader will be in a better position than the characters in "Fox Trot."

This book is to be used as an introductory-level text for students wishing to learn electronic troubleshooting. The author is planning to prepare a series of additional textbooks on this subject. These books will present topically specific subjects, such as computers, consumer products, and communications devices. The set of books is planned to be used for those wishing to learn more of the specifics related to troubleshooting and repair of these types of devices.

The author of this book, Joel Goldberg, has spent most of his life in the field of electronics. He began by using surplus electronic equipment available after World War II as the basis for learning about electronics. This experimentation led to employment at Michigan State University, where he learned much about servicing and installing electronic devices. In addition, he worked for the university radio department and station as both a service/installation technician and an equipment operator.

FOX TROT **by Bill Amend**

After he left the university he worked in the distributive field of wholesale electronics for nine years. This experience included sale, servicing, and diagnosis of electronic parts and equipment. During this period he managed, owned, and operated a wholesale distributorship of electronic parts.

Dr. Goldberg returned to Michigan State University, where he earned a bachelor's degree in industrial education. This was followed by three years of teaching electronics at the high-school level. While teaching, he continued his education at Wayne State University. This culminated with his earning a doctorate in industrial education. While completing his degree work he started his teaching career at Macomb Community College, in Warren, Michigan.

Dr. Goldberg has now completed his 27th year at the college. He has taught in its electronic engineering technology program and is currently teaching in the electronic communications service program. He has written eight other textbooks related to the field of electronics and electronics servicing, including a widely accepted basic theory text. In addition, his articles have been published in professional magazines.

For more than 19 years Dr. Goldberg has served as a consultant/evaluator and a member of the Team Chair Corps for the North Central Association of Schools and Colleges, Commission on Institutions of Higher Education. He presented materials related to employers' needs for electronics education program graduates at the 1990 Trends in Occupational Education Conference, and at the 1991 conference he spoke on occupational program assessment.

His hobbies have also involved him in public service activities. While attending Michigan State University, he obtained an amateur radio license with the station call letters W8HIU. This license is still held and is used daily. Dr. Goldberg currently is the deputy emergency coordinator for the Amateur Radio Public Service Corps of Oakland County, Michigan, under whose auspices amateur radio operators provide supplemental communications for all public service units during civil emergencies.

His diverse background has given him extensive knowledge of the electronic servicing field and has prepared him well as an author. His desire to share his knowledge is evidenced by the number of materials he has published.

The task of putting together a book requires the assistance of many people. The author and publisher wish to thank the following reviewers for their valuable insights and constructive comments: Robert L. Adams, Charles R. Alexander, Louis D. Bean, Stanley M. Bejma, Stanley P. Creitz, Donnin Custer, Ronald J. Davis, Michael Durren, Audrie E. Hall, Leigh A. Hargis, Wendall Johnson, William F. Maddy, Larry Oliver, Lee Rosenthal, Gerald Schickman, Eugene R. Shackley, and William H. Watson.

SECTION 1

Rules for Troubleshooting

CHAPTER 1

Basic Electronic Rules

INTRODUCTION

Electrical and electronic repair and service has grown into a worldwide business. This is due in part to the rapid growth of the use of computers, cellular telephones, and personal paging equipment. There is an ever-increasing demand for competent people to work in this business. Those who are considered experts in electrical and electronic service are those who have developed a solid foundation for successful diagnosis and repair. This book is designed to present a format that will serve as the basis for success. One of its major themes is the development of methods of approaching service-related problems. Another major theme is that the basic rules do not change; only applications of these basic rules change as technology changes. Successful service personnel have learned to apply the basic rules with all types of electrical and electronic equipment as repair procedures are followed.

OBJECTIVES

Upon completion of this chapter, the reader/student should:

 understand the need to learn the basic electrical and electronic rules;

 recognize the need to relate the function of blocks used in equipment to operational units;

 recognize the three major levels of repair as unit, module, and component;

 understand the need to know typical repair histories of components as they relate to failures;

 recognize the difference between signal processing and current flow in systems; and

 recognize that the basic rules for electricity and electronics seldom change—only their applications change.

KEY WORDS AND PHRASES 

board level repair: This type of repair involves replacing a specific component, or set of components, that have failed on a circuit board or module.

bracketing: This is a method of identifying the area of trouble in any unit. Brackets are placed at either end of the suspected area. As tests are conducted, one of the brackets is moved until only one section is suspected to be the problem area.

functional block: A functional block is a unit that performs a specific function in an electronic device. The block may contain one or more individual circuits. Often the complexity of the circuits makes it more convenient to recognize the block and its action in the circuit.

Kirchhoff's law: There are two forms of Kirchhoff's law. One is related to voltage in a closed loop, or series circuit. This law states that the voltage drops in a closed loop electrical circuit will add to equal zero. Another way of stating this law is that the voltage drops that develop across the loads in a series circuit will add to equal the applied voltage.

The second version of Kirchhoff's law applies to current flow at any junction in an electrical circuit. The law states that the current entering a junction, or node, is equal to the current flowing out of that same junction.

modular level repair: This refers to identifying a malfunctioning module, or board, in a device and replacing it.

Ohm's law: The relationship among circuit voltage, current, and resistance is described in Ohm's law. According to this law, one volt applied to a circuit produces one ampere of current flow through one unit, or ohm, of resistance.

parallel circuit: This is a type of electrical circuit; it has two or more paths for current flow.

series circuit: This is a type of electrical circuit; it has only one path for current flow.

unit level repair: This refers to identifying a defective, or nonfunctional, unit and replacing the entire unit as a method of repair.

Watt's law: This is the relationship among voltage, current, and electrical power. Watt stated that the electrical power in any circuit is equal to the products of voltage and current in that circuit.

WHY THE NEED FOR ELECTRONIC SERVICE PERSONNEL?

One dramatic change of recent years is the major increase in the application of electronics in our daily lives. Most of the products in use today are either controlled or operated by electronic components. A great many of these devices use miniature computers, or microprocessors, to control their actions. Unfortunately, the manufacturing world has found that consumers are not willing to pay the price for foolproof and fail-safe products, so each of these devices has the possibility of failing.

The fact that many of our electronic products will fail has developed into a "bad news, good news" type of situation. The bad news is that the products will fail at some point during their operating life. The good news is that this possibility of failure has created an electronic service industry as well as the occupation of electronic service repair technician. The development of sophisticated electronic units demands that service technicians be able to perform rapid diagnosis and repair of these units. As equipment becomes more complex, the knowledge of additional sections and their specific functions is necessary. The other major factor is that these highly sophisticated pieces of equipment have replaced workers, thus shifting the emphasis from the person to the machine. One nonfunctioning machine can shut down a complete manufacturing or data-processing system. This is why efficient and rapid diagnosis and repair are necessary.

A great many of these units are constructed in modular form, so service technicians require more than knowledge of the basics of electronics to repair this modular type of equipment. The ability to rapidly diagnose these complex units requires a thorough understanding of how the units normally function. Many of these units contain integrated circuits (ICs). These integrated circuits contain miniature blocks, or subunits, that cannot be repaired individually. Service-related information about these integrated circuits often does not provide the actual electronic schematic diagram for them. Instead, the service literature shows a block diagram of the IC, identifying the individual blocks that form the unit. This is one of the major reasons why it is so important to understand how individual blocks function. Successful service technicians also must know how these individual blocks function and how to recognize when any of them have failed.

Consider the technician's feeling of elation when a successful diagnosis and repair is made on a piece of electrical or electronic equipment. Consider the feeling if this repair is accomplished in a minimum amount of time. This is possible, even very probable. There is one catch, however; the technician who is doing the servicing must base the diagnosis and repair on a set of very solid systematic procedures.

The ability to develop this set of procedures is the theme of this book. Almost anyone who is knowledgeable about electrical and electronic theories should be able to repair and return to operating condition any device that uses electricity. The more experienced repair/service technicians can normally do this in a very rapid and efficient manner. How they developed the skills and knowledge to be able to do this is, again, what this book is all about. To be both efficient and productive, better service/repair technicians have developed a set of procedures that will enable them to quickly diagnose and repair almost any type of equipment. Of a secondary nature, but equally important, is that once this technique is learned, it can be applied to almost every service and repair situation, regardless of the type of device requiring the repair. It may also be used for the diagnosis and repair of nonelectronic devices, since the format for the procedure is common to all types of devices.

The efforts required to perform almost any task can be learned by memorizing specific processes. This type of learning, however, does not permit transference of this knowledge to the performance of any other task, since it lacks flexibility. Too often the learner is expected to know that a specific type of failure is indicative of the failure of a specific component in the circuit. This cause-to-reaction type of learning is difficult to transfer from one system to another. As technology and electronic circuits change, the service technician needs to have the ability to transfer basic knowledge from one system to another.

In the world of electronic service and repair, memorizing a specific solution to one problem may not always be productive. I know service technicians who know which component to check when a specific symptom occurs in a specific piece of electronic equipment. While this knowledge is helpful, it does limit their ability to transfer evaluation procedures to other types of equipment. A far better method of approaching any type of electronic service problem is to be able to use all of your capabilities to do these things:

1. recognize that a problem exists;
2. identify the area in the unit where the problem exists;
3. recognize the type of electronic circuit used;
4. consider the rules that apply to that circuit;
5. apply those rules in a manner that provides specific information;
6. use the results of these tests to evaluate and consider how and where to make additional tests; and
7. continue to repeat steps 3, 4, 5, and 6 until the problem area is located or the repair completed.

These steps do not appear to be very difficult to perform. In reality, they are very easy to do. The most difficult part is to remember to do each of them in an organized manner, such as the order shown above. Too often the service/repair technician attempts to skip some steps in order to repair the malfunctioning unit quickly. This haste actually may double or triple the cost and time to make the repair.

Every repair should be approached using the basic rules of electricity and electronics. These rules, presented by Ohm, Kirchhoff, and Watt, do not change with their use. Only the applications of these rules change. This is why it is so important to know these basic laws and to be able to apply them as various electronic products are serviced.

Several major steps are required to successfully service and repair electronic products. Actually, these steps can be applied to almost any type of service-related activity. As you read this book each of these will be identified and discussed. Learning these and following them in the order they are presented will provide a method for successful analysis and repair of most electronic products.

The process of diagnosis and repair is illustrated in Figure 1–1. Most electronic devices are formed from the assembly of several individual units. Each unit has a specific function. Each of the units can be broken down into several additional levels. These include a stage, a circuit, and ultimately a component part. The process of successful servicing is approached by recognizing these levels and then following the steps outlined in this figure. First, identify the function of the unit. Then identify the specific stage in which a problem is considered to be located. This is followed by the location of a specific circuit and, finally, the specific component(s) that have caused the failure. This approach is the one used by successful electronic servicers. It may also be used in fields other than electronics.

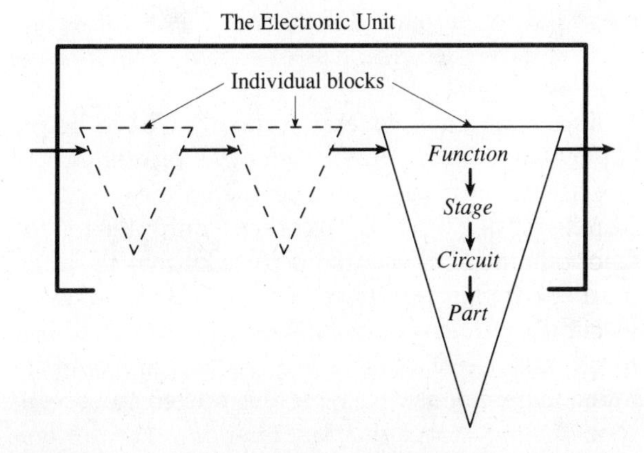

Figure 1–1 The steps involved in the process of troubleshooting or repairing an electronic device.

Repair of defective electronic units is a major industry in the United States. Much of the equipment used in the home, business, or industry is expensive. Replacement costs are often more than the user can afford. When a piece of electronic equipment fails, one of two major steps can be taken to permit the operator to return to a functional role—either replace the whole unit with another one known to be in working order or request repair by a service technician or an independent service company. Time is very costly to most companies. Hidden factors, such as the cost of preparing purchase orders, paying the invoices, shipping, and installation must be factored into the overall cost of any replacement. If the unit is to be repaired instead of replaced, the service call and the specific replacement parts, while often expensive, may return the unit to service much more rapidly than the length of time required to order, ship, deliver, and set up the new equipment. In addition, many of these nonoperable units can be repaired at minimal cost to the owner. Often the malfunctioning components are located on a replaceable module or board. The decision to repair or replace should not be made until after an initial analysis of the service/repair problem is made. There have been stories about companies replacing properly working units because the operator was not aware of a simple thing such as a disconnected cable. Often the initial analysis by a qualified service technician will locate these problems and minimize the need to purchase new equipment.

There are three basic levels of repair for electronic equipment on the market today: unit, module, or board, and component. **Unit service/repair** often means replacement of the entire unit with another one. This may be the simplest method of addressing the problem. This is particularly true when the replacement cost of the unit is low, when the unit's replacements have additional and desired features, when the original unit has become obsolete or slow, or when the replacement unit is readily available. In addition, today's replacement parts market—with the necessary profit margins for the servicer—may make it more expensive to replace a part than to replace the entire unit. One example of this is the cost to the consumer of the television cathode ray tube, or picture tube, which often exceeds the replacement cost of the entire set in today's very competitive consumer marketplace.

Modular, or **board**, **service/repair** has become one of the more popular methods of servicing units today. This is particularly true in the computer field. A great many of the control and operational segments of the computer are contained on plug-in boards or modules. The simplest method of service/repair of these is to replace them with another unit known to be working properly. The final level of service/repair is replacement of a specific component in the unit. This level requires the use of testing equipment to locate the individual component, or components, that have failed. This process also requires a greater depth of knowledge to efficiently remove and replace the defective component. In addition, the technician either has to have a stock of replacement components or know where to obtain them rapidly.

Several of the larger service/repair facilities use a two-level type of repair service. Level one—basically a "get the equipment up and running quickly" philosophy—means replacing a major board or module in the unit. Once this is accomplished, the service/repair technician will often return the suspect board or module to the home office's major repair facility, when level-two repair takes place. The major repair facility has the depth of equipment required to diagnose the problem at the component level. Decisions are made at this level concerning the economics of repairing or destroying the board or module. This two-level process is used because the customer's main concern is to return the equipment to productivity as quickly as possible. Unless they are very inquisitive, customers are not really interested in knowing which specific component has failed, or even why it failed.

Each level of service/repair is important. Usually the equipment owner's or operator's initial contact is with a service/repair technician, who must be able to recognize that a problem exists, locate the problem area in the unit, and then offer the proper solution to the problem. Reducing the downtime (time the equipment is out of service) is often more important than the cost of the repair. After all, if the equipment is not operating, it cannot be used. An unused piece of equipment does not earn any income for its owner. If it was purchased on a time/lease arrangement, it is costing the company money to stand idle. The operator must consider the additional costs of a nonproductive piece of equipment.

The first major question asked by both service/repair technicians and those who own or operate the equipment is, why repair? The answer to this question is very subjective. Many factors influence the decision whether to repair or to replace a defective unit. Each of the levels of service/repair must be considered when making this decision. Often the repair is considered a simple one. It could entail replacing a cable or a power cord on the unit. It also could be the replacement of a plug-in module or board. Either of these steps can be accomplished in a very short time. Another factor to consider is whether the repair parts are available. One example of this is the failure of a print head on a computer's printer system. At the time, the manufacturer did not have any replacement printer heads in stock. Delivery was expected in "two to three months." The printer was needed immediately and could not wait for a replacement printer head. So the decision to replace the printer rather than wait for a replacement part was made. While this was a rather expensive approach, the loss of income would have been even more costly.

Since that time, several companies have begun advertising the repair of defective printer heads. Other companies specialize in the repair of specific boards, disk drives, and other computer-related components. These companies are known as "aftermarket" repair facilities. They are not associated with the original manufacturer of the unit; their sole purpose is to efficiently repair specific parts of major units. One may ask why there is a need to use this type of service organization. First of all, these companies are experts in the repair of specific modules, boards, and components. They often have a supply of repaired units available and can ship one out to the customer the same day. In keeping with the concept that downtime is expensive and nonproductive, the ability to obtain a certified replacement repair part is very desirable.

Let us expand each of the seven steps previously described to see why they must be done in a specific order and why each of them is necessary for the effective and productive service/repair technician.

RECOGNITION THAT A PROBLEM EXISTS

It is almost impossible to consider a unit nonfunctional unless you have a basic understanding of how the unit is supposed to function when it is working properly. Let us start with a very basic unit—a circuit used to operate, or turn off and on, a light assembly fixture. A schematic diagram of this circuit is shown in Figure 1–2. In this circuit a source of electrical power is required. This source is considered the local electrical power company and its distribution system. Let us assume that when the switch in this circuit controls the light normally, the lamp in the fixture will be illuminated. In this situation, placing the switch in the "on" position will make the light glow and create illumination. However, when the switch is placed in the "on" position, nothing noticeable occurs. Since the light does not illuminate, a problem does indeed exist.

IDENTIFICATION OF THE PROBLEM AREA

In almost every service/repair situation, you should be able to use a block diagram or a schematic diagram. The need for this is evident for many complex circuits. Often you can visualize and analyze the simple circuits by creating a mental picture of the circuit. If this does not work, then draw it on a piece of paper. In this circuit, there are several places where the problem could exist. Actually, each component making up this circuit should be under immediate suspicion. A system of testing must be used that will efficiently reduce the area of the problem from the total circuit to an individual component or section. This is the type of thinking and planning performed by the successful service technician.

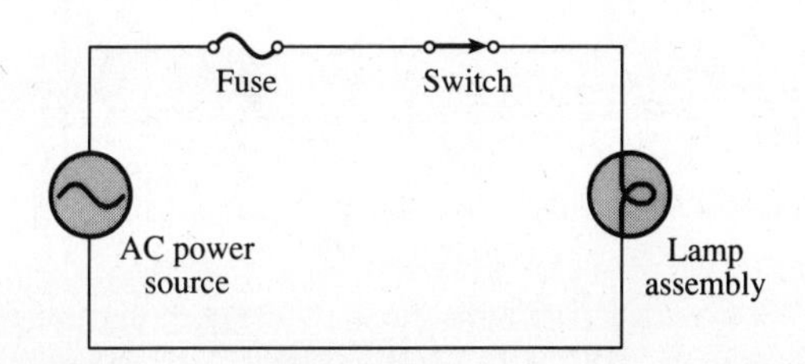

Figure 1–2 **A power source, a fuse, a power switch, and a lamp assembly make up this electrical circuit.**

RECOGNITION OF THE TYPE OF ELECTRONIC CIRCUIT

Only three circuits are used for electrical and electronic wiring. These are the **series circuit**, the **parallel circuit**, and the use of a combination of both series and parallel components in a **series/parallel circuit**. Using the basic rules for circuits, the circuit shown in Figure 1–2 is a series circuit, because the major characteristic of a series circuit is a single path for electron flow. This circuit meets this criterion.

CONSIDER THE RULES FOR THIS TYPE OF CIRCUIT

What are the basic rules for this circuit? Rules, or laws, presented by Ohm, Kirchhoff, and Watt apply to all electrical and electronic circuits. In the series circuit, rules affecting current flow, voltage drop, and total resistance will apply. These rules are:

1. Current flow is the same value throughout the circuit.
2. Voltage drops that occur must add to equal the source voltage.
3. The sum of individual resistance values will equal the total circuit resistance.

APPLY THE RULES IN A SPECIFIC AND EFFICIENT MANNER

The basic electrical and electronic circuit consists of a source of energy, a load using the source power, and lines—or wires—used to connect the load to the source. In all circuit testing, the service technician must be able to eliminate those components or

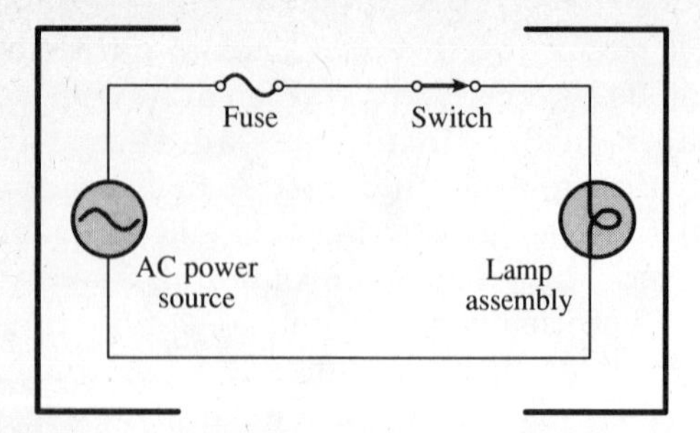

Figure 1–3 An initial set of brackets surround the complete system as the starting point for analysis.

units that are functioning properly to identify the area of the problem. This process is one of elimination of the functioning sections so that the technician can pay attention to the nonfunctioning areas of the set, or unit. An efficient method of testing will help in accomplishing this. The ability to develop a set of efficient skills will assist technicians in becoming efficient and, it is hoped, high-salaried employees.

A proper method of analysis is to develop a set of checks, or tests, to reduce the problem to one area. In Figure 1–3 the same circuit is shown with a set of brackets around the entire system. This is done to identify the problem area and to show on the schematic diagram all the major components in the section under suspicion. The technician in this situation will (1) identify a specific type of testing equipment and (2) identify a place in which to make an initial test. There are four components in this circuit: source of power, fuse, switch, and lamp. The testing equipment used for analysis of this circuit is one capable of indicating the presence (or lack of presence) of voltage. Typical testers are usually either a voltmeter or a test lamp.

Efficiency of testing will tell us that the first test to make in a circuit of this type is at, or near, the middle of the circuit, as illustrated in Figure 1–4. The reason for this is that this test will reduce the area under suspicion to one half of the entire circuit. If you find the presence of voltage at the left side of the switch, then you can assume correctly that both the power source and the fuse are functional. An additional test is now required; this is shown in Figure 1–5. Voltage was found to be correct at the center of the system, so the left-hand bracket is moved from its original position to the point of the test.

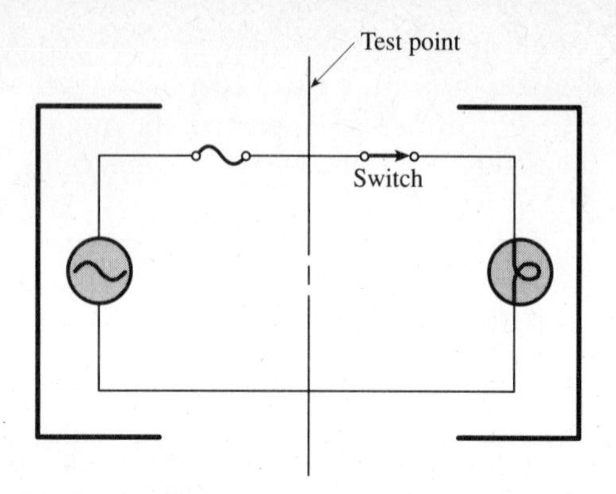

Figure 1–4 The first test point in a system of this type is at its center.

USE INITIAL TEST RESULTS TO DETERMINE SITES OF ADDITIONAL TESTS

The technician's attention is then directed to the balance of the circuit: the switch and the lamp. A test should be made at some point between these two components. This is done using the same rule as stated earlier: split the remaining system in half to make an efficient test. The left-hand bracket would then be moved from its position at the center of the circuit, where the test was made, to its new position

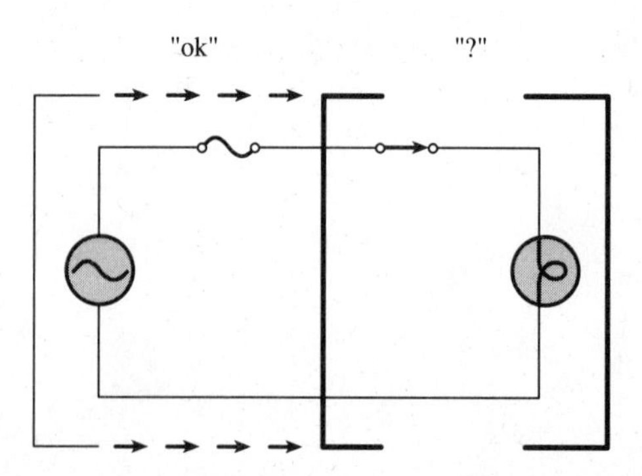

Figure 1–5 When the initial test indicates voltage at the test point, the left-hand bracket is moved to the point of the test. This eliminates half of the circuit.

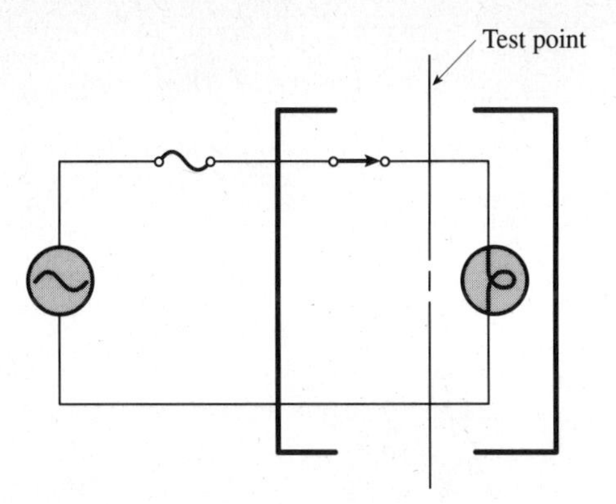

Figure 1–6 **A final test is made at the halfway point for the balance of the circuit.**

between the remaining two components, as shown in Figure 1–6.

CONTINUE STEPS UNTIL REPAIR IS COMPLETED

If voltage is present at this test point when the switch is in the "on" position, then the problem area is identified as the lamp. Otherwise, if the presence of voltage is not indicated when the test is made between the switch in the "on" position and the lamp, the problem area is localized to the switch.

In a more complex circuit, the steps identified in this section would be repeated until only one area is still under suspicion of malfunctioning. In each of the above steps, the area of suspicion was reduced to one half its previous size. In a system with many more sections, this type of approach is a very efficient method of analysis and diagnosis.

You probably think that the method described in this section is very time-consuming and inefficient. After all, don't we all know that the major cause of failure in a lamp circuit is the bulb itself? Although this is true, the purpose of this book and the total concept involved in it is the development of a method of thinking and approaching an electrical or electronic problem to effect a repair. The process of jumping to a conclusion will not work in every type of service problem. Suppose the lamp described in this example is housed deep inside a major unit and the time required to remove it exceeds one hour! If the repair technician's immediate thought is to replace the lamp, thinking that the normal problem is a burned-out lamp, the time involved to test this theory would not be very productive. This would be very true if the problem did not turn out to be the lamp, but was elsewhere in the circuit. Therefore, the development of a basic set of operating rules for problem solving is essential. In some service situations, following the basic rules may appear as though one is taking the long way around to arrive at the cause of the problem. Using this process will aid in the development of a set of service-related skills that will ultimately provide you with a way to approach most service-related problems successfully.

REPAIR HISTORY RELATED TO FUNCTIONAL BLOCKS

Each type of electrical or electronic circuit has its own service or operational history. Not only is this true for the acceptable manner in which the circuit is supposed to function, it also is true for how the circuits normally malfunction. The typical history will often aid the service/repair technician in making an initial diagnosis of the problem. Many of the more sophisticated systems even keep a log of repairs. A review of this "personal" history will often aid in the diagnosis and repair process. As some of the more common types of failures that occur in each general type of circuit or system are reviewed, you should keep in mind that knowledge of how the system is supposed to function and how the individual building blocks in the system are used is of major importance. In addition, knowing why certain sections fail more often than others can aid you in the diagnosis process.

BLOCK DIAGRAM DIAGNOSIS

Systems have become more complex and the actual components in most of the electrical and electronic devices currently in use have been reduced to just a few integrated circuits. This is cost-effective for the manufacturer of the device; however, it does make the service/repair technician's work more difficult.

Visual inspections of the units have less and less meaning under these conditions. One is almost forced to use a block diagram. The need to relate functions of the system to individual **functional blocks** in the system has become very important to the service technician. It has become necessary to rely on diagrams of the system. The most helpful of these are block diagrams that relate or describe the function of each of the building blocks that make up the system. If the operator or the service/repair technician is able to describe the action of the device as it malfunctions, then use of a diagram that describes the functions of each segment can be very helpful in localizing the problem area. This type of diagnosis is also used when the repair can be accomplished by replacing a board or module in the system. These, too, have become more commonplace in systems. The electrical and electronic systems of today consist of several building blocks, fairly basic units that make up the operating system of the unit. Today's state of the art for electrical and electronic units is to use integrated circuit design and construction. Often two or more discrete functional sections of the unit are contained in a single integrated circuit. A knowledge of functional blocks is necessary to be able to service the units. Of major importance to anyone wishing to be successful in the service/repair field is a knowledge of how each of the fundamental types of building blocks really functions.

Some electrical and electronic equipment manufacturers are now including basic service literature with delivery of their equipment. Much of the literature is limited to schematics and block diagrams. The complexity of many of these units requires the use of block diagrams as well as schematic ones. The block diagram provides information about the function of each of the many building blocks in the system. The ability to be able to comprehend the specific electrical or electronic circuitry inside the block in these systems is of little importance. The major requirement is knowledge of the function of the blocks and how electrical current and electronic signals are processed in the system.

An example of the "block" system is shown in Figure 1–7. Notice the difference between this type of diagram and the diagrams shown in the last five figures. This type of diagram does not display the entire circuit. It has broken the system into discrete blocks, and each block has a specific function in the system. Highly complex circuits and systems often use this type of display to illustrate either signal flow processing or the sequence of events that occurs during operation of the system. The block diagram approach eliminates the need to follow current flow in the system. It is an excellent method of showing the building blocks in the system. Using the block diagrams, the technician is able to develop a system for approaching the analysis and identification of the area of the problem in the system. Often arrows, as indicated in this diagram, are used to show the flow of the process involved in the system.

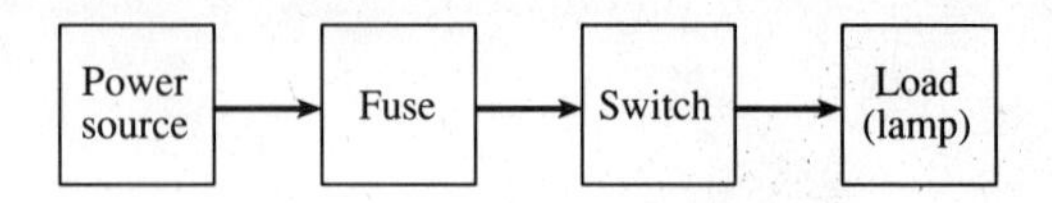

Figure 1–7 Block diagram for the lamp assembly lighting circuit.

One example of the need to know the basic function of the building blocks is when the term amplifier appears as a part of the title of a specific block in a system. An amplifier, by definition, is a device that uses a small amount of power to control a larger amount of power. Recognition of this definition should indicate to the service/repair technician that the input power to this block is relatively small, while the output of the block should provide an increase in the power level, or a power gain. If the amplifier has the title "voltage amplifier," then you should expect to measure an increase in the signal level voltage at the output of the block. If this occurs, this block is functioning properly. If the measurement does not produce the voltage gain, the block has a problem and you need to spend time locating the specific problem.

The most likely cause of any failure in electrical and electronic circuits is heat. All components used in these circuits generate some form of heat. The ability to keep the components operating at normal room-temperature levels goes a long way in extending the life of the components. Often, air-circulating fans are used for this purpose. In some industrial and broadcasting units, the heat-generating components are encased in jackets. Water is circulated through the jackets to keep the components at nominal operating temperatures and to extend their operating life. Water cooling is only of major importance when the devices have the potential to operate at extremely high temperatures. Normal cooling procedures use less drastic measures, such as heat sinks or fans, to maintain component "cool."

Vacuum Tube Circuits Vacuum tubes are still in use in several types of electrical and electronic circuits. Vacuum tubes are used as the cathode-ray tube found in television systems and in computer displays. In addition, they are still used in industry, the military, and in broadcasting. The highest incidence of vacuum tube failure is its filament. The filament of the vacuum tube is considered a part of the power supply block, or section, of the unit. A failure of the unit containing vacuum tubes should indicate to the service/repair technician that an initial area to check is the power supply block. Details concerning how to approach diagnosis and analysis of power supply blocks is found in a later chapter.

A block diagram for a typical power supply distribution system is shown in Figure 1–8. In this very traditional system, the power source is used to supply the necessary operating power to the DC power supply. In addition, a second block is connected directly to the power source. This block is used as the source for voltage and current required for the filament circuits of the tubes. Once the power source is developed, the arrow indicates that it is used to supply the power required to operate the filaments on all vacuum tubes in the system. This is another example of the use of the block diagram to show system function.

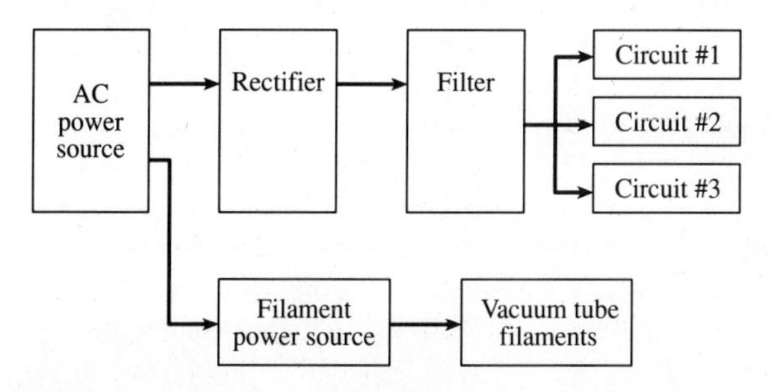

Figure 1–8 Block diagram for a power supply for vacuum tubes and other electronic devices.

Simple Solid-state Circuits Failures in simple solid-state devices may be limited to the diode. This is more difficult to pinpoint, since diodes are used in many different sections and blocks of functional devices. The importance of knowledge of how the system is supposed to function cannot be overestimated. Knowledge of system operation as it relates to specific functional blocks will often assist in locating a nonoperating block.

Complex Solid-state Circuits More complex solid-state devices, such as transistors, SCRs, triacs, etc., also fail. Historically, these failures create failures in specific sections, or blocks, of the system. In many service-related situations, one of the best methods of locating these problem devices is with the use of an oscilloscope, which allows you to trace the electronic signal through the unit. Knowledge of how the specific sections are supposed to function and reference to a block diagram of the unit will often aid in locating a problem in a specific section, thus requiring further analysis.

Single-function Integrated Circuits Integrated circuits do indeed fail. The only true methods of determining which one of several integrated circuits has failed are the visual method—looking for an overheated component in the same section as the integrated circuit—or using an oscilloscope or signal tracing device to locate an area of failure. This, of course, can only be accomplished after you have used a block diagram to localize the area of the problem.

Multifunction Integrated Circuits Historically, only one section of a multifunction integrated circuit will fail. This failure will force the service/repair technician to replace the entire unit, since no one has been able to get inside these integrated circuits to make a repair. Here, too, either an oscilloscope or a signal tracing device must be used to locate the problem. As with all other devices, this requires the use of a block diagram to assist in locating the specific problem area.

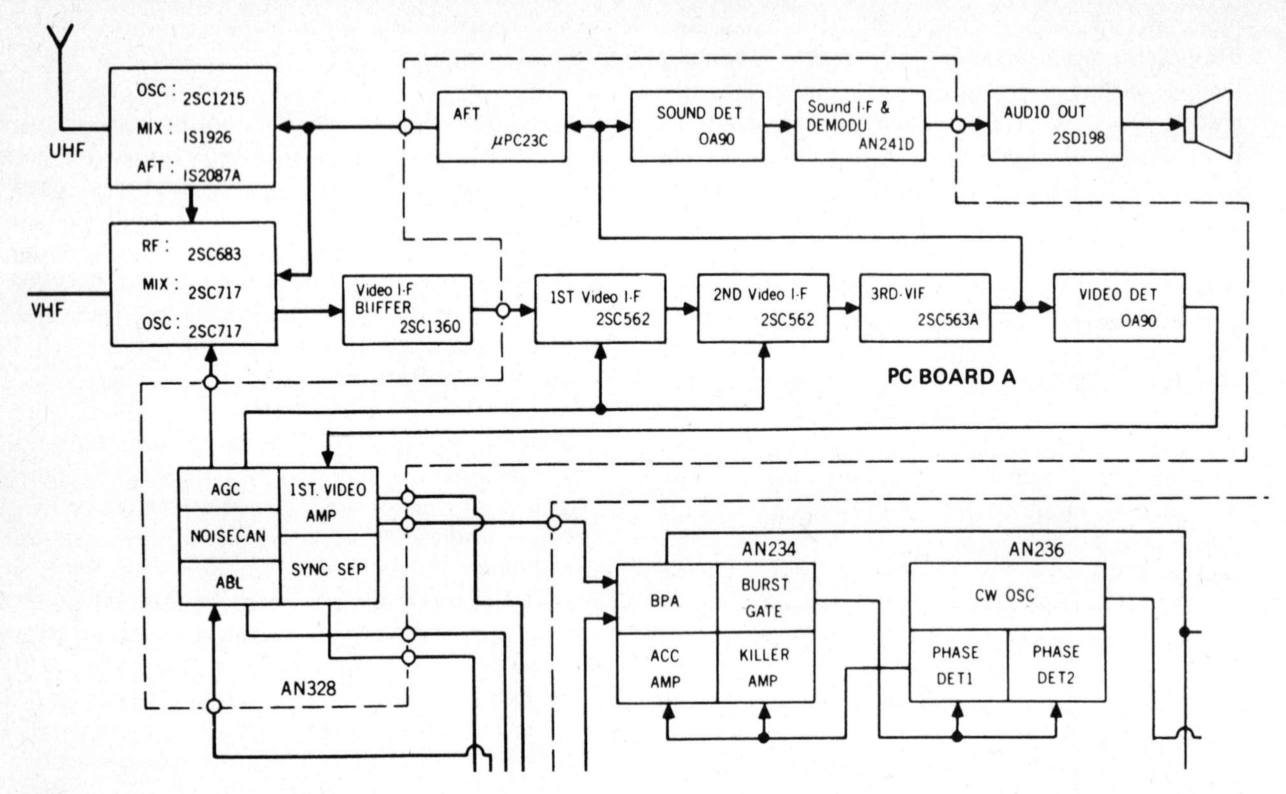

Figure 1–9 **Partial block diagram indicating specific functions is used instead of individual components with complex circuits.**

Look at Figure 1–9. It is typical of information found on the schematic diagrams currently in use. The specific electronic circuitry is too complex to display; instead, the manufacturer provides functional block information for the service technician. In this type of complex integrated circuit, the display of blocks makes more sense than does including the specific electronic circuits that make up each of the individual circuits inside the integrated circuit.

THE NEED TO KNOW THE BASIC RULES

Of major significance to service/repair technicians and to those wishing to enter this field is the fact that the basic rules have not changed very much since they were presented to the world. The original presentation of **Ohm's law** is rumored to have been incorrect. German physicist Georg Simon Ohm is said to have made the proper corrections to it, and we now use it to diagnose most electrical and electronic circuits. The same is true for the rules established by German physicist Robert **Kirchhoff** and Scottish inventor James **Watt**. It is important to note that we are not discussing the mathematical analysis of these laws, only their applications.

Service technicians must have a total understanding of the basic laws. Ohm described the relationship of voltage, current, and resistance in a basic electrical circuit. Those technicians who are successful recall these relationships while performing service and repair techniques. Often the use of Ohm's law becomes a "background" activity for technicians. This means that they use the law without really stopping work to consider how it is applied. For example, the diagnosis of a circuit similar to that shown earlier in Figure 1–6 would use Ohm's law. If the load is inoperative, then the possibility that the power source is malfunctioning is one area of the diagnostic procedure. Ohm's law states that the current flow in the circuit is dependent upon the amount of voltage used to develop a flow of current. If the voltage source is nonfunctional, then little or no current would flow from the source through the wires to the load. If the source voltage is too low, then insufficient current would flow in the circuit to make the lamp light. The result is that the load does not operate. The technician does not actually complete each of these steps as the problem is diagnosed. They remain in the background and are used without the technician's really thinking about them. If they had not been learned initially, the technician would be unable to apply them to reach a rapid and correct diagnosis of the problem.

Rules such as Ohm's law may be used in any of several different ways. Fortunately for all of us, these rules do not change with their applications. The applications do change, however, and this is good. It does help keep us up to date about how the various systems function. Return to the function described earlier as "amplifier" and consider the various common uses of it. In electronic units you can often find voltage amplifiers, buffer amplifiers, power output amplifiers, and current amplifiers. Each of these describes a use of the amplifier circuit. It is also possible to describe amplifiers as (using strictly solid-state terminology) common emitter, common base, and common collector (or emitter follower) amplifiers. Each of these has a specific set of characteristics that should be a part of the service/repair technician's basic knowledge.

Some of the previously described terms may be combined into a single description of the section of block of the unit. For example, you could find a common emitter, intermediate frequency voltage amplifier in a radio or television receiver. Each of these basic terms, "common emitter," "intermediate frequency," and "voltage amplifier," is in general use. The experienced technician would have to recognize that this amplifier is used in communications receivers covering the whole radio frequency range. The specific range of these frequencies is identified as intermediate frequencies since they fall in an area of the radio spectrum that is between the actual broadcast signal frequency and the frequency processed by the amplifier in the receiver. This circuit amplifies signals that are normally found in that range. It will provide a voltage gain when comparing its output signal to its input signal. Whether the application is found in a home AM or FM radio, an auto radio, a shortwave receiver, a cellular telephone, a pocket pager, or a satellite television receiving system is not important. What is important is the ability to identify the application and what to expect to find when you are performing diagnostic work on the unit. The recognition of this type of thinking is critical to the success of those wishing to be outstanding electrical and electronic service technicians.

As you have seen, knowledge of the basic terminology is important. Also important is how the circuits are used in a variety of applications. The impact of this description should show the need to understand fully the function of each block in a system. It also should assist you in recognizing knowledge about the various types of applications found in electrical and electronic devices.

BASIC RULE APPLICATION

You can use several processes to localize a problem area. The specific process often will depend upon the type of circuit or system being analyzed. Two of the more common processes refer to the analysis of either the processing of the electronic signal through the system or the path taken by the electron current as it flows through the circuit. The choice of process will often depend not only on the kind of circuit but also on the complexity of the system.

SIGNAL FLOW SYSTEMS

In most of the electronic schematics in common use today, the signal flow path is similar to the way we read a book. The signal is shown as entering the schematic diagram at its upper left-hand corner. Signal flow then continues in a straight line across the page to the upper right-hand corner of the page. It then drops down to the next horizontal line and continues across it in a similar manner. The com-

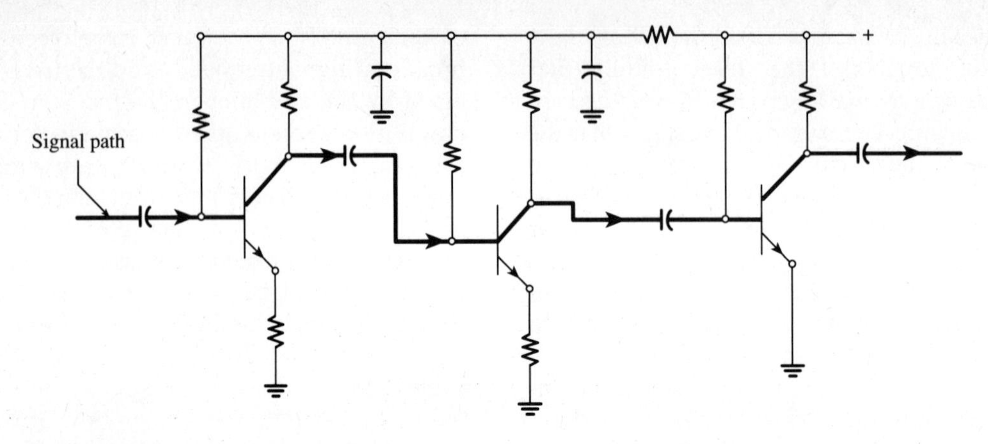

Figure 1–10 One method of showing signal flow is the use of a heavy, dark line on the schematic diagram.

plexity of the system dictates how many lines are required for the entire system. An illustration of this is shown in Figure 1–10. In this illustration the heavy line indicates the path of the signal in the system. Note that the output of the system is located at the right-hand side of the page at the end of the signal flow path diagram. This placement is typical for display of the schematic diagram.

A similar format is used to display the processing blocks of the system. Figure 1–11 shows this method. The flow path is also from the upper left of the page to the upper right. It then returns to the left-hand side of the second line and flows to the right-hand side of that line. This continues until the end of the path is reached.

CURRENT FLOW SYSTEMS

Current flow is not always indicated on a schematic diagram, and it is not normally found on the block diagram. The experienced service technician must know how to trace the current flow in each section of the unit. There are two methods of describing current flow. One of these is known as "conventional" current flow. In this system the current flow is from the positive (+) terminal of the power source, through the specific components in the system, and ultimately to the negative (−) terminal of the power source. Engineers often use this type of analysis. The second method of describing current flow describes the flow of electrons in the circuit. It

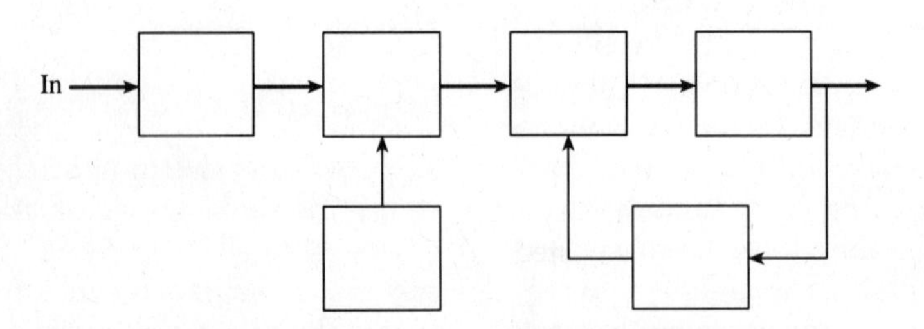

Figure 1–11 Signal flow paths are often indicated with the use of a heavy line to display the specific block-to-block path.

is known as the "electron" current flow method. In the electron flow method the direction of current is reversed. Electrons are negative particles, and their movement typically originates from an area of heavy concentration, or the negative terminal, of the power source. The movement of electrons flows from the negative terminal, through a conductive path, to the load. From the load the electrons continue through another conductive path back to the power source. They complete their journey at the positive terminal of the power source. Either of the two current flow theories is correct; the chosen one depends upon who is describing the circuit. In most instances, electronic technicians will use the electron flow method to describe current flow.

In almost any electrical or electronic circuit, the main point to consider is that, when the current is moving through the load of the circuit, some work will be performed by the load. The specific direction of the current flow may not make any difference to the operation of the load. Since conventional current flow describes the movement of positive charges, and electron flow describes the movement of negative charges, both are accomplishing the same thing. Again, while current is able to flow through the load, some amount of work will be accomplished by the load and the circuit is probably functional.

The service/repair technician must be able to follow the current flow path to analyze the circuit. The rules expounded by Ohm, Kirchhoff, and Watt will apply to most circuits. Analysis of the circuit using these laws will normally aid in identifying the nonfunctioning components in the circuit.

THE NEED FOR PLANNING

One of the most important steps in the repair process is thinking and planning your approach. Too often, the novice service technician will want to appear to be "busy" and start to make some tests without developing a plan first. This may lead to wasted efforts. Tests may be made that have little or no relation to the resolution of the problem. Sometimes these tests may even provide information that will open a testing path that is totally wrong.

This first step of stopping and thinking about the problem is important. Too often, as children, we were expected to be busy in school. A student who sat in the chair and did not appear to be busy was "wasting time." This concept is absolutely not true for those in the service field. You must be able to take the time to think about the problem and plan an effective approach to analyze the problem.

One aspect of the planning step is identification of what you should find when making a test. If, for example, you are testing a voltage amplifier, then you should expect to find a larger signal at the output of the circuit when compared to the circuit's input signal amplitude. In addition to the expected result, the service technician must identify the type of testing equipment needed to make a qualified analysis of the circuit and must be able to adjust the test equipment properly to observe the results of the test. Then, based on the discussion of breaking the large system into smaller segments, this same service technician must know where to make the next test.

At some point in the analysis process, the service technician will stop to review the results of the various tests and plan the necessary tests required to continue to localize the problem. There is nothing wrong with this approach. In fact, it is a very good idea to do just this during the analysis process. Too often you will be so wrapped up in making the tests that you will ignore the analysis portion of the process. One of the best tools we, as humans have, is our minds. When we are able to use our minds and the knowledge contained in them to analyze, we can successfully solve most of our problems.

The funnel-shaped information system illustrated in Figure 1–12 describes the process of electronic troubleshooting. The steps involved in this process are:

1. Identification of the problem area. The techniques available for this include use of the manufacturer's service literature; your senses of sight, sound, and feel; and the ability to use your knowledge of electronics.
2. Knowledge of the function of discrete blocks is required for this step. Basic block functions do not change.

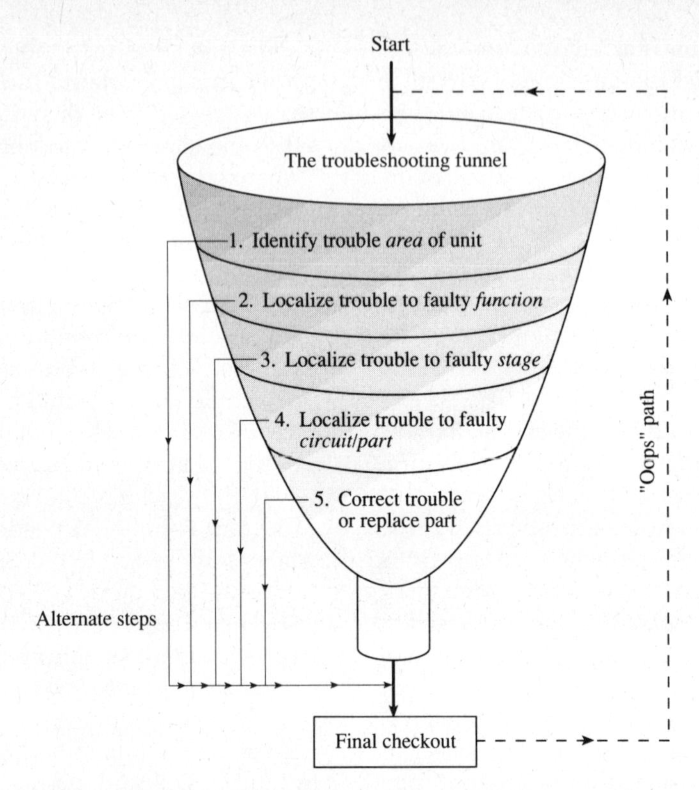

Figure 1–12 The information funnel illustrates the process of starting with the largest section and reducing the area of suspicion to a minimum size.

3. Once the block is identified, the following step is to locate the problem stage in that specific block.
4. Continuing on, the next phase is identification of a specific circuit and part that has created the problem.
5. The next to last step in this process is to repair or replace the problem component.
6. The final step is to check out the repair to ensure that it resolved the problem.

An "oops" path is also identified in this figure to show that the process described here is not always perfect; it is possible for the service technician to make an incorrect judgment. This could result in failure to correct the malfunction in the unit. When this occurs, the service technician must then return to earlier steps to locate the problem area and repair the unit.

REVIEW

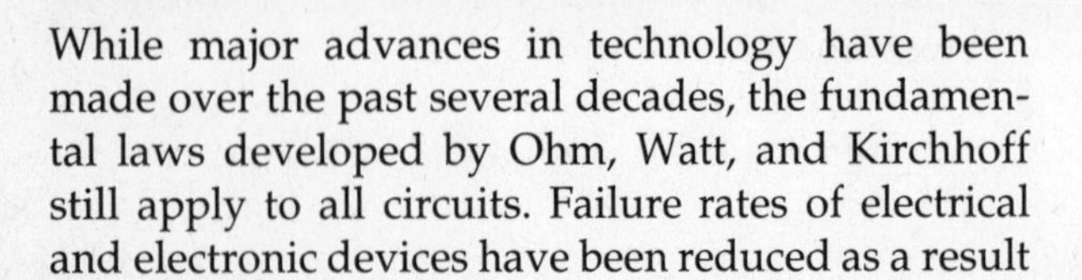

While major advances in technology have been made over the past several decades, the fundamental laws developed by Ohm, Watt, and Kirchhoff still apply to all circuits. Failure rates of electrical and electronic devices have been reduced as a result of better manufacturing quality control. In addition, many devices are constructed using solid-state and integrated circuit technology. This has created the need to recognize blocks or modules in the development of schematic diagrams and service literature. Those involved in the service industry require knowledge of functional blocks, signal flow processing, and component function. There are two fundamental levels of service personnel currently in the industry. One of these is the field service person. This person is sent to the site of the problem, where an initial level of diagnosis is performed. Knowledge of how the system functions is necessary. The field service person normally services at the board or module level. Thus, the process involves returning the system to normal operation as quickly as possible. Usual procedure for this type of service includes location and replacement of one of the plug-in boards in the system.

The next level of service is normally conducted at a service/repair center. The boards and modules that have been replaced are placed into a test jig for diagnosis, and the specific component is then located and replaced. The board is tested for its ability to function properly and then returned to service. Technicians working at this level require a much more sophisticated level of troubleshooting knowledge and techniques.

Both levels of diagnosis require knowledge of the basic rules of electricity and electronics and the ability to problem solve. Service personnel need to be able to recognize that a problem exists, diagnose and locate the area of the problem, and make the necessary repairs effectively.

REVIEW QUESTIONS

1. What has occurred in the past several years that requires a change in thinking for the servicing of electrical and electronic products?
2. Explain the need to understand the function of blocks in electrical and electronic devices.
3. Why should you develop a system of approach to any repair?
4. Identify and explain the three basic levels of repair for electrical and electronic equipment.
5. Why is it necessary to determine whether the unit should be repaired or replaced?
6. Explain why it is necessary to know how the system normally functions as an initial step in the repair process.
7. Identify the three basic types of electrical circuits.
8. Identify and explain each of the seven basic steps for effective repair described in this chapter.
9. Explain why taking a shortcut during service work is not recommended.
10. Why should you know about typical failures for specific sections, blocks, or components for successful servicing?
11. What is block diagram diagnosis?
12. Where are most vacuum tubes used in today's electronic devices?
13. Why is it necessary to know, understand, and apply the basic electrical and electronic rules during the servicing process?
14. Why is the use of blocks important for signal flow display in the service literature?
15. What is the typical direction of signal flow in service literature?
16. What is the typical display for current flow in service literature?
17. Why should you take the time and expend the energy to plan how to approach a service-related problem?
18. Explain why it is important to anticipate a proper signal or voltage value at a test point during the testing process.
19. What is the difference between conventional current flow and electron current flow?
20. What difference, if any, is there between conventional and electron current flow for the servicing process?

CHAPTER 2

Applications of the Basic Rules

INTRODUCTION

The basic theories of electricity and electronics consist of some simple rules. These are known as Ohm's, Watt's, and Kirchhoff's laws. Each of these laws can be expressed by use of mathematical formulas. They also can be explained in terminology that is familiar to all of us without the need for a high level of mathematical analysis. The purpose of this chapter is to do just that—explain these laws in simple terms. The beginning portion of the chapter will review these basic laws and their applications. Examples of each of the laws will be given to assist you in understanding them.

OBJECTIVES

Upon completion of this chapter, the reader/student should:

1 recognize Ohm's law and its applications;

2 recognize Watt's law and its applications;

3 recognize Kirchhoff's voltage law and its applications;

4 recognize Kirchhoff's current law and its applications;

5 understand the importance of the above laws in the field of electrical and electronic service;

6 understand the application of the terms "ground" and "common" in electrical and electronic work;

7 understand how values of voltage, current, power, and resistance are measured; and

8 understand the terms "junction" and "node."

KEY WORDS AND PHRASES 

current: In the context of this book, current is defined as the flow of electrons in a complete circuit. Electron flow goes from the negative (−) terminal of the power source through the circuit's load to the positive (+) terminal of the power source. Current is measured in units of the ampere, or amp (A). In some technical manuals, the letter I is used to indicate the intensity of current flow. The letter A indicates the quantity of current in amperes.

load: All electrical devices are designed to use some amount of electrical power. Load describes the device that uses this power to perform some work. The load can be any device, ranging from a simple lamp to a very sophisticated computer-controlled machine. Loads are often indicated on the schematic diagram as resistance values in circuits instead of by their physical description.

power: Electrical power is created when the application of voltage to a circuit produces a flow of current. Electrical power is required for the performance of some form of work. The values of both voltage and current are combined in a formula to indicate the electrical power. The letter P is used to indicate electrical power. The letter W indicates the power level in units of the watt.

resistance: Resistance is the opposition to the flow of current in any circuit. Normally all components have some resistance value. Resistance is measured in units of the ohm (Ω). The letter R is used to show resistive units.

voltage: Voltage is the electrical pressure developed by the power source in the circuit. Voltage is also called pressure, or electromotive force (EMF). Application of a voltage creates a flow of electrical current when the source is connected to a load in an electrical circuit. Another term used to describe voltage is "potential," since voltage has the potential ability to create an electron flow in an electrical circuit. Voltage has the potential to cause some work to be performed when it is applied to the circuit. Voltage is measured in units of the volt (V). In some technical manuals, the letter E is used to indicate electromotive force. The letter V is used to describe quantities of voltage in a circuit. This book uses E for electromotive force and V for values of voltage in a circuit or system.

BASIC TERMINOLOGY

OHM'S LAW

Ohm's law is the fundamental rule used to explain the relationship of **voltage**, **current**, and **resistance** in a simple electrical circuit. Ohm stated that when one volt of electricity is applied to a circuit that produces one ampere of current flow, then the opposition to this current is equal to one ohm of resistance. This provides the mathematical analysis of the circuit. The three mathematical relationships are:

$E = I \times R$
(voltage = current times resistance)

$I = E \div R$
(amperes = voltage divided by resistance)

$R = E \div I$
(resistance = voltage divided by current)

A block diagram for a simple electrical circuit is shown in Figure 2–1. It consists of a source of electrical energy, two lines, or wires used to carry the electrical current, and a **load**. The purpose of the load is to use this current to accomplish some form

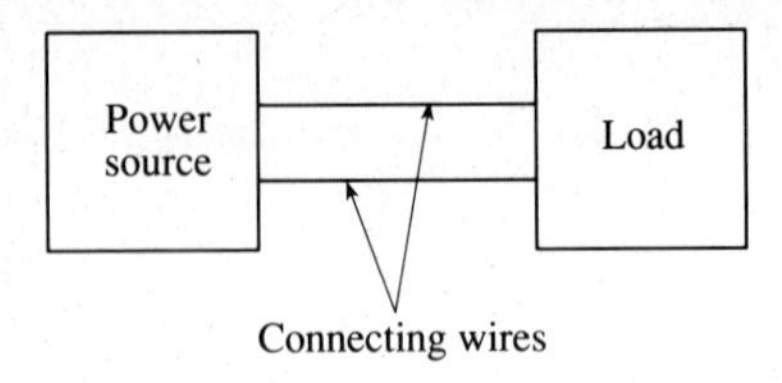

Figure 2–1 **The block diagram for a basic electrical circuit contains a power source, a load, and two connecting lines, or wires.**

of work. These components are the minimum required for an electrical circuit. The **power** source creates a difference in electrical potential between its two terminals. This potential difference is necessary to create a flow of electrons. When the load is connected to the power source with the wires between the two units, then current will flow throughout the circuit. This flow of current will create some form of work in the load.

Figure 2–2 shows the same circuit using commonly used graphic symbols. The power source is illustrated as either a direct current (DC) source or an alternating current (AC) source. In addition to the graphic symbols, a value of voltage is required for a complete description of the power source. Normally, this is illustrated with a numeric value. Both an AC and a DC source are illustrated on the left-hand side of the circuit. The AC source is the left-hand symbol of the two, and the DC source is to its right.

The load is shown on the right side of the schematic diagram. The universal symbol for a load is the same symbol used for a resistor. While this does not provide a true description of the load, it does indicate the presence of a device that uses electrical energy. The specific symbol displayed is not critical. The most important thing to realize is that when a current is flowing through any load, the load is actually performing some work. The specific type of work can vary from the simplest lamp circuit to a complex electronic computer.

In the circuit shown in Figure 2–3, the component that creates the potential for electron flow is the power source. Since this is identified as the voltage potential, current will flow only when the source is connected to a load. When the circuit is complete, electron current flow starts at the negative terminal of the power source. It continues through the load and then returns to the positive terminal of the power source.

Circuit analysis can be accomplished by approximation once the basic circuit configuration is understood. One does not require the use of a calculator to determine what the general value of voltage, current, or resistance should be in the circuit. Often you can approximate the values in a circuit. This, of course, depends upon your ability to use Ohm's formulas and to approximate values. Circuit values are not always the same as those provided by the manufacturer. In many circuits the values are close to the original. A measurement of the circuit value that is close to the noted value is an indicator that the circuit is properly functioning.

Here is an example of the above situation: A circuit contains a 4.7 kΩ resistance in series with a

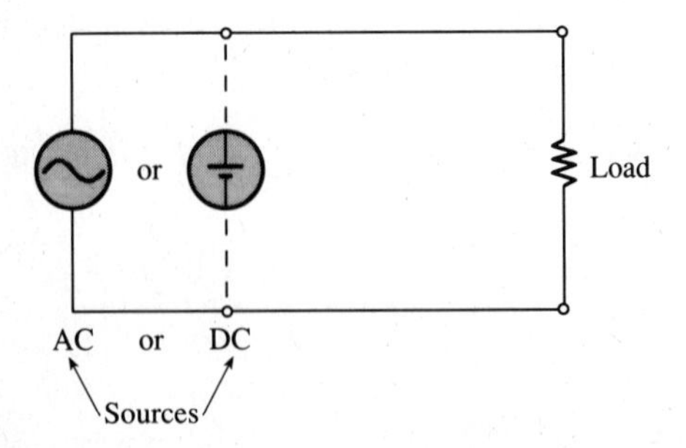

Figure 2–2 **The basic electrical circuit drawn using schematic symbols instead of block symbols. Both an AC and a DC power source are shown.**

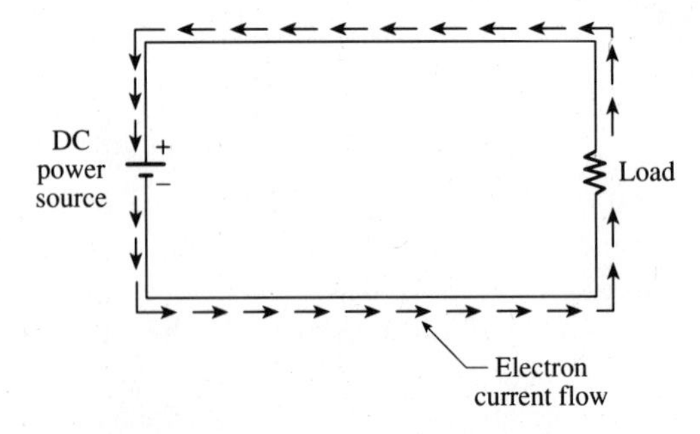

Figure 2–3 **Electron flow in the basic circuit is from the power source's negative terminal through the load and then to the source's positive terminal.**

10 kΩ resistance. Total resistance is actually 14.7 kΩ. This value is very close to 15 kΩ, so why not use the 15 kΩ value to mentally calculate some circuit values? The two values of resistance have a 1:2 ratio: the 4.7 kΩ value is just less than half of the 10 kΩ resistance. If the 4.7 kΩ value were "adjusted" upward to 5 kΩ, then the 1:2 ratio would be correct. Since this ratio exists, the voltage drops across the two components should be in the same form. If a 24 V source were applied to this circuit, the 4.7 kΩ resistance would develop a voltage drop of just under one-third of the 24 V total, or just slightly less than 8 V.

You do not need to be an expert mathematician to determine this. All that is necessary is an application of the basic rules for voltage drops in the circuit as they apply to resistance values. In truth, the actual voltage drop measured across the 4.7 kΩ resistance would be 7.67 V *if* both of these resistances had 0 percent tolerances and measured their actual design values. This seldom happens in actual practice, so why not round off the numbers to determine a "close" value? The measurement of any voltage value close to 8.0 V would indicate that the circuit was operating in the correct manner.

A series circuit is illustrated in Figure 2–4. This circuit provides the values for both the power source and the load resistance. Using the values of 100 V and 100 Ω, the Ohm's law formula will provide the current flow in this series circuit. In this example the current formula is:

$$I = E \div R$$

Substituting numeric values:

$$I = 100 \div 100 = 1 \text{ A}$$

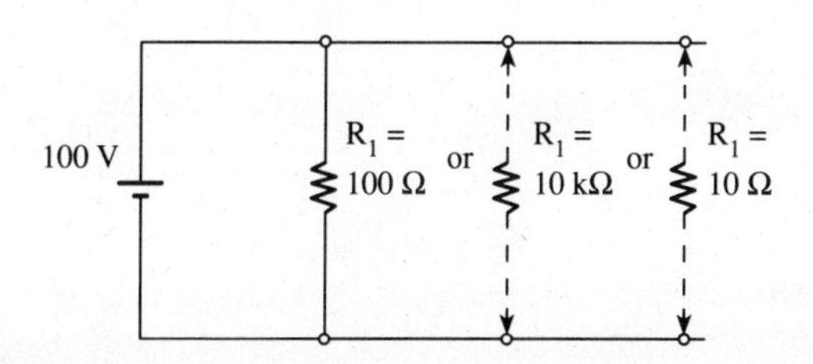

Figure 2–4 Series electrical circuit with a source and a load resistor. Three values of load are shown for analysis in the text.

Therefore, the total current in this circuit is 1 A when a 100 Ω resistance is used for the load value. If the value of voltage remains constant in a series circuit, the current is controlled by the amount of opposition, or resistance, to its flow. In any series circuit when the resistance is increased, the current flow will decrease. It really is a simple statement. However, the service technician may forget this rule when attempting to localize a problem to a specific section or component in the circuit. Just remember, when the resistance increases in the series circuit, total current will decrease.

Figure 2–4 also has a second load resistance value shown. This value is 10 kΩ. When the higher resistance value of 10 kΩ is used instead of the original 100 Ω resistor, the operating conditions in the circuit change. The circuit resistance has now increased to 100 times its original value. When you consider the relationship of voltage and resistance and Ohm's formula, then you would expect a reduction in the amount of current flow. This can be shown by use of the formula:

$$I = E \div R$$

$$I = 100 \div 10{,}000 = 0.01 \text{ A}$$

When the load resistance is less than its original value, the amount of current will increase. Load resistance value is 10 Ω. The expected result is an increase in the amount of current flow under these conditions:

$$I = E \div R$$

$$I = 100 \div 10 = 10 \text{ A}$$

Both examples use the premise that the power source can supply an unlimited amount of current. In actual installations this often is not true. Manufacturers will design or select power supplies for equipment having adequate current capabilities for normal use. Usually there is some amount of additional capacity, often 10 to 20 percent in addition to the amount required for normal design operation. Normally, there will be a reduction in the level of voltage when current demands of the load exceed the capability of the power source.

Using the circuit shown in Figure 2–4, consider what will probably occur when the system load demands more current than the power source is capable of producing. System design developed a unit capable of producing 1 A of current when 100 V was applied to a 100 Ω load. Should the load resistance decrease in value to 25 Ω due to a partial malfunction, load current will attempt to increase to 2.0 A. This is not possible, due to the 1.0 A design limitation of the power source; therefore the source voltage will decrease under this new current demand condition. Using the Ohm's law formula:

$$E = I \times R$$

$$E = 2 \times 25 = 50 \text{ V}$$

Source voltage, as shown on the graph in Figure 2–5, has decreased to one half of its design value due to the load demand in excess of the power source's capability. Also shown on the graph is the value of source voltage when the load's ohmic value is reduced to 10 Ω. In this situation, a current of 2.5 A will attempt to flow. If this current flows, then the source voltage is reduced to a lower value of 25 V.

This type of condition may be observed in equipment being repaired. The reduction in source voltage is usually due to a decrease in one component's ohmic value. Location and replacement of this malfunctioning component usually will correct the problem and restore the equipment to its normal use. The graph shown in Figure 2–5 illustrates the conditions encountered in a system when the load requirements exceed the design of the power source. The ultimate excess current condition occurs when the load has a value of zero ohms. When this happens, the source voltage also will be reduced to zero.

You must remember that a reduction in the ohmic value of any component will result in an increase in current flow in the circuit. Under these conditions, a reduction in the value of the load's resistance does not indicate that less work is being done. What it does indicate is that the work has increased due to the change in the ohmic value of the load.

WATT'S LAW

Watt identified the relationship of voltage and current to electrical power. In one sense, electrical power is related to the performance of work. We use electrical power, or energy, to operate most electrical devices. Earlier, the term "voltage" was described as an electrical potential. Voltage applied to a circuit will create a flow of electrons. When the electrons are in motion through a load, then the load is performing some work. This is true whether the load is a battery-operated clock or a very sophisticated computer or a complicated piece of machinery. Therefore, both voltage and current must be functional to perform electrical work.

Watt's law simply states that the total amount of power in any system or circuit is the product of applied voltage and the amount of current that flows. The formula for this is:

$$P = E \times I$$

Figure 2–6 is an example of how the power formula is used. In this circuit, 100 V is applied to a 100 Ω load. This creates a 1 A current flow in the circuit. Using Watt's law:

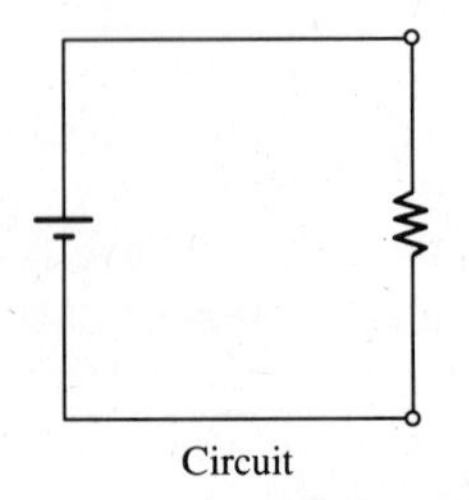

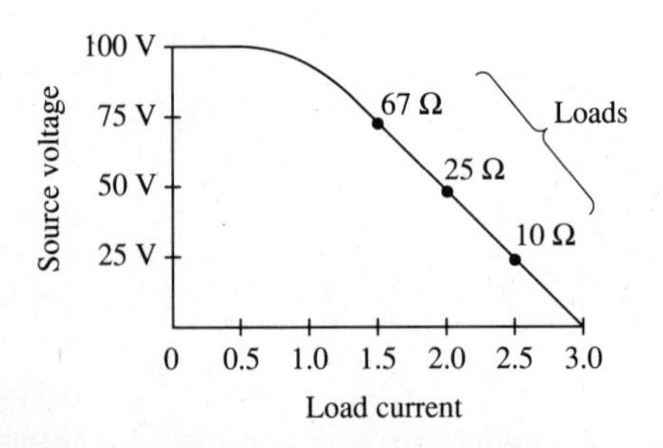

Figure 2–5 Illustration of the drop in available voltage due to circuit-loading conditions beyond the power source's capacity.

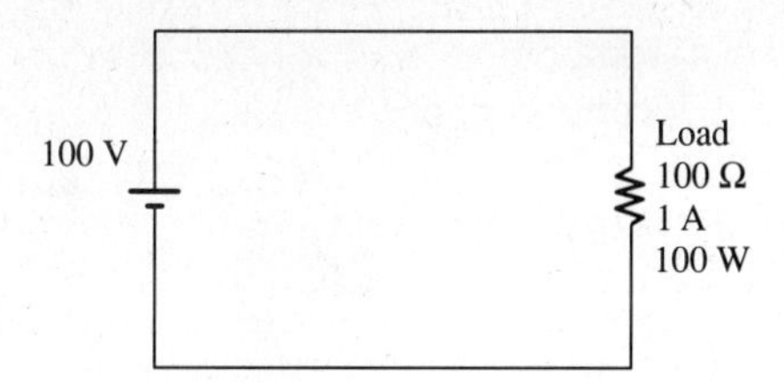

Figure 2–6 Application of the power formula in a basic electrical circuit.

$$P = E \times I$$

$$P = 100 \times 1 = 100 \text{ W}$$

A second method of using this circuit is to determine the flow of current in a circuit when both the source voltage and the power values are known:

$$I = P \div E$$

$$I = 100 \div 100 = 1 \text{ A}$$

This method may be used by determining the quantity of current flow required for an electrical device when both the power rating and the operating voltage are known.

Of equal importance to the service technician is the concept of electron flow creating heat in the circuit or device. Each component or system has a value of internal resistance. This resistance creates opposition to the flow of electrons. In a mechanical system, opposition to the performance of work is called friction. One result of overcoming this opposition is the creation of heat. This happens in electrical and electronic circuits and systems. The opposition to electron flow creates heat in the circuit. Heat is often one of the greatest enemies of electronic components. The amount of heat generated by electron current flow is determined by using Watt's law. The value of power as determined by using Watt's Law is also the rate at which heat is generated in a device.

Figure 2–7 illustrates the heating concept of Watt's law. In this circuit 25 V will create a current flow of 0.005 A.

$$I = E \div R; \quad I = 25 \div 5000 = 0.005 \text{ A}$$

Using Watt's law:

$$P = E \times I; \quad P = 25 \times 0.005 = 0.125 \text{ W}$$

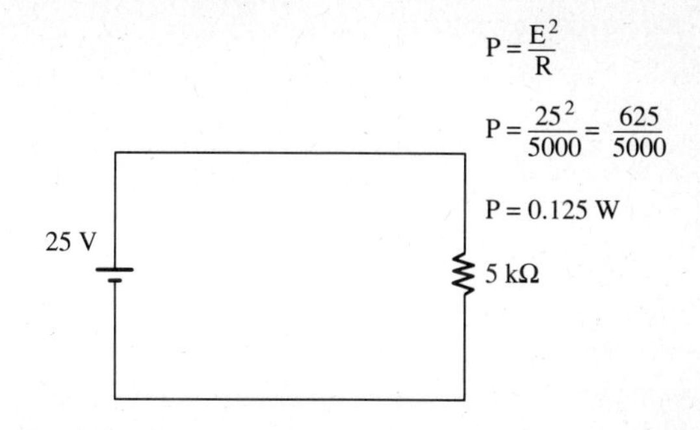

Figure 2–7 Current flow creates 0.125 W of heat in this circuit.

In this circuit the heating effect on the load resistance is equal to 0.125 W or 1/8 W. Most electronic devices used as loads will have a power rating in addition to their component value rating. Design will dictate the power rating's ability to handle a specific amount of heat without destructing. In this circuit the load's power rating will be a minimum of 0.125 W. In practice, the power rating should be higher than the minimum required for the load.

If the source voltage in this same circuit should be increased to 75 V, then the power rating for the load also would have to be increased. Power requirements increase in a nonlinear, or squared, manner with the increase of source voltage. For example:

$$I = E \div R; \quad I = 75 \div 5000 = 0.015 \text{ A}$$

$$P = E \times I; \quad P = 75 \times 0.015 = 1.125 \text{ W}$$

If the load's power rating had remained at 0.125 W, then a secondary action would have occurred. One method of describing this action is that the load will overheat and destruct. Either the load will heat to the point where it becomes discolored or charred, or the excess heat will cause it to break into two pieces. In either of these conditions, the load will no longer be able to sustain its design current and the circuit will malfunction. The experienced service technician is able to detect the odor of a burned or overheated component while examining the unit. If the failure occurred during operation, you might see some smoke coming from the unit. This, too, would indicate an overcurrent condition.

EXCESS HEAT EFFECTS

When semiconductors are overheated they have a tendency to self-destruct. This develops a condition called thermal runaway. Heat in the device will reduce its internal resistance. As was described earlier, a reduction in the resistance of a circuit while maintaining the applied voltage will permit additional quantities of current flow. Additional quantities of heat are generated as more current flows in the circuit. This action and reaction continues until the semiconductor becomes so hot that it fails. Under normal operating conditions thermal runaway is not allowed to occur.

Other electrical and electronic devices also may fail when they are overheated. The example of a resistor charring or breaking into two is one of the results of excess heat. Connections on units that have been assembled with solder may fail due to the melting of some of their soldered connections. Failure may occur inside some of these units and may not be apparent to the observer. At times the wire windings on the voice coils of loudspeakers will melt and open due to excess current flow. This can also happen with the wires on transformer windings due to excess current flow and heat because of a failure elsewhere in the system. When this type of failure occurs, the technician must rely on knowledge of how the specific blocks are supposed to operate to localize the area of failure.

COMPONENT COOLING METHODS

In electronic design and construction, the process of maintaining an acceptable operational temperature for components is done in one of two ways. One of these is to use a device known as a heat sink. The heat sink often is the metal chassis constructed as a part of the unit's construction. It may also be a finned device attached to the device that effectively increases its surface area. One such heat sink is illustrated in Figure 2–8. Semiconductor devices, such as the power transistors seen on this heat sink, are often attached to the metal frame or chassis of the unit. Heat generated by electron action in the semiconductor is conducted away from the body of the semiconductor through the metal frame or chassis. Device cooling is accomplished since the combined surface area of the metal and semiconductor is much greater than the surface area of the semiconductor alone. The heat will dissipate over the larger area and assist in keeping the semiconductor from an excess heat failure. A word of caution must be mentioned here. The frame, or chassis, or heat sink, of many electronic devices is often connected to circuit common. One of the elements of some semiconductors may be attached to its case or metal body. This element is often the collector of the transistor. If the transistor is an NPN type, its collector is usually connected to the most positive point in its circuit. These semiconductors must use an insulator between their case and the metal frame or chassis used as the heat radiating heat sink. Use of the insulator does not affect normal operation of the device. It isolates the transistor's case and collector lead from circuit common, permitting its normal operation.

Figure 2–8 Additional heat buildup during power transistor operation can be dissipated by the use of heat sinks. (*Photo by J. Goldberg*)

Another point to remember when working with solid-state devices that use heat sinks is that these sinks should *not* be touched when the circuit is under power until a voltmeter is used to check the presence of operating voltage. A person who is unaware of voltage on a heat sink may receive an electrical shock when it is touched and another part of the body is touching circuit common. The result could be very serious—the same as receiving an electrical charge from any power source. You could be knocked down or hurt your knuckles when you touch what you thought was a harmless heat sink.

A second method of cooling electronic components is the use of a fan inside the case of the unit. The design of the case is such that the fan will draw outside air into the unit. Additional openings elsewhere in the unit's case act as exhaust vents for the heated air. The movement of the air will reduce the internal temperature of the unit. This air flow assists in keeping the semiconductors in the unit from overheating and possibly destructing.

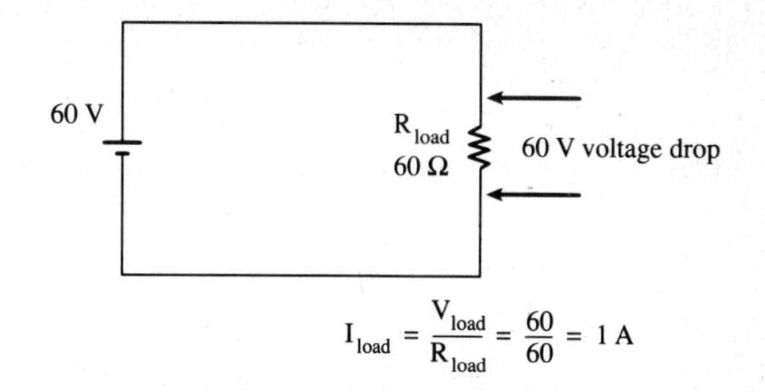

Figure 2–9 Voltage drop across the load is equal to the source voltage in this basic circuit.

KIRCHHOFF'S LAWS

The third major set of laws to be presented in this chapter are those identified as Kirchhoff's laws. One of these laws is related to voltage drops, while the other is related to current flow. Both of the laws may be discussed in mathematical terms. They also can be presented in a nonmathematical manner. This second method is the way in which Kirchhoff's laws will be described here.

Kirchhoff's voltage law applies to the voltages present in all circuits. A difference in electrical potential develops across a component when current flows in an electrical circuit. This difference is commonly called a voltage drop. If the circuit contains one source and one load, the voltage drop across the load is equal to the value of the applied voltage. The ohmic value, or resistance, of the load does not influence the value of the voltage drop. The ohmic value of the load will only affect the circuit current; this is illustrated in Figure 2–9. The calculated value for load current in this circuit when the load is 60 Ω is 1 A. When the ohmic value of the load is increased to 120 Ω, the current will be reduced to 0.5 A. The value of voltage developed across the load by the application of the source voltage will remain 60 V in either case.

A different set of conditions occurs when more than one load resistance is connected as shown in Figure 2–10. This circuit has three different values of load resistance connected in a series configuration. Voltage drops are developed across each of these three loads. The sum of the individual voltage drops in this type of circuit will equal the applied source voltage.

A rule to remember is that the voltage drops in the series circuit are directly proportional to the individual ohmic values of the load resistances. The largest voltage drop will be developed across the largest ohmic value load. If this is true, then the smallest voltage drop will be developed across the lowest ohmic value load. One method of quickly determining the value of the voltage drop if it is not provided on the service literature is to identify percentages of the total ohmic value. In this circuit load R_3 has a value that is equal to one half of the total amount of load resistance in this circuit. The circuit contains three resistances of 100 Ω, 200 Ω, and 300 Ω. The sum of their individual ohmic values is 600 Ω. The 300 Ω value is one half of the total resistance in this circuit. Considering that the voltage drops are directly proportional to the proportional ratio of the ohmic values, a value of one half of the total resistance should create a voltage drop that is one half of the applied voltage. The voltage drop developed across load R_1 should be equal to one-sixth of the applied voltage since this ohmic value

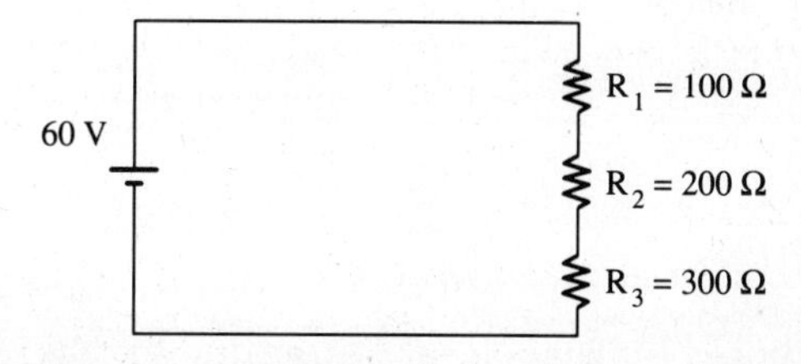

Figure 2–10 Voltage drops across each component add to equal the total of the applied source voltage in this series circuit.

of 100 Ω is one-sixth of the total resistance value of 600 Ω in this circuit.

The concept of voltage drops illustrated in Figure 2–10 will apply to any series circuit regardless of the number of loads or resistances in the circuit. This is also true when the loads are other than true resistances. The circuit shown in Figure 2–11 contains two resistances and a block identified as a transistor. The term "transistor" was introduced by combining the terms "transfer" and "resistor" into this new term. In any service-related activity, the technician should recognize that the transistor will act as a resistor in the circuit. (This is also true for other types of conductive devices, such as vacuum tubes.) Service technicians should analyze the circuit and treat the transistor as a resistance; this will make circuit analysis much easier. In a great many amplifier circuits, the transistor will exhibit close to one half of the total circuit resistance. This, then, should indicate to the service technician that close to one half of the applied voltage should be measured across the terminals of the transistor.

In this circuit, the value of R_1 is 5 kΩ and the value for R_2 is 100 Ω. Initial analysis of this circuit should be made before the use of any test equipment. This approach provides a set of expected results for the technician. Then, when actual measurements are made, the technician will be able to use the expected results to tell whether the voltages are correct or incorrect. The expected ohmic value of the transistor will be close to one half of the total circuit resistance. In this circuit this value will approximate 5 kΩ. The expected voltage drops for this circuit will be: E_{R1} = 9 V to 10 V; E_{R2} = less than 1 V; and $E_{transistor}$ = 9 V to 10 V. In other words, almost all of the applied 20 V will be measured across the series combination of the transistor and resistor R_1 since these have the largest percentage of resistance in the circuit. The 100 Ω value is such a small percentage of the total 10,100 Ω resistance in this circuit that its voltage drop will be less than one volt. Measurement of these values will assist the technician in determining whether the circuit is functioning properly. Measurement also will indicate whether further measurements in this circuit are required.

When measuring circuit current in this circuit, the basic rule for series circuits applies. Circuit current is equal throughout the circuit. This is true regardless of the individual values of load resistances used for proper circuit operation.

Kirchhoff's current law is also applied to all circuits. The rule Kirchhoff presented for current also can be explained in very simple terms. It states that the electrical current entering any junction is equal to the current leaving the same junction. This statement may be expanded to include the word "sum." The addition of this word then modifies the statement to read: The sum of the currents entering a junction is equal to the sum of the currents leaving the same junction. In Figure 2–12 an electrical junction is identified by the dot. Current flow direction from C into the junction would then flow out of the junction into lines A and B. In this situation, there is only one current at C, which would be a total value for this partial circuit. Therefore, its sum and its value are the same. At the junction the current will divide into the two available paths of A and B. The specific values for these currents will be determined by the values of resistance in each path. The

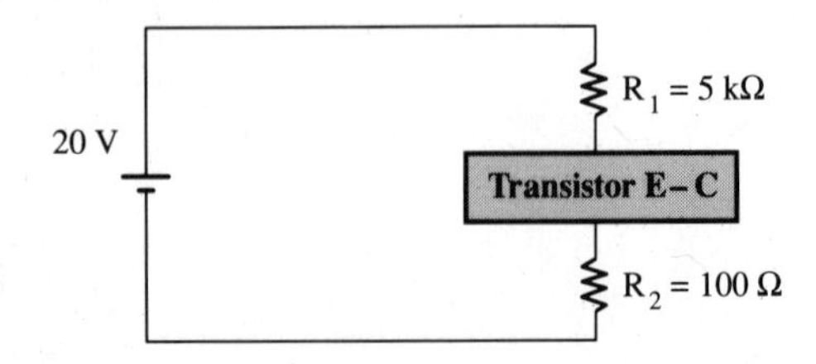

Figure 2–11 The rule for voltage drops in the series circuit applies regardless of the types of loads used in the circuit. A transistor acts as a resistance load in this circuit.

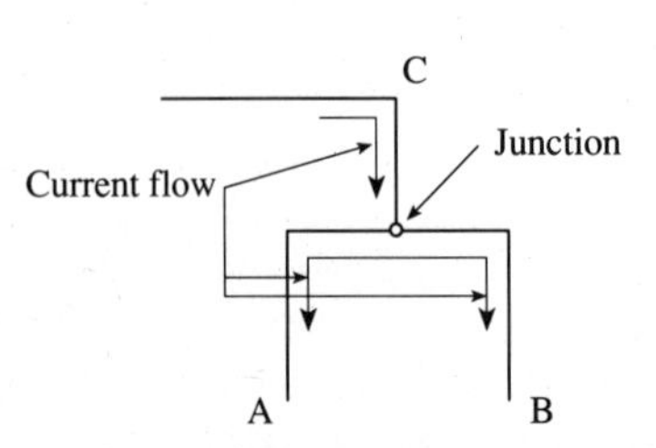

Figure 2–12 The total current entering a junction is equal to the total current leaving the same junction.

sums of the current in each path will be equal to the sum of the current at C.

Should the direction of current flow be reversed, the same rules will apply. The sums of currents through A and B that enter this junction will be equal to the sum of the currents leaving the junction at C.

Analysis of a series circuit will indicate the same type of results for current flow. Figure 2–13(a) is a series circuit. Each junction is indicated by a letter in this circuit. Measurement of current at point C should provide the same value as current measured at point D. This is true because a junction exists between these two points. The same rule is true when current is measured at point E and compared to the current at point F. The sum of the current entering either of these junctions is equal to the sum of the current leaving the individual junction. This is why current is the same in all parts of the series circuit.

Part (b) of Figure 2–13 is a parallel circuit. Junctions are indicated at each of the dots on the diagram. The sum of the current entering junction H is equal to the sum of the currents at points E and F in this circuit. Also true is that the sum of the currents at D and G are equal to the sum of the current leaving junction C. The basic rule here is that the total current entering a junction must be equal to the total current leaving that same junction. This is true regardless of the type of circuit involved. The individual branch resistive values will determine the value of current in those branches.

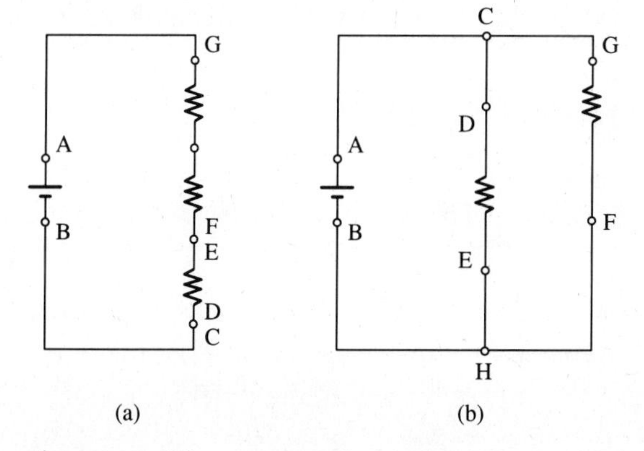

Figure 2–13 Kirchhoff's current law applies for both series circuits (a) and parallel circuits (b).

TROUBLESHOOTING TERMINOLOGY

Some terms are unique to the area of troubleshooting electrical and electronic devices. Some of the terms in use by service technicians include common, ground, voltage drop, current flow, and junction. Proper understanding and use of these terms will aid in the development of good work habits for those entering into the service profession.

COMMON

The term "common" indicates a point of reference for measurements in a circuit. In many electrical and electronic circuits, one connection from the power source is used as a point of reference for all voltage measurements. The negative terminal of the power source is most often the one used for this reference. This is usually considered to have a voltage value of zero volts. Keep in mind that there are exceptions to this and the service technician must refer to the service literature for the polarity used in the specific equipment being analyzed.

Voltage values may be either positive or negative. Systems with their negative power source terminal connected to circuit common have positive value voltages. This type of circuit is illustrated in Figure 2–14(a). In this circuit the negative lead of the power source is connected directly to circuit common. It is

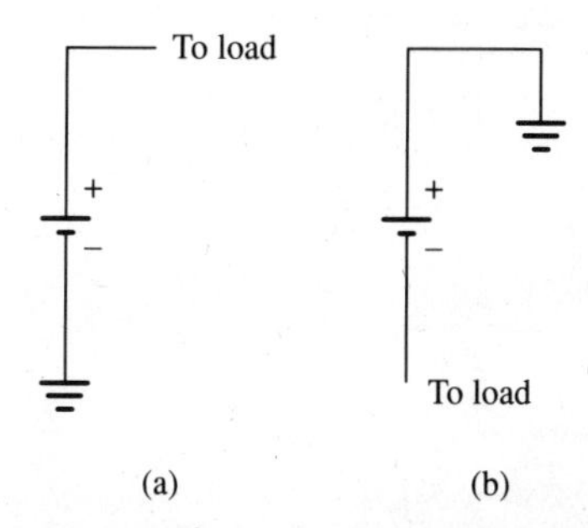

Figure 2–14 The direction of conventional current flow is shown by use of the (+) signs at the end of the arrows, and electron current flow is displayed by the use of a (−) sign.

the most common type of circuit used in electronic equipment. When the positive terminal of the power source is used as circuit common, as shown in Figure 2–14(b), the voltages are negative in value. This type of power source is found in systems using PNP types of transistors.

In some systems neither the positive nor the negative power source leads are connected to circuit common. Systems using this technique may have both positive and negative polarity voltages present in individual circuits. Two of these are shown in Figure 2–15. Part (a) illustrates a system with a single power source connected to a voltage-dividing network. Neither of the leads of the power source is connected to circuit common. The center connection of the two series resistances is used as the common point for the load circuit. The ohmic values of the two resistances usually are equal, thus the two output voltages are also equal in value. One of the output voltages will be considered as positive and the other one as the negative power source.

Part (b) of this figure has the same type of output polarity configuration. The difference between this circuit and the previous one is that this one uses two individual power sources connected in series. The connection between the two power sources is considered circuit common. Both circuits will provide the required dual-polarity power source. One application for this dual-polarity power source is found with operational amplifier circuits.

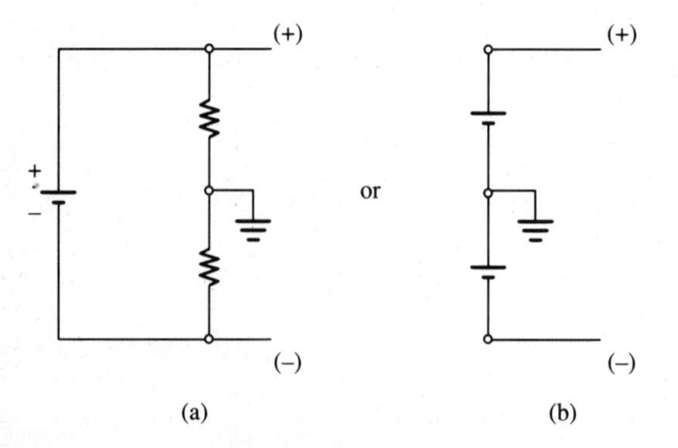

Figure 2–15 Two methods of obtaining both a positive and a negative voltage source.

GROUND

The term "ground" has evolved from an original application of electricity. Electrical codes in most communities currently require one of the two wires used in homes to be connected directly to the earth, or ground. Many early electrical and electronic circuits also connected one of the wires used for electrical power directly to the earth. This wire was considered the negative, or reference, wire in the system. It was supposed to be at a zero volt potential. This concept has evolved into the use of the term "ground" as a reference for zero volts and a point of measurement.

Many electronic circuits in current use are not connected to any external source of electrical energy. These devices may use chemical cells, or batteries, for their power source. Others may use solar cells. In automotive applications the negative terminal of the battery is connected to the body and frame of the vehicle. There is no connection to any other power source. Each of these has a common voltage reference point. Although the term "ground" is still used to identify this point, technically, it should not be. Today's technician must understand that the terms "ground" and "common" are used interchangeably by those working in electrical and electronic servicing.

Possibly the reason for this is a carryover from older days and types of training in the electrical and electronic field. The concept of a point of reference for measurements is still valid, regardless of which term is used. Some of the symbols used for common and ground are shown in the Appendix of this book.

VOLTAGE MEASUREMENTS

These measurements are normally made by placing the negative test lead of the voltmeter on the circuit common and the other test lead at some predetermined point in the circuit. Measurements are then "read" as some value of voltage at the point of the test. There is an understanding that the measurement is actually determining the difference in electrical potential, or voltage, measured between circuit common and the point of the test. While this long explanation is not often expressed, this is the true measurement of a voltage value in a circuit. A typical way of expressing this measurement is simply to state that "the voltage at this point is 250 volts."

VOLTAGE DROP

Voltage drop is used to indicate a difference in electrical potential, or pressure, from one lead on a component to the other lead. It is possible to measure the voltage drop developed across a specific component in the circuit. This is done by placing one of the test leads of the voltmeter on one wire or connection on the component. The other test lead is then placed on the other wire or connection on the component. The measurement is then considered to be "across" the component. A description of this will indicate the specific type of measurement. For example, the statement, "Resistor R_1 has a voltage drop of 25 volts," would indicate that the actual measurement was made from one lead of this resistor to the other lead of the resistor. This is the same as stating that this is the difference in electrical potential at each terminal of the resistor.

CURRENT FLOW

Current flow describes the movement of electrical charges within an electrical circuit. The charges may be considered as either positive or negative. If they are considered positive, then the direction of current flow is from the positive terminal of the power source through the load to the negative terminal of the power source. This is shown in Figure 2–16 by use of (+) signs. When the charges are negative, the flow is in the opposite direction. Current flow from positive to negative is known as "conventional" current flow. When the flow is from negative to positive, it is called "electron" current flow. This is shown in Figure 2–16 by the use of (−) signs.

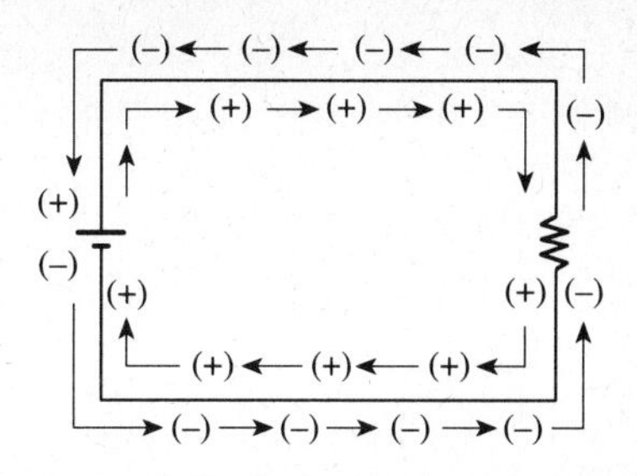

Figure 2–16 **Electron current flow is illustrated with arrows having a (−) sign, and conventional current flow is illustrated with the (+) sign in front of the arrows.**

JUNCTION

Junction is the connection of two or more components in any electrical or electronic circuit. This connection is also known as a node. The simplest junction is one where the leads of two components are connected to each other. A complex junction may have three or more component leads connected to each other. The rules for current flow as identified by Kirchhoff apply under either of these conditions. In some engineering books, the term "node" is used as a substitute for the term "junction".

REVIEW

There are three basic sets of laws that relate to electrical and electronic circuits. These are the ones presented by Ohm, Watt, and Kirchhoff. Each of these basic laws must be used to successfully service any electrical or electronic device. Ohm's law describes the relationship of voltage, current, and resistance in any electrical or electronic circuit. Watt's law describes the relationship of voltage, current, and power in the circuit. Kirchhoff described the concept of voltage drops in a series circuit and current flow at any junction in a circuit.

Correct use of these laws improves the abilities of the service technician as diagnosis and repair procedures are developed and employed. The fundamental rule for all successful servicing procedures is that these basic laws have not changed since their introduction. Circuit applications have become more sophisticated in the past years, but this has not affected the ability to use the basic rules of Ohm, Watt, and Kirchhoff for servicing products using electricity and electronics.

Ohm's law is used for circuit analysis. When the value of the source voltage is kept at a constant level, the current flow in the circuit depends on the ohmic value of the load. If the source voltage value is changed and the ohmic value of the load is constant, current flow will change.

Watt's law describes the amount of power in the circuit. This law is often used to determine the power rating of individual components in a circuit. Heat is generated by the flow of current in a circuit. This heat must be controlled, or the possibility of component failure is very high. Cooling systems are used for this purpose.

In his voltage law, Kirchhoff described the relationship of voltage drops in a series circuit. He determined that there is a direct relationship of the ohmic value of series-connected components and their respective voltage drops. He also found that the sum of the individual voltage drops developed in a series circuit will equal the value of the applied source voltage.

Kirchhoff's current law explains the relationship of current flow at any junction, or node, in a circuit. The current entering any junction is equal to the current leaving this same junction. This law applies to junctions in both series and parallel circuits.

The point of reference in any circuit is technically known as circuit common. Many texts refer to this connection as circuit ground. Those who have been working in the field of electronic service may also refer to circuit common as circuit ground. You should be able to recognize both uses of this term.

REVIEW QUESTIONS

1. Explain these terms: voltage, current, resistance, and power.
2. What conditions, or combination of conditions, would cause a resistor to char and break in half?
3. Explain why voltage measurements are usually made when one of the two test leads is connected to circuit common.
4. Describe two methods of temperature maintenance used in electrical and electronic circuits.
5. Why is the load considered a resistance in the circuit?
6. You encounter a blown fuse in a circuit. Is this a result of excess current or too little current?
7. State the major components of a basic electrical circuit.
8. Ignoring physical size, is it possible to install a resistor rated at 1 W in a circuit to replace an original resistance having a power rating value of 1/2 W? Explain your answer using electrical terminology.
9. Again, ignoring physical size, is it possible to install a resistor rated at 1/2 W in a circuit to replace an original resistance having a power rating value of 2 W? Explain your answer using electrical terminology.
10. Three resistors are connected in a series circuit having a source voltage of 25 V. Their individual values are 10 kΩ, 4.5 kΩ, and 15 Ω. Indicate the approximate voltage drops that develop across each of them and state your reasons for your answers. Do *not* use a mathematical computation for your explanation.

Describe what you would expect to find in each of the following circumstances:

11. One component in a parallel circuit does not operate. What effect does this have on circuit voltage and current values?
12. Current flow is less than expected in a parallel circuit. What component(s) should be suspected?
13. Where in a series circuit would you look to locate a short circuited component?
14. The voltage drop across any one component in

a series circuit is incorrect. What component(s) should be suspected?

15. The voltage measured across any load in a parallel circuit is incorrect. What component(s) or blocks should be suspected?
16. Source voltage in a series circuit is held at a constant value and the load resistance increases. What is the effect on the other components in the circuit?
17. Load in a series circuit is kept constant and the source voltage decreases. What is the effect on the other components in the circuit?
18. Source voltage in a parallel circuit is correct, but current flow much higher than normal is measured. Where should you look for potential problems?
19. The expected brilliance of a lamp used in a series circuit is reduced. What components can cause this problem?
20. The filaments of indicator lamps in a circuit burn out very quickly. What sections of the circuit should be checked?

CHAPTER 3

Path Analysis

INTRODUCTION

Each task we learn requires some form of education or training, either formal or informal. Often the education or training is used as a basis for further learning. The inexperienced traveler will have difficulty in determining a route by which to arrive at a destination. Learning how to use a road map makes travel planning much easier. The time required to locate an acceptable route will be longer the first time or two the map is used. Route identification and planning time is usually much shorter after some experience is gained. This same approach should be used for servicing electrical and electronic products. Most of these products have some form of road map available for service technicians. In the field of electricity and electronics, these roadmaps are known as block diagrams or schematic diagrams.

OBJECTIVES

Upon completion of this chapter, the reader/student should:

recognize the difference between signal paths and current paths;

recognize the specific path types used in electrical and electronic equipment; and

understand and apply the rules for specific path types as they apply to diagnosis and repair.

KEY WORDS AND PHRASES 

current path: The path used for current flow in one or more circuits in an electronic device.

feedback path: This system is used to control the signal's output level. It processes a portion of the output signal and uses it to control the input to the circuit.

joining path: In this system, the signals or currents of two or more paths meet and join into one common path.

linear path: The path for signal or current that follows one signal line; also known as an "in-line" or "series" path.

signal path: The path used for signal processing inside an electronic device.

splitting path: In this system, the signal or current flow divides into two or more paths. Normally this system has one common input block, or section, and then the signal or current divides into two or more paths.

switching path: In this system, an electrical or mechanical switch is used to select one or more discrete paths. It can have one common input and several selected switched outputs. It also can have several more inputs with one selected by the switch.

BLOCK DIAGRAMS

Block diagrams indicate the operational flow of information or data as it is processed through the system. A typical block diagram is shown in Figure 3–1. This block diagram does not contain voltage information or information about the specific component circuitry in the system. Block diagrams are useful for following the flow of information and control signals through a variety of units. Block diagrams are very useful when the device is complex and individual integrated circuits (ICs) contain a multitude of functional subunits. One example of this is the integrated circuit illustrated in Figure 3–2, known as a "video jungle." There are several individual blocks in this IC. The schematic circuitry is too complex to understand easily; it is much easier for the service technician to use a block diagram approach to understand the functions of this IC chip. When the service technician recognizes the function of the block, then the operation of the system is better understood. Also, the service technician has some knowledge of what to expect to observe at the output of the specific block.

Schematic diagrams are useful for the display of the current flow paths in the system. Schematic diagrams normally provide voltage, current, and signal shape information to the servicer. A typical schematic diagram is shown in Figure 3–3. The schematic diagram often provides **signal path flow** as well as information about the current flow paths. A high-quality schematic diagram will include voltages at various points in the circuit, current values, and typical electrical signal waveform information. Many of these are shown in Figure 3–3.

Recognition of the type of processing that is being done in any unit is one of the early steps in the diagnosis process. This recognition normally starts as the identification of the signal processing paths. One of the major signal path systems is then used to localize the problem area. After the problem area is identified, the current flow path is analyzed. The rules for localizing the problem area or component work in a manner that is similar for both the signal path and the **current path**. The service technician must be able to recognize the type of path used in the system.

The fundamental rule for troubleshooting is to reduce the area of the problem to one section or component in the system. This is accomplished in an

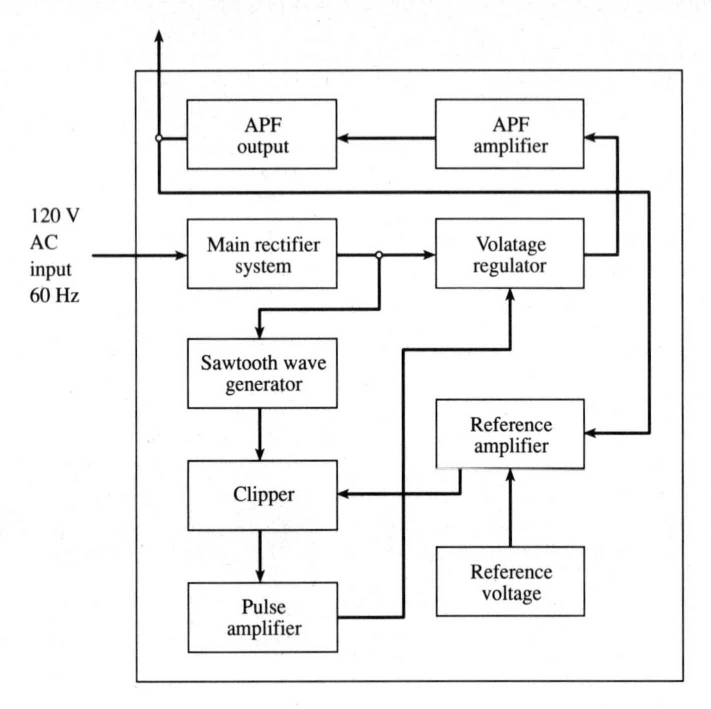

Figure 3–1 Typical block diagram used to indicate function in an electronic device. The complexity of the internal circuit requires the use of the block diagram.

orderly manner, following some very basic procedures. The basic concept used by the expert technician is to first eliminate those sections or blocks that seem to be functioning properly. After all, there is little sense in attempting to troubleshoot a section that is operating in an expected, or normal, manner. It makes a lot more sense to spend time on those sections of the unit that do not appear to be functioning properly.

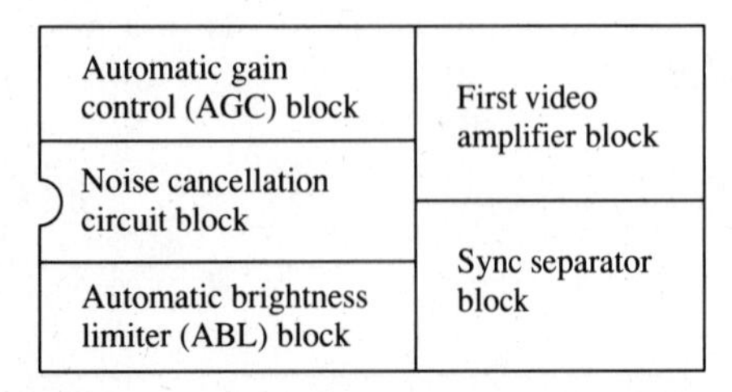

Figure 3–2 This very complex integrated circuit is used in TV receivers and is known as a "video jungle".

Following this procedure, the technician will identify the type of signal flow system used in the specific unit. The next step after this is application of the specific rule for that type of system to reduce the area of the problem. These rules are actually a series of steps to follow to reduce the suspected area to a single block or section of the entire unit. The importance of these steps cannot be overstated; they are necessary to develop a systematic method of approaching the service problem. This approach will seem slow at first. But, ultimately, with practice and experience, it will produce both a rapid and the correct analysis of the system.

One of the best methods of learning the procedure is to start by developing a plan on a sheet of paper. Use this plan and check off each step as it is completed. Make some notes adjacent to each step of the plan indicating the type of test to make, the type of

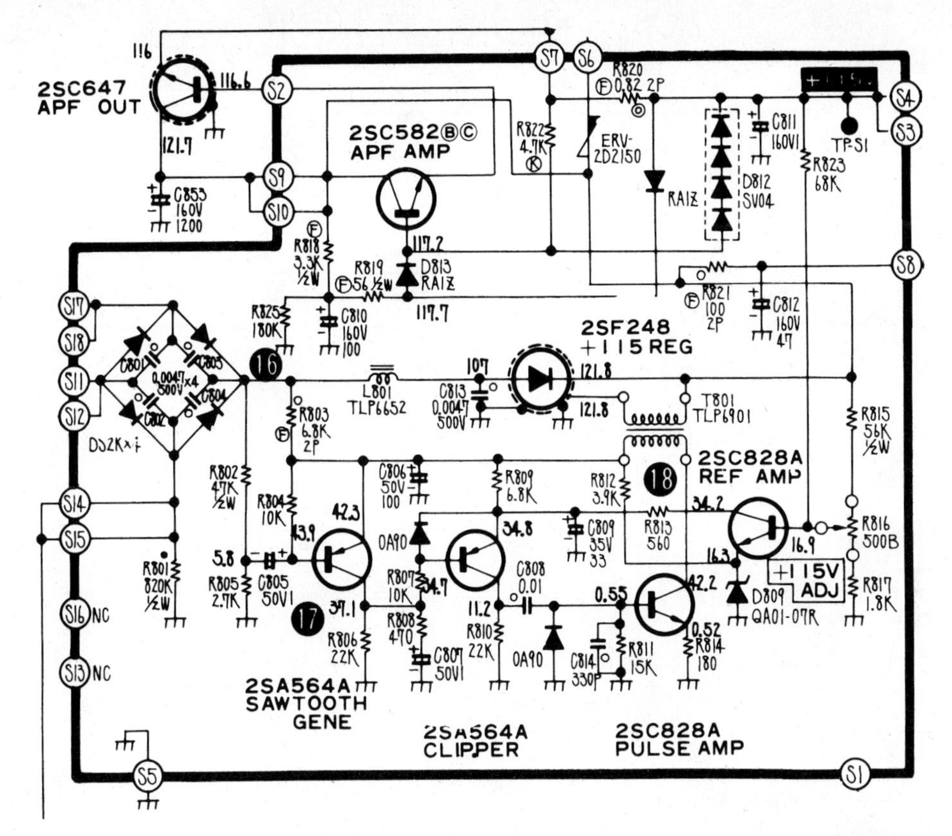

Figure 3–3 Schematic diagram for the circuit shown in Figure 3–1. Its complexity makes circuit analysis easier when the block diagram is used.

testing equipment to use, and what to expect to find at that point if the system is functioning properly. A sample of this worksheet is shown in Figure 3–4. Start to complete the planning sheet with basic information about the unit. Include a description of the unit, the initial problem, and the type of signal flow system used. You need this information if you are to start developing the skills required for successful product servicing. The other information needed for this planning sheet often requires thought and time to complete. Initially, this will be time-consuming, but once you have developed this type of thinking you will need less time to complete. Often it will be possible to develop the skills necessary to accomplish this without the use of the paper form. After many planning sheets have been filled out, the thought process will be learned. Then overall planning time will be reduced. Keep in mind that most of us required some practice before we learned to walk well. This is also true for those learning the field of electronic servicing.

The column headings for this sheet should be easy to understand. Start by identifying the equipment being analyzed and the problem. Continue with the type of signal flow path system used in the unit. Locate an initial test point. Identify the type of test equipment to be used for the test and what to observe if the system is functioning correctly. If there is a malfunction, then an improper value at the test point will aid in locating the problem area. The right-hand column should not be ignored. The feedback column, as used here, is the place to indicate whether the analysis efforts were in the best possible direction. It also may be used to offer a better step

PLANNING SHEET

Equipment: ______________________________ Date: ____________________

Model: ____________________________ Serial Number: ____________________

Statement of the Problem: ____

____

____

Signal Flow System: ____

TEST POINT	TEST EQUIPMENT	EXPECTED RESULT	FEEDBACK

Figure 3–4 Planning sheet for use in the development of a logical approach to service, diagnosis, and repair.

to accomplish the analysis when this type of system is analyzed in the future. Most of us can learn by analyzing our mistakes. Your ability to recognize that you may have made a mistake is critical to your ability to learn from the experience. This type of learning will aid in the development of a system for efficient and correct analysis of the malfunctioning system.

The specific path systems used for analysis are identified by use of their respective names. The path systems presented in this chapter are identified as **linear**, **splitting**, **joining**, **switching**, **feedback**, or some combination. Recognizing them is important to the service technician. The rules to follow identifying the specific type of path will make diagnosis much easier. All analysis in this manner requires a valid test signal to the input of the system. This should be included on the planning sheet as one of the initial steps.

LINEAR SIGNAL PATH SYSTEMS

One of the path systems is the linear, or in-line, type. This path system is the same as the series circuit. There is only one path for the signal to follow. A block diagram for this type of system is shown in Figure 3–5(a). The input to the system is block A and the output is block F. A system like this may consist of a few blocks or it may contain many blocks. A standard method of analysis can be used regardless of the number of individual blocks.

One method of accomplishing the steps described in each of these processes is to use a system of brackets. The experienced service technician also does this. The difference between an experienced technician and an inexperienced student is that the student should perform these steps using paper and pencil. The experienced technician often performs the same steps using only mental processes. The set of brackets initially will be around the entire system, as shown in the figure. Each test will indicate how the brackets should be repositioned, and they will always be moved closer to each other. In addition, each test will indicate which one of the two brackets should be moved. The ideal result is to move the brackets so they finally are around only one block in the system. This is done by eliminating properly functioning blocks at each step of the test procedure.

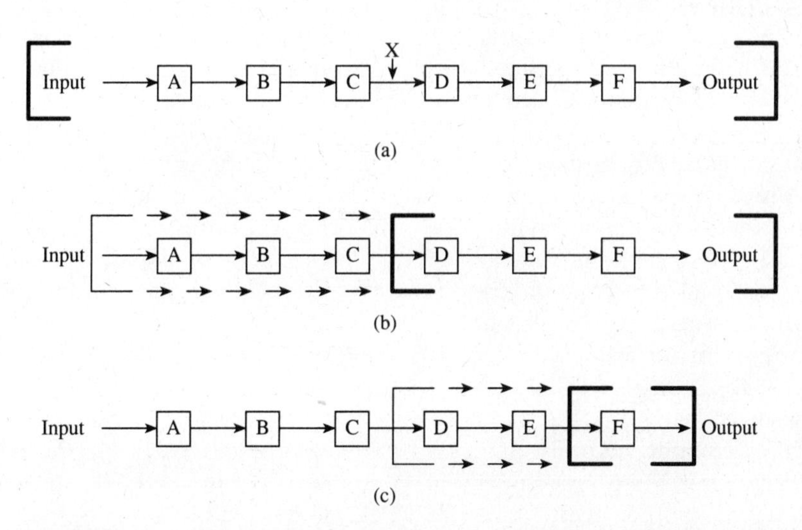

Figure 3–5 **(a) Block diagram for a linear, or in-line, signal flow system showing brackets placed at the input and the output of the system.**
(b) A proper signal at the test point moves the left-hand bracket to the test point.
(c) The final set of brackets is located around the suspected block in the unit.

In the linear system the most efficient method of approach is known as the "half-split" process. A general statement for this rule is to make a test at, or near, the center of the area under suspicion.

Linear System Path Analysis

1. The initial set of brackets is placed around the entire system. This means that one bracket is placed at the system input and the other bracket is placed at the system output. The first test determines whether the input to the system is correct. This will be at the input to block A on the diagram. If there is a lack of signal at this point, the problem area has not been located. After all, if no signal is present at the input, it will be impossible for the system to create an output signal. Return to the planning sheet for a second plan to analyze the troubleshooting process. If there is an adequate signal at the input, then proceed to the next step.
2. Next locate a point at or near the center of the system by making the second test. It is not necessary to find the exact center of the system; any point that is close to the middle is acceptable for this test. The advantage of this approach is that it will eliminate one half of the entire system with a single test. This point is located between blocks C and D.
3. There will be one of two choices to make after this test. If the proper signal is present at this point, then the half of the unit between the original bracket and the test point is functioning properly. There is little need to perform any additional testing in this area. The bracket at the input is now moved to the test point. This is shown in Figure 3–5(b). A single test has reduced the area of suspicion by half.
4. The next test will again split the remaining system blocks in half. Since there are only three blocks left to analyze, another choice must be made. An experienced service technician would make this test between blocks E and F. If the test produces valid results, then the circuit up to this point is working properly. If the signal is missing or distorted, then the problem is before this point.

   Often, the final block is the one that will fail. The reason for this is that output blocks are usually the hardest-working blocks in any system. These blocks require the largest quantities of current and often generate the most heat. They will have a tendency to fail before other blocks. This is another bit of knowledge that experience provides. This move is shown in Figure 3–5(c).
5. Once the block or section of the system is located, the process is continued using current flow analysis. The process for current flow analysis is explained in detail later in this chapter.

The analysis of this system was accomplished using three tests. Often the inexperienced person wishing to locate the area of the problem will use a less efficient method. This method starts at either the input or the output of the system. A test is made between each of the blocks to locate the specific area of the problem. If the problem area in this system were the output block F, then six tests would have to be made to determine which one was malfunctioning.

You could locate the problem block after the first test if you were to make the initial test at the output between blocks E and F. This would be a very lucky analysis. It would be more efficient if that was the problem block. You should not attempt to service in this manner since it does not reflect good working habits. This method depends upon luck rather than skill development. After all, if block F were not the problem block, then this approach would not have been successful.

Figure 3–6(a) displays the same block diagram as the one that was used for the previous example. Let us examine the approach used if the test between blocks C and D indicates a lack of proper signal.

1. The right-hand bracket will now be moved from its original position to the point of the test. This is shown in Figure 3–6(b).
2. The second test will be made at a point halfway between the input and the original test point. Whether it should be between blocks A and B or between blocks B and C is arbitrary. For the purpose of this discussion, make it between blocks B and C.
3. If the test produces a proper output signal at this

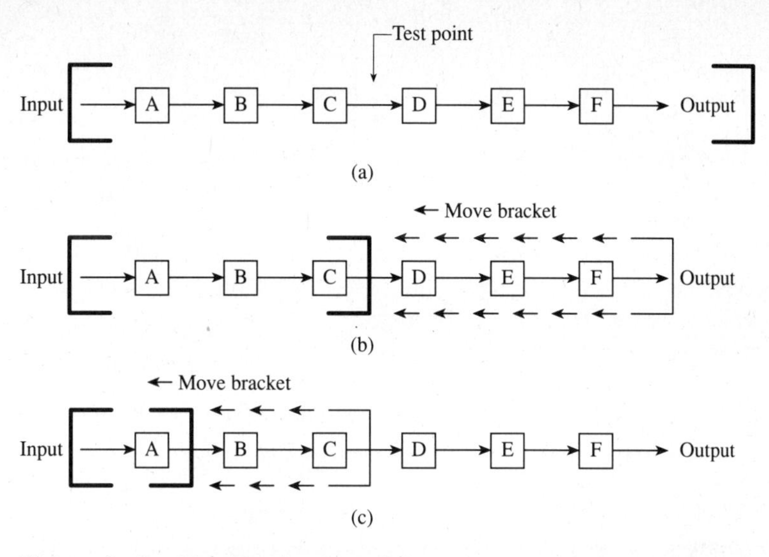

Figure 3–6 **(a) Identification of the center point of the system in order to make the first test. (b) Movement of the right-hand bracket due to a lack of signal at the second test point. (c) The final set of brackets is located around the suspected block in the unit.**

point, then the problem block is either block A or block B. The right-hand bracket is moved from between blocks C and D to between blocks B and C. A test at the output of block A will identify which of the two is the specific problem block.

4. If the test indicates that the proper signal is not present, then the right-hand bracket is moved to this point. The area of the problem is between input block A and the test point between blocks B and C. An additional test must now be made to locate the specific problem area or block.
5. The right-hand bracket is positioned between blocks A and B using the half-split method. This test will aid in the determination of which of the two blocks are not functioning properly.

SPLITTING SIGNAL PATH SYSTEMS

The splitting signal path system is also known as a diverging, or separating, path system. These path systems are often used in television receiving equipment. One application is the separation of audio and video signals. Another application is the separation of monaural and stereo signals in a frequency modulation (FM) receiver. A simple block diagram for the splitting signal path system is shown in Figure 3–7(a). Block A contains all of the signals present in the system. Its output is designed to separate these signals into blocks B and C. The circuitry in these blocks will either accept or reject the specific signal.

Splitting Signal Path Analysis

1. First determine whether the input to the system at point A is functional. Brackets are placed to the left of the input and around both output blocks. This is shown in Figure 3–7(a).
2. Check one of the outputs to this system. If it is normal, then the problem is in the second output stage because the splitting signal must be functioning properly to create either of the two output signals. If it is malfunctioning, then both output signals will be incorrect or missing. The bracket will then be moved from the left side of the input at block A to the input to the remaining block. If block B is tested and found to be normal, then block C contains the problem.

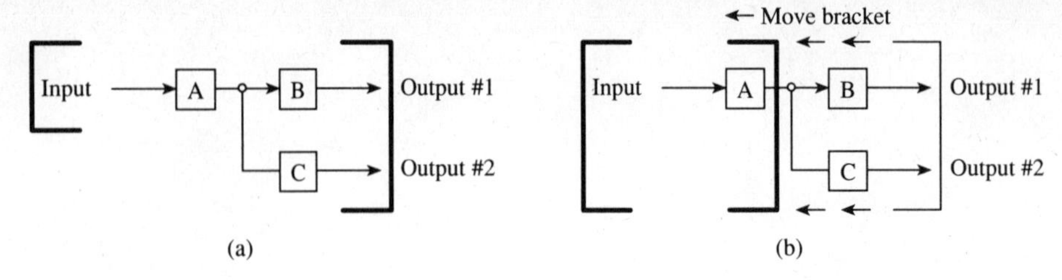

Figure 3–7 **(a) The splitting signal path system has one input block and multiple output blocks. (b) One test will determine which bracket is moved and its placement for the next test.**

3. If neither of the output block signals is present or proper, then the brackets on the right are moved as shown in Figure 3–7(b). This will then identify the problem block as block A in the system. The reasoning behind this statement is that the only block that is common to both of the output blocks is block A. Since both output blocks are working properly, the problem area has to be the one block providing a signal to both of the other blocks. In this example it is block A.

JOINING SIGNAL PATH SYSTEMS

The joining, or converging, signal path system is the reverse of the splitting type of signal path system. The block diagram for the joining signal path system is shown in Figure 3–8. There are two input blocks in this type of system. The output of each input block is then directed to a common block, where the two signals are processed. The result of this processing is created at the output of the common block. Blocks A and B in this illustration are both considered input blocks. The output of this system is obtained from the output of block C.

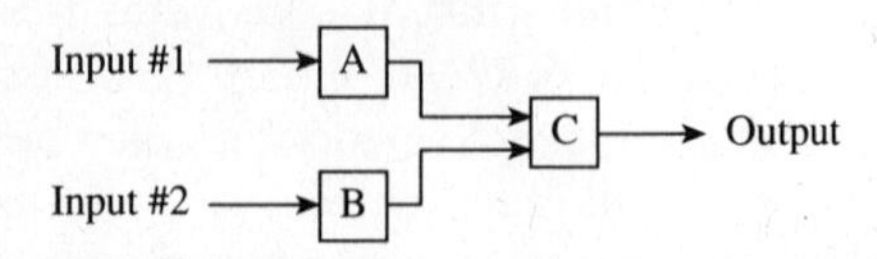

Figure 3–8 **The block diagram for a joining system has two or more input blocks and only one output block.**

There are two types of joining signal path systems. One of these is called the *alternative* system and the other is the *summative* system. The alternative system is similar to OR logic (explained below) blocks found in digital computer systems. With this system, either input A or input B will produce the desired output.

The summative system must have both types of signals present before the correct output is obtained. Also used in digital computing systems, this system is called AND logic. Another application for this system is used in most types of radio frequency receiving systems. It is the core of the heterodyne converter, or mixer, stage of these receivers.

The circuit shown in Figure 3–9 is used as a summative type of joining system. It must have signals at both input A and input B to create a valid output signal. In this example a signal from a broadcast station must be present at the input of the RF amplifier block. A second signal must be created by the local oscillator block. Both signals are delivered, or injected, into the mixer block. The electronics in this block will mathematically process the two signals. The result of this is actually four signals. One signal is the original RF signal. A second one is the original oscillator's signal frequency. A third is the sum of the frequencies of the two signals, and the fourth is the mathematical difference between the two signal frequencies. A circuit located between the output of the mixer block and the next stage is used to select one of these as the desired signal frequency for further processing. This type of circuit is used in most types of devices that are capable of receiving and

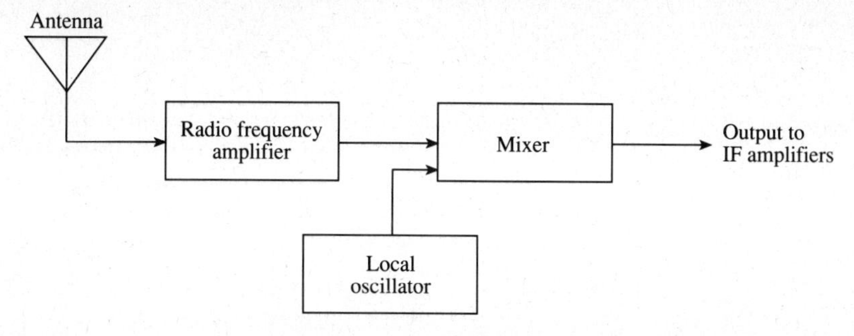

Figure 3–9 The summative type of joining system requires both inputs to be "on" to achieve the proper output signal.

processing radio frequency signals. These include AM and FM radios, shortwave receivers, satellite receivers, TV receivers, pocket pagers, and cellular telephones.

Alternative types of signal processing blocks function in a different manner. These blocks only require a single signal at one of the input blocks. A block of this type could be considered an electronic switching block. If a signal is present at block A, then it is processed through block C and becomes the output signal. If a signal is present at block B, it is then processed through block C and forms the output signal for this system. In the world of logic, this system is called an OR system, since either input A OR input B will create the desired output. This type of joining signal path system is illustrated by Figure 3–10.

One requirement when servicing the joining block system is to be able to identify which type of processing system is being used. This will require knowledge of the circuit being analyzed, its theory of operation, and how it functions in the system. Diagnosis may be performed once this is decided.

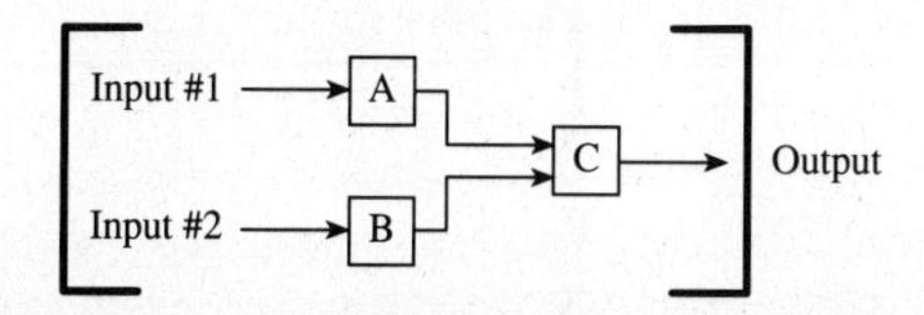

Figure 3–10 In this alternative type of joining system, only one block has to be "on" to provide the correct output signal.

Joining Signal Path Analysis

When checking a system that uses summative signal processing:

1. Place brackets around the complete system.
2. Check the output from each stage. If one of these is abnormal, then move the right-hand bracket to this point. This will identify the problem area as that specific stage.
3. If the outputs of both stages A and B are acceptable, then move the left-hand bracket to the left of the block in which the signals meet. This is the area of the problem.

When checking a system that uses alternative signal processing:

1. Place the brackets around the entire system.
2. Make the first test at the output of block A. If it is acceptable and the output is abnormal, then the problem is in the joining stage. Move the left-hand bracket to the input side of this stage because the problem is localized.
3. If the output of block A is abnormal while the output of block B is normal, then move the right-hand bracket to the output side of block A. The block containing the problem is now located.

SWITCHING SIGNAL PATH SYSTEMS

Another type of signal path circuit is the switching path system. This system may be linear, joining, or splitting in design. There is one basic rule for anal-

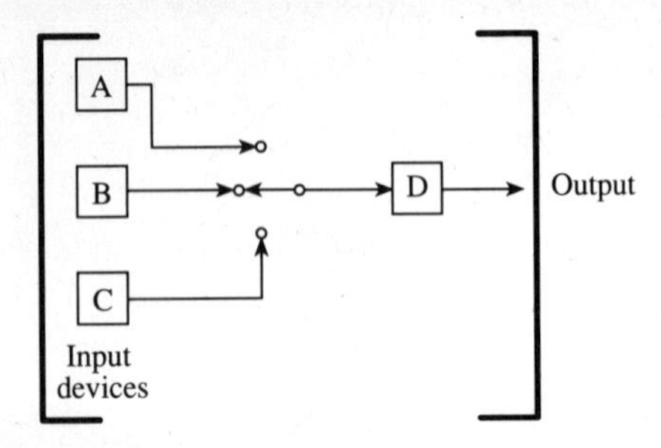

Figure 3–11 The switching type of signal processing system uses an electrical or a mechanical switch to select one input.

ysis of the switching type of system—use the switch as an initial test device. A system similar to that shown in Figure 3–11 is typical for switching signal path systems. This system has three inputs. Any of these can be selected by rotating the selector switch. This is true for mechanical switches as well as for electronic switching circuits. Each of the inputs is individually selected by the switching circuit and then processed into a common block shown as block D in this diagram.

Switching Signal Path Analysis

1. Place brackets around the entire system as an initial step.
2. Move the switch to a second position. If the problem persists, then move the left-hand bracket from its original position to the area of the switch. This will localize the problem to either the switch or block D in this diagram.
3. If the problem is no longer present, then move the right-hand bracket to the left of the block in which the trouble appeared. Switching to another block will provide the test necessary to locate the area of the problem.
4. It is not necessary to check all the switch positions. If one position is functional, then the probability of the other positions being functional is very likely. The steps described above will provide sufficient information to localize the problem.

FEEDBACK SIGNAL PATH SYSTEMS

A feedback type of circuit is normally a control circuit. Feedback circuits are used to control machine motion. They are also used to control automatically the level, or volume, of audio and radio frequency signals. The block diagram shown in Figure 3–12 is a typical feedback system. One example of this system is the automatic gain control circuit used in most consumer receiving systems. Automatic gain control, or AGC, will electronically adjust the amount of gain in the intermediate frequency amplifier (IF amplifier) sections of the receiver. The design of the system determines a specific amplitude of signal voltage at the output of the IF amplifier block and the input to the demodulator block. This is required to fully demodulate the IF signal and to

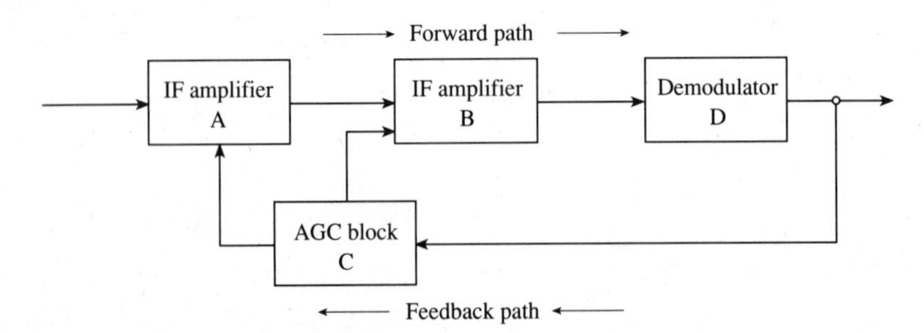

Figure 3–12 Block diagram for a feedback system used in communications receivers. A portion of the output is returned to the input to provide a correction voltage.

provide sufficient signal to the audio amplifier section of the device.

Electronic circuitry in the feedback block changes its level of bias voltage when the output of the IF amplifier is below the design standard. This change in bias voltage will provide an increase in gain in the IF amplifier stage of the receiver. The result of this is an increase in the signal level at the input to the demodulator block. This action is continuous and is constantly repeating and rechecking for an adequate level of signal. If the amount of signal is too high, then the feedback block's electronics will provide a bias voltage that will create a reduction in the gain in the IF amplifier block.

Feedback is a reverse signal processing type of system. It samples the output of the block and sends a portion of the signal back to the input of the block in the form of a control signal voltage.

Feedback Signal System Analysis

1. Change or modify the feedback path and observe the results of this change. The initial set of brackets will be placed around the entire system, as shown in Figure 3–13.
2. If the output of the system is changed when the feedback path is modified, the system is functioning properly. This will eliminate both the forward path and the feedback path.
3. When the feedback path is changed and there is little or no effect on the output, the problem is located in the forward path. The bracket on the left side of the system should be moved to the left of the feedback block.

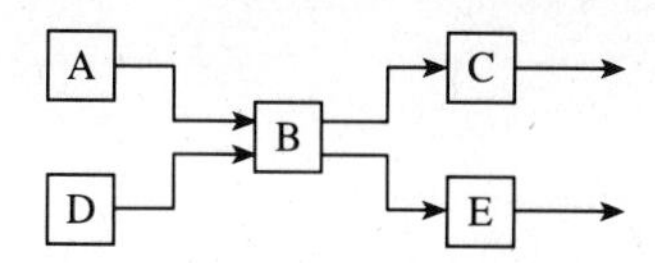

Figure 3–14 **Multiple-path system uses two or more of the individual systems. This one has both a joining and a separating system.**

Keep in mind that the forward path is often an amplifier system during the process of testing the feedback system. The amount of amplification is determined by the amount of feedback provided by the system. If the feedback path is temporarily isolated or removed from the system, then the remaining system can be handled as if it were a linear system. The rules for the linear system would then be applied by localizing the problem to one block or section of the unit.

MULTIPLE-PATH SIGNAL SYSTEMS

Some electrical and electronic systems use more than one type of signal path system. One such system is shown in Figure 3–14. This system has a joining section where blocks A and D join at block B. It also has a separating section in which the signal from block B is separated into blocks C and E. Systems such as these require more than one set of tests. The system must initially be reduced to either a joining one or a splitting one. Diagnosis of the system in this manner will identify the problem area. These complex blocks must first be divided into either a

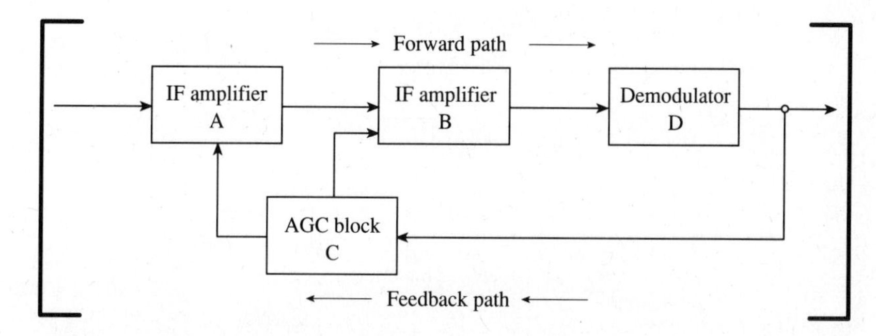

Figure 3–13 **Brackets are placed around the entire system as an initial step in analysis.**

joining or a separating system for analysis. Follow the rules for either the joining and splitting system as an initial step. Eliminate one of these systems or sections of the unit and then use the rules for the other system to determine which specific block is at fault.

REVIEW

There are two types of paths to consider when approaching a diagnosis for repair problem. One of these is the signal processing path and the other is the current flow path. Recognizing and applying the rules for both are important in the field of repair and service. In many applications the same rules apply to both circuits.

A competent service technician will know which of the rules to apply and how to apply them for the most efficient method of diagnosis and repair. The basic path types include linear, or in-line; splitting, or separating; joining, or connecting; switching; feedback; and multiple, or combination, path. Each of these has its own set of rules. They should be learned and applied as the need exists. Using them will decrease your downtime and make you much more efficient.

REVIEW QUESTIONS

1. Name the signal path system shown in Figure 3–15.

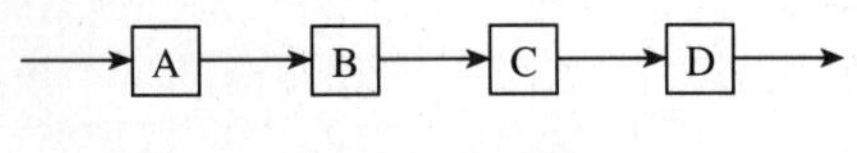

Figure 3–15

2. What is the significant identifying component in the system shown in Figure 3–15?
3. State the rules for troubleshooting as they apply to the signal path system in Figure 3–15.
4. Name the signal path system shown in Figure 3–16.

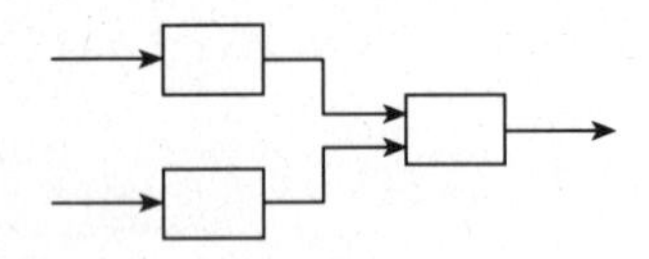

Figure 3–16

5. What is the significant identifying component in the system shown in Figure 3–16?
6. State the rules for troubleshooting as they apply to the signal path system in Figure 3–16.
7. Name the signal path system shown in Figure 3–17.
8. What is the significant identifying component in the system shown in Figure 3–17.

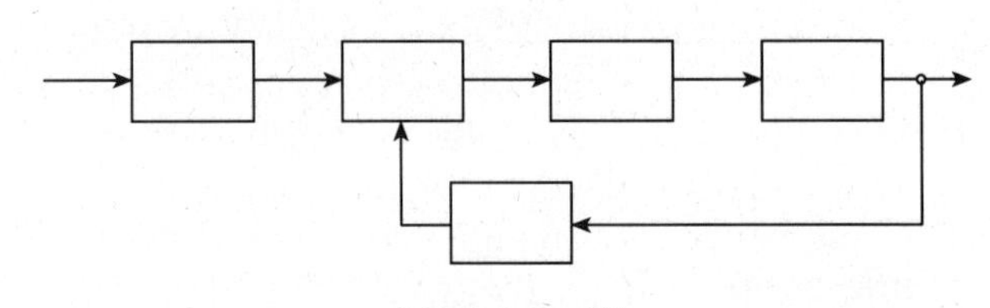

Figure 3–17

9. State the rules for troubleshooting as they apply to the signal path system in Figure 3–17.
10. Name the signal path system shown in Figure 3–18.

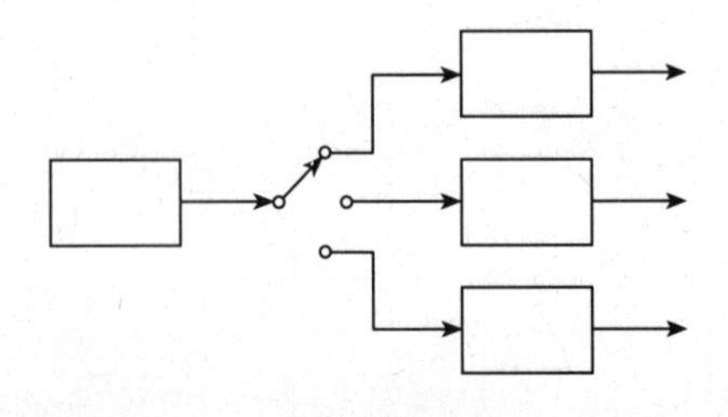

Figure 3–18

11. What is the significant identifying component in the system shown in Figure 3–18?
12. State the rules for troubleshooting as they apply to the signal path system in Figure 3–18.
13. Name the signal path system shown in Figure 3–19.
14. What is the significant identifying component in the system shown in Figure 3–19?

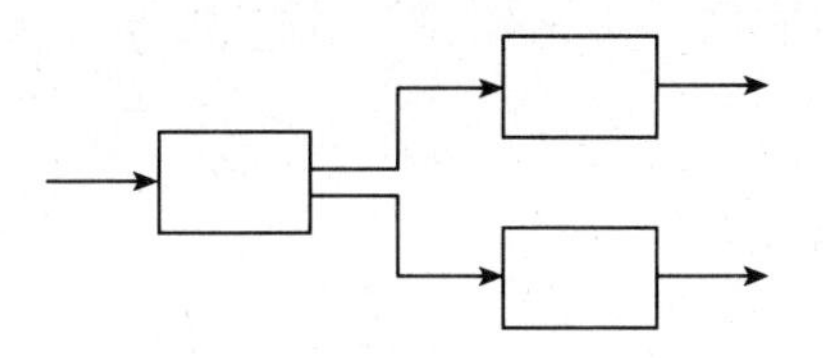

Figure 3–19

15. State the rules for troubleshooting as they apply to the signal path system in Figure 3–19.
16. Name the signal path system shown in Figure 3–20.

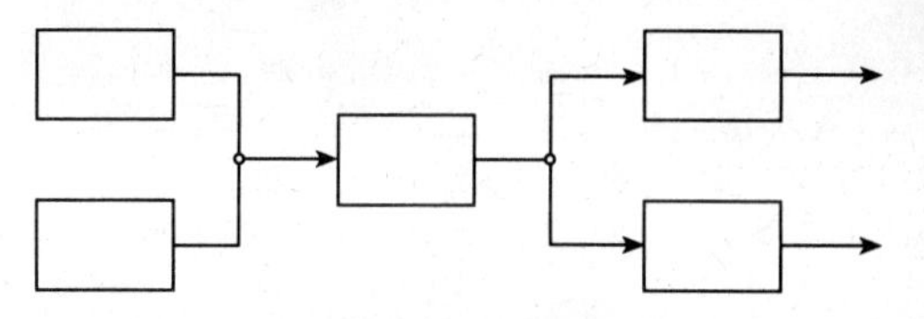

Figure 3–20

17. What is the significant identifying component in the system shown in Figure 3–20?
18. State the rules for troubleshooting as they apply to the signal path system in Figure 3–20.
19. How do the rules for resolving Problem 12 differ from those required to resolve Problem 7?
20. The identification of signal and current processing paths is important for the service person. Explain why this is true.

CHAPTER 4

Component Theory

INTRODUCTION

Individuals wishing to learn about electronics and the repair of electronic devices need to understand the theory of operation of each basic device. These fundamental units include resistors, capacitors, inductors, transformers, and some of the more basic solid-state devices such as diodes and transistors. Each device has some effect on an electronic circuit. If this were not true, then the manufacturer would not install them in circuits. The addition and installation of unnecessary components does not improve circuit operation; all that is accomplished is increasing the product's manufacturing cost.

Each of these components plays an important part in the operation of the unit. The expert technician recognizes the function of each component and the effect on the circuit or system if the part should fail. Some of you who read this book may feel that you have the required knowledge about components. I encourage you to review this chapter to refresh your memory and help find a more efficient repair method.

OBJECTIVES

Upon completion of this chapter, the reader/student should:

1. be able to recognize basic electronic components;
2. understand the operation of these basic components;

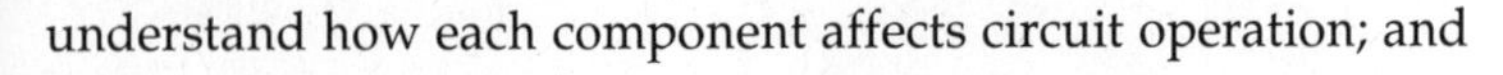

3. understand how each component affects circuit operation; and
4. be able to use this knowledge for efficient diagnosis and repair of electronic components.

KEY WORDS AND PHRASES 

active value components: These are components whose electrical characteristics change during operating conditions.

bias point: This term is used to describe the point of operation of an electronic device.

bypass (or decoupling): This describes one function of the capacitor. In this function any signal or voltage variations are minimized in the circuit.

capacitor: An electrical device composed of two sets of metalized plates and a separating, or dielectric, material. The purpose of the capacitor is to maintain circuit voltage levels.

coupling: The process of transferring signal information from one stage to another in an electronic device.

dielectric: This describes the insulating material used to separate the plates of a capacitor.

diode: This is a two-element electronic device. One element is considered positive and the other negative. When a voltage is applied to the elements of the diode, it will act as either a very low resistance or a very high resistance. The specific amount of resistance depends upon the polarity of the applied voltage.

inductor: This is an electronic device made up of a coil of wire. Its function in an electrical circuit is to maintain circuit current levels.

static value components: These are components whose electrical characteristics remain constant under varying operating conditions.

transformer: An inductive device containing two or more insulated windings, or coils, of wire. Energy is transferred from one coil to the other. In this manner voltage or current levels can be modified, depending upon the characteristics of the windings of wire.

transistor: This is a three-element electronic device. It is used as an active device in electronic circuits.

vacuum tube: A type of active electronic device that functions only when its internal elements are in a vacuum.

STATIC AND ACTIVE DEVICES

In the world of electronics, many people recognize three basic fixed, or **static, value components**. The term "fixed value" is used to indicate that the values of these components remain constant as they are used in circuits. Three components are classified as having static values. The resistor, the **capacitor**, and the **inductor** make up this set of components. Another set of components is classified as active. These are the **diode**, the **transistor**, and components that operate as **vacuum tubes**. Even though today's world is mostly solid state, one must be able to recognize how vacuum tubes operate. They are still used as the display device in computer terminals and television receivers as well as in industrial devices where large voltages and currents are required.

This chapter will discuss how the components function in circuits rather than how they are constructed and marked with values. I am basing this on the concept that the student/reader will have some background in the basics of electronic theory before reading this book.

STATIC DEVICES

RESISTORS

The fundamental purposes of any resistor are twofold: to limit current flow in the circuit and to provide a voltage drop across its terminals. These purposes are consistent and are not dependent upon the resistor's ohmic value or its power rating. The operation of a resistor in the circuit is shown in Figure 4–1. This circuit consists of two resistors, R_1 and R_{LOAD}. The value of R_1 is 10 kΩ and the value for R_{LOAD} is 5 kΩ. R_{LOAD} represents a working load device for the circuit. The source voltage for this circuit is 24 V; the load requires an operating voltage of 8 V. Since the system (only part of the entire system is illustrated) power source is 24 V, a voltage-dropping resistor is required. This resistor is identified as R_1 in the diagram.

The previous example identified the source voltage as 24 V and the voltage drop across R_{LOAD} as 8 V. If this is true, then the voltage drop developed across R_1 was determined to be 16 V. This can be proven by using Kirchhoff's voltage law. This law states that the voltage drops in a closed loop, or series circuit, when added will equal zero. Using the same circuit and assigning polarities to each of the three components, as seen in Figure 4–2, is proof of this law. You are able to start at any point in the circuit to prove this. In this example, start at the midpoint between the two resistors and count in a counterclockwise manner. The first component, R_1, is the negative end of this resistor. As you proceed from this negative end of the resistor to the other end of the same resistor, you do not know the value of voltage. It is identified as $-V_{R1}$. Moving to the power source, the value of voltage is +24 V. Finally the voltage at the bottom end of R_{LOAD} is −8 V. When these three values are added:

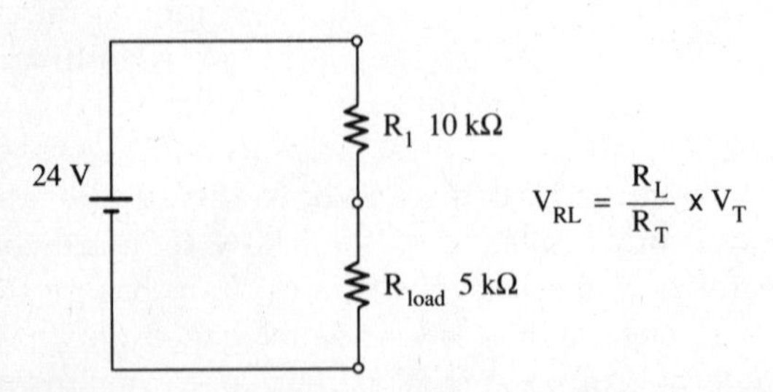

Figure 4–1 In this two-resistance circuit resistor, R_1 provides a voltage drop for the load.

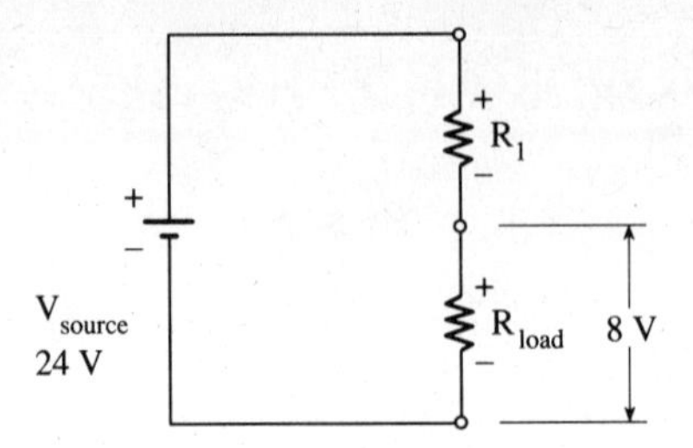

Figure 4–2 Polarities are assigned to each component and used to verify Kirchhoff's voltage law.

$$-V_{R1} + 24 - 8 = 0$$

solving for V_{R1},

$$V_{R1} = 24 - 8 = 16$$

$$\text{thus, } +24 - 8 - 16 = 0$$

This establishes the correctness of the circuit voltage values using Kirchhoff's voltage law.

The difference between the 24 V source and the required 8 V for the load is 16 V. The value of the resistor required to provide this 16 V drop must be determined. When the value of the load current is known, this is a relatively easy project. All that is necessary is to use Ohm's law to find the resistance of the voltage dropping resistor, R_1:

Given: voltage drop across R_1 = 16 V

current flow through R_{LOAD} = 0.002 A

Using Ohm's law and the previously known information, the value of the unknown resistance is:

$$I_{LOAD} = 0.002 \text{ A}$$

$$I_{R1} = I_{LOAD} = 0.002 \text{ A}$$

$$R_1 = \frac{V_{R1}}{I_{R1}}$$

$$R_1 = \frac{16}{0.002} = 8000\ \Omega$$

The purpose of the resistor R_1 in this circuit is to reduce the operating voltage from its value of 24 V to the required value of 8 V for proper circuit operation. Thus, the resistor is used to establish a voltage drop for this circuit.

The second purpose of the resistor is to control circuit current. In this same example, if R_1 was not in the circuit, as seen in Figure 4–3, the current flow would be:

$$I = \frac{V}{R} = \frac{24\ V}{5000\ \Omega} = 0.0048\ A$$

Previously it was determined that the current flow in this circuit with a total of 15 kΩ was 0.0016 A. The resistor, R_1, has the purpose of limiting current flow in this circuit to the desired value of 0.0016 amperes.

The issue of the power rating of the resistor in the example has not yet been discussed. The power rating, as mentioned earlier, has no direct relationship to the resistor's ohmic value. The power rating is equally as important as is the amount of resistance in the circuit. When the power rating is less than the resistor's ability to dissipate heat, the resistor may fail. Carbon-composition resistors can fail and often will *decrease* in their ohmic value. Film- or wire-wound resistors have replaced most of the carbon-composition types in today's electronic equipment. Their failure results in an increase in the marked value of resistance.

Using the circuit shown in Figure 4–4, the value of current is determined by use of Ohm's law:

$$I = \frac{V}{R} = \frac{500\ V}{25{,}000\ \Omega} = 0.02\ A$$

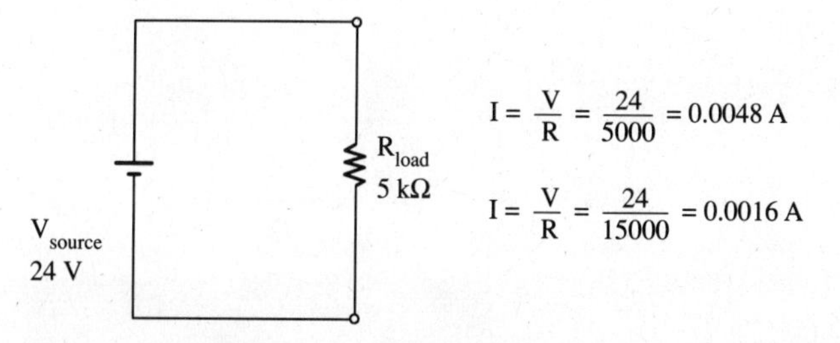

Figure 4–3 Removal of R_1 places the source voltage across the load and increases circuit current.

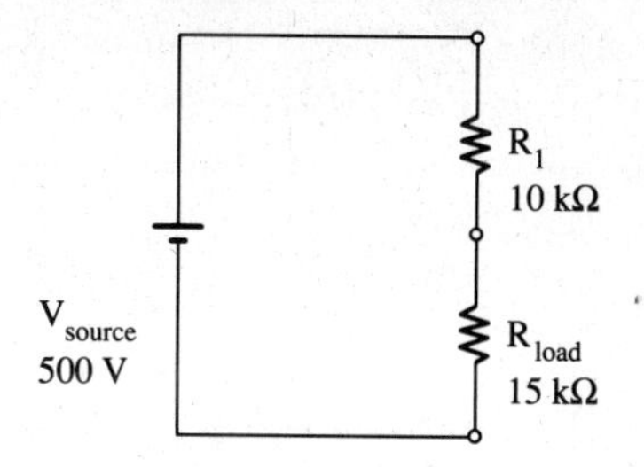

Figure 4–4 Current in this circuit is determined by using Ohm's law.

The power rating for resistor R_1 is determined by using Watt's law:

$$P = V \times I = 500\ V \times 0.02\ A = 10\ W$$

Good design practice would use a resistor having a minimum power rating of at least 12.5 W.

It is possible for the value of the load resistance to decrease during operation of the circuit. The load, represented by the resistor symbol, could be any electronic device having resistive qualities. If the load's internal resistance decreased to 2000 Ω, then the current in the circuit would increase. This additional current flow would also create additional heating and require a larger wattage rating for resistor R_1:

$$I = \frac{V}{R} = \frac{500\ V}{12{,}000\ \Omega} = 0.04\ A\ (\text{rounded})$$

The power rating for resistor R_1 would have to be:

$$P = V \times I = 500\ V \times 0.04\ A = 20\ W$$

Thus, the decrease in the resistive value of the load would require a resistor having a rating of at least 20 W instead of its original value of 10 W.

Resistor Failures Often, when a resistor exhibits signs of excessive heating in a series circuit, the most obvious component to replace is that resistor. In many situations, the excessive heating of the resistor is the result of the failure of another component in the circuit. This is often true even when a transistor is the load device in the circuit. The experienced service technician will always check for the components that *caused* the failure of the resistor. It may be good practice to replace the resistor that showed signs of overheating while servicing the circuit. Resistors are relatively inexpensive and easily replaced. Doing this may save hours of additional labor due to recall and additional servicing in the future.

CAPACITORS

The capacitor has several purposes in the circuit: to attempt to maintain the circuit voltage, to block the DC operating voltage and permit the varying signal voltage to pass, and to act as a regulated charge and discharge timer in a circuit. The first place to start when discussing capacitors is to examine their construction.

The basic construction of the capacitor is two sets of metallic surfaces, or plates. The simplest capacitor has two plates; others may have several individual plates connected in parallel, making up two sets of multiple plates. Some of these are illustrated in Figure 4–5. The two plate sets are separated from each other by a **dielectric** insulating material. This dielectric material may be air, oil, ceramic, paper, or almost any other nonconductive material. Many of today's capacitors use mylar, polyester materials, and tantalum for the dielectric material.

Capacitor action consists of charging and discharging of the plates. When the plates are charging, capacitor action attempts to keep the circuit voltage from rising. When the plates are discharging, the function is to attempt to maintain circuit voltage values. The action of attempting to maintain circuit voltage levels is similar in many ways to a sponge. When the sponge is dry and placed in a liquid, it will absorb the liquid. When the sponge is saturated, it will give up its liquid.

This type of capacitor activity is illustrated in Figure 4–6. The upper portion of this figure is a block diagram of a half-wave rectifier power supply. The lower half shows the schematic symbol for the same supply. Waveforms as observed on an oscilloscope are shown between these two diagrams. The AC source is shown as a sine wave. The rectifier cuts off one half of the wave, creating the half-wave form shown. The addition of the capacitor creates the output waveform for the capacitor and load.

One wire from the capacitor is connected to circuit common. When the voltage is rising, the capacitor's plates will charge to the value of the applied voltage. The charging process is not instantaneous, and there is a slight delay between the time the voltage is rising and when the capacitor's plates are fully charged. When the voltage is falling, the capacitor will give up its charge. Once the capacitor's plates are charged, an attempt to maintain the voltage between the plates occurs. Since one lead of the capacitor is connected to the zero volt common line in this circuit, the capacitor will attempt to maintain the level of positive voltage at the output of the rectifier section of the power source. Capacitor action

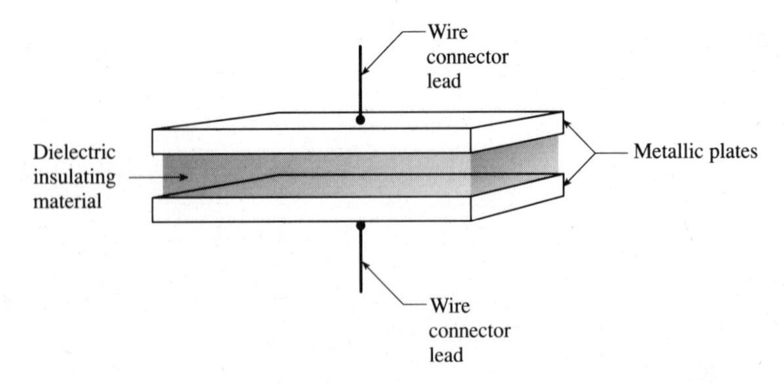

Figure 4–5 Capacitor construction includes two electrically isolated metal surfaces, or plates, and an insulating material between them.

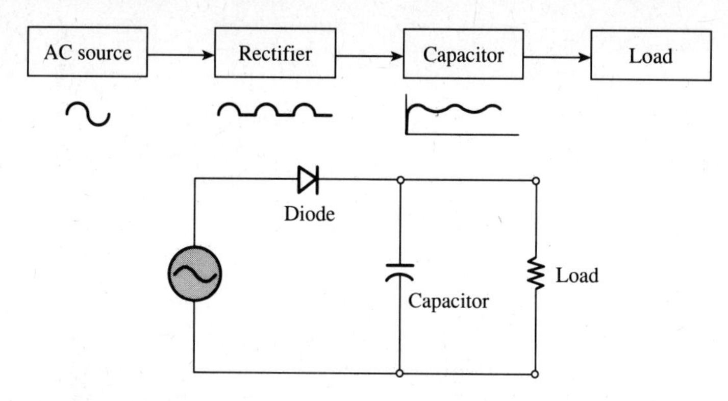

Figure 4–6 Power source used to convert AC into direct current for load operation. A block diagram, circuit waveforms, and the schematic diagram are discussed in the text.

occurs when the voltage from the rectifier changes in amplitude, or value. It is connected in parallel with the load, and one side of the load is connected to circuit common. The charge on the capacitor will actually operate the load during this portion of the AC cycle. This is one illustration of how the capacitor attempts to maintain circuit voltage.

A second application for the capacitor in a circuit is commonly called **bypass**, or **decoupling**, action. The need for bypass, or decoupling, capacitors is illustrated in Figure 4–7. This circuit consists of two parallel series paths. One, made up of resistors R_1 and R_2, is used to establish the operating point, or **bias point**, for this transistor amplifier circuit. The other path consists of resistor R_3, the collector-to-base connection of the transistor Q_1, and the emitter resistor R_4. Transistor action affects current flow in the second path. Since the transistor circuit is connected directly to the power source, the current flow in the power source is also affected. The output voltage will vary if the power source is not regulated. This variation of the power supply output voltage will affect all other individual circuits connected to it.

The method of minimizing this action is shown in Figure 4–8. Each of the capacitors in this schematic diagram is identified by the letter "C." Two of the capacitors, C_1 and C_3, are connected to the plus (+)

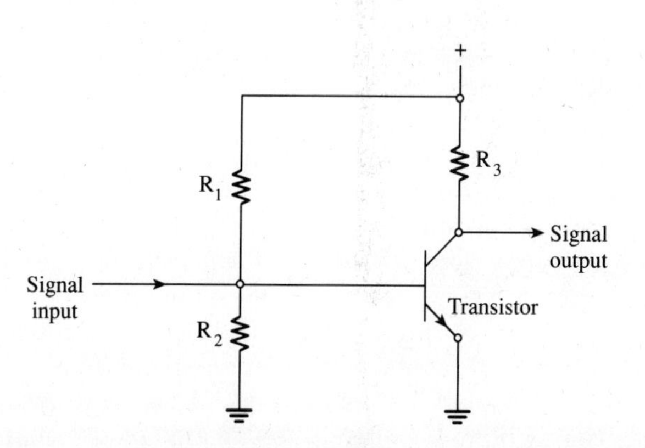

Figure 4–7 Transistor amplifier circuit. Circuit operation will cause the current to vary and reflect back into the power source.

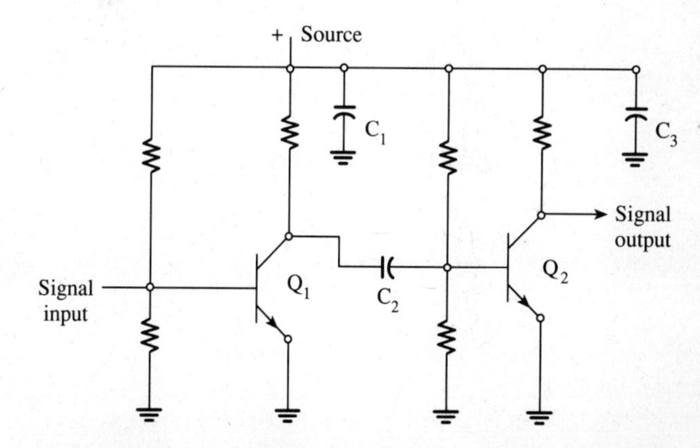

Figure 4–8 Decoupling, or bypass, capacitors C_1 and C_3 are used to maintain circuit voltage between stages of this circuit.

terminal of the power source. Transistors Q_1 and Q_2 are used as amplifiers in this circuit diagram. As the transistors operate, their individual action will affect the power supply voltage. This action is minimized by the addition of the two bypass, or decoupling, capacitors, C_1 and C_3. Any attempt to create a variation of the power supply voltage by transistor Q_1 is counteracted by the addition of C_1, the bypass capacitor. The same is true for the second amplifier circuit. Capacitor C_3 is used to oppose any voltage variations in this circuit. The function of the third capacitor, C_2, is discussed below. These two illustrations show how capacitors are used to maintain circuit voltage levels.

Figure 4–9 is similar to the previous diagrams. The major difference is the addition of resistor R_4 and the placement of capacitor C_1. Both of these components are parallel connected between circuit common and the emitter element of the transistor. Transistor action will cause a change in the amplitude of the voltage between the emitter element and circuit common when a resistor is connected between these two points. This may be an unwanted action in the circuit. The bypass, or decoupling, capacitor is used at this point to minimize any variations in the design voltage at the emitter of the transistor.

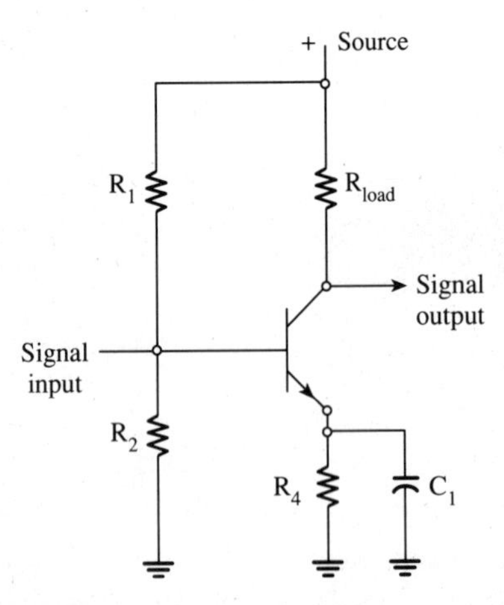

Figure 4–9 **Use of resistor R_4 and capacitor C_1 raises the operational voltage at the emitter and holds it at a constant value.**

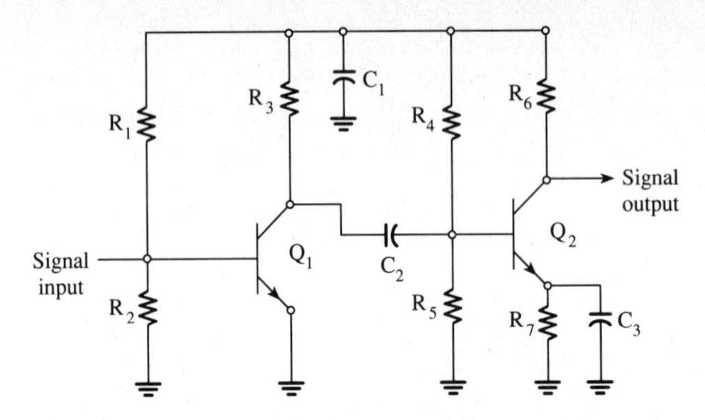

Figure 4–10 **Capacitor C_2 is used to transfer signal voltage between stages of this amplifier circuit.**

Figure 4–10 is used to illustrate how the capacitor can transfer, or couple, a signal voltage between two stages of an electronic circuit. This is accomplished by the use of C_2. This capacitor is connected from the collector of transistor Q_1 to the base of transistor Q_2. In this type of use, neither of the capacitor's plates is connected to circuit common. Changes in the amplitude of the operating voltage at the collector of transistor Q_1 are transferred to the base of transistor Q_2. The value of the operating voltage at the collector is not permitted to transfer, or couple, to the next stage of this circuit. This is one example of the **coupling** action of the capacitor.

The third use of the capacitor is as a timing device. The time required to charge and discharge the capacitor is used in oscillator and wave-shaping circuits. The capacitor's charge-discharge time will control the frequency of an oscillator circuit. This time can also be used to shape the signal wave for circuits such as integrators and differentiators.

Capacitor Failures Capacitor failures fall into three categories: open, short, or partial short circuits. Wire leads from the plates of the capacitor are used to connect it to an electronic circuit. At times, the connection at one of the plates will fail. This will be a complete open, and there will not be any capacitor action. The open is best located by use of an instrument to measure capacitance values.

The second failure may be considered a dielectric failure. The insulating material between the plates fails, due to a puncture of the material because of too high an applied voltage or to aging of the dielectric. When the plates touch each other, the internal resistance of the capacitor is close to zero ohms. This short circuit path does not permit current flow to the load. Short circuits can be located by use of an ohmmeter.

The third failure is the most difficult to locate. The value of the dielectric material can change due to excess heat or its aging. This will create a partial short circuit inside the capacitor. One term for this is "leakage." Internal leakage will reduce the ability of the capacitor to be fully charged, which, in turn, cuts down its efficiency. One of the better methods of identifying this problem is the use of a test instrument called a capacitor analyzer. Information about this type of test instrument is presented in a later chapter.

INDUCTORS

Inductors are basically coils of wire. Their size, shape, number of turns of wire, wire diameter, and type of core material influence the specific value of inductance. The purpose of a single inductor is to attempt to maintain circuit current. This is different from the purpose of the capacitor, which is to maintain circuit voltage. When current flows through the inductor, a magnetic field is established around it. The size of this field is directly related to the amount of current flowing in the conductor. When the amount of current flowing in the conductor changes, its field is in motion and it opposes any changes in its size. This is due to the electrical properties of the field. In many ways an inductor acts in a manner similar to that of a flywheel. The energy in the flywheel attempts to keep it in motion at a constant speed. Inductance tries to keep circuit current at a constant value.

One example of the use of the inductor is shown in Figure 4–11. The inductor is identified as L_1 in this diagram. It is connected in series with the diode rectifier and the load. Capacitors are parallel connected between the positive and negative lines in this circuit. The capacitors will attempt to maintain circuit voltage, and the inductor attempts to maintain circuit current. Together they smooth out any variations of the DC voltage created by D_1, the rectifier diode.

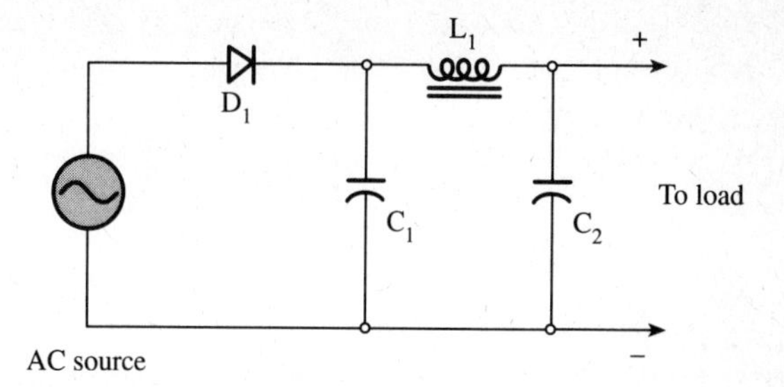

Figure 4–11 **Inductor L_1 is used to maintain circuit current in this power supply circuit.**

Transformers The **transformer** is an application of the inductor. The basic transformer consists of two or more windings of wire on a single core, or frame. The process of electromagnetic induction transfers energy from the input, or primary, of the transformer, to the output, or secondary, of the same unit. Transformers are used to isolate, increase, and decrease the secondary voltage. This is accomplished by the relationship of the number of turns of wire on the primary winding and the number of turns of wire on the secondary winding.

Transformers can also have multiple windings. Many electronic devices manufactured for world markets are designed to operate on either 120 V or 240 V AC sources. Primary windings on these devices can be switched, or wired, in either parallel or series, depending upon the specific source voltage available. This is illustrated by the two diagrams in Figure 4–12. Some units must be wired to make this change, while other units have built-in switching circuits. Many power transformers have multiple secondary windings. The specific number of windings and available voltage levels depend on the specific needs of the individual device.

Transformer Failure Failures in transformers are often due to excessive current demands. These result in overheating of the transformer windings. Often the insulation on two or more windings may melt and cause an internal short circuit. If this does not happen, the connection on the transformer be-

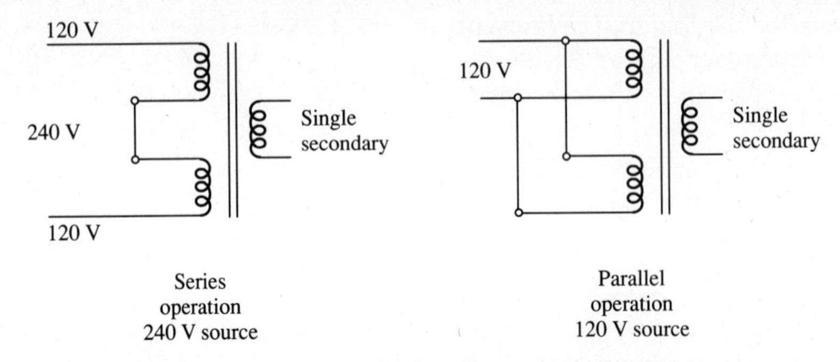

Figure 4–12 Transformer primaries can be wired in series or in parallel, depending upon the value of the operating voltage source.

tween the internal winding wire and the lead to the external circuit has been known to fail. In either case, the output of the transformer is less than its expected value.

When the failure is due to excessive current, the cause must be located. Often the source of this problem is in one of the circuits in the unit. Most likely it will be a component failure that has created a lower than normal resistance path in one circuit. Normal troubleshooting procedures are used to locate this type of problem.

When the failure is an open connection, a voltage test of the terminals of the transformer will locate it quickly. Once the problem has been identified, an ohmmeter will help to locate the specific connection. These often may be repaired by resoldering the connection. If this does not resolve the problem, locating the broken wire connection to the terminal and splicing a wire to it and to the terminal will effect the repair.

ACTIVE DEVICES

Active devices, or **active value components**, are those capable of changing their design characteristics as they operate. The more common active devices include semiconductor diodes, semiconductor transistors, and vacuum tubes. The purpose of this chapter is to review some of the basics involved; therefore the text material will not cover all active devices. You should remember that all active devices normally act as variable resistances in electronic circuits. If this knowledge is used for circuit analysis, the ability to locate and identify malfunctioning components is easier.

SEMICONDUCTOR DIODES

Semiconductor diodes are used in many different electronic circuits. Some people refer to the diode as a two-position switch in a circuit. Others consider them to be resistances. For purposes of this section of the book, let us compare diode action and resistance. Figure 4–13 shows the diode's schematic symbol. The diode is fundamentally a two-element semiconductor. One element is considered to be positive and is known as the anode. The other element is considered to be negative and it is known as the cathode.

When a voltage is applied to the elements, the diode acts as a resistor in any circuit. If the voltage at the anode of the diode is more positive than the voltage applied to its cathode, the diode's internal resistance will be very low. This permits relatively large quantities of current to flow through it. If the

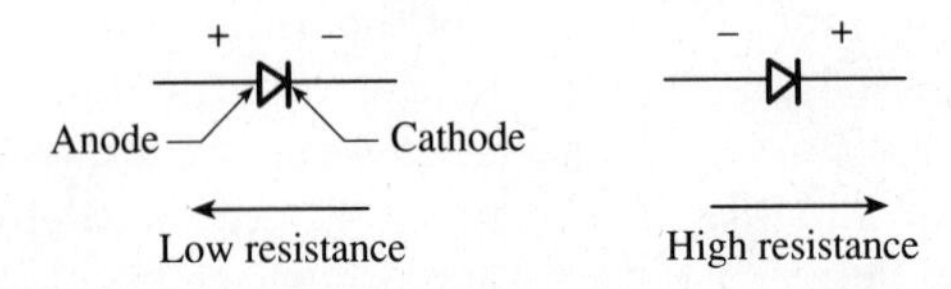

Figure 4–13 The semiconductor diode acts as a resistance in the circuit. Its value is dependent upon the polarity of the applied voltage.

diode is made of silicon semiconductor material, the voltage drop developed across its terminals in this condition is between 0.5 and 0.7 V. When the diode is constructed with germanium semiconductor material, the voltage drop is less, between 0.2 and 0.3 V. The diode is said to be forward biased under these voltage polarity conditions.

The right-hand side of the figure shows the diode with reversed applied voltages. The anode has a negative voltage applied to it, and the cathode's voltage is more positive. The internal resistance of the diode is very high under these conditions. It is said to be reverse biased. The current flow, if any, is minimal, and the voltage drop developed across its terminals is very high. The diode could be considered to act as an open circuit when it is reversed biased.

Figure 4–14 shows one very common application of the diode. This circuit is a half-wave rectifier circuit and is used to convert AC voltages into pulsating direct current. The AC source reverses its polarity during its cycle. The polarity of the voltage applied to the circuit and the diode also reverses during each cycle of the AC source.

When the AC source is applying a positive voltage to the anode of the diode, the diode is forward biased. The voltage drop across the diode's elements is less than 1.0 V, and the majority of the applied voltage develops across the load. The load will operate under this condition. When the AC source polarity is reversed and a negative voltage is applied to the anode of the diode, the diode is reverse biased. Its internal resistance is very high, and little, if any, voltage can develop across the load resistor. During this half cycle the load does not operate.

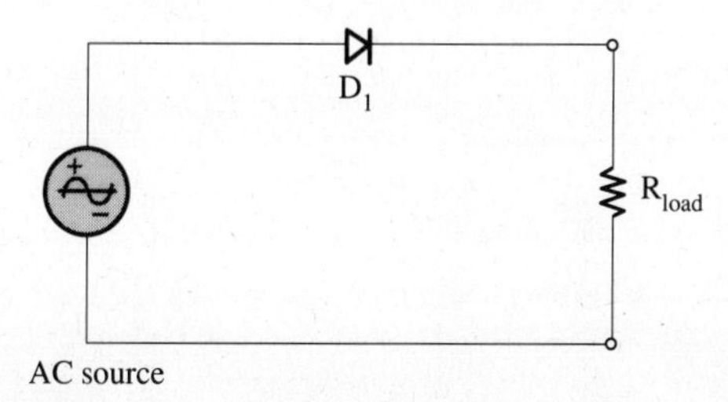

Figure 4–14 The semiconductor diode is used to convert alternating current into pulsating direct current.

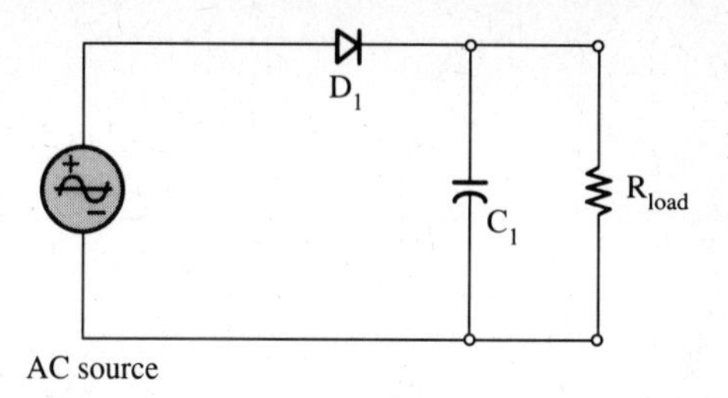

Figure 4–15 Using a capacitor in parallel with the load will minimize variations in the output voltage from the rectifier diode.

When a capacitor, such as the one shown in Figure 4–15, is placed in parallel with the load, capacitor action will occur during the period when the diode is reverse biased. The voltage discharge from the capacitor will "fill in" the missing source voltage and permit the load to operate during each half of the AC source voltage cycle.

Diode failures do occur. Often the diode will open and not permit current to flow during any portion of the applied AC source cycle. Diodes also develop short circuits. This places a very low resistance across the circuit and usually causes protective circuit breakers to trip or fuses to fail. The short-circuited diode is often caused by the failure of some component in the unit.

Diodes can be tested with an ohmmeter. Many of the digital multimeters available today are capable of testing diodes. They place a voltage across the terminals of the diode and can indicate the electrical condition of the diode.

SEMICONDUCTOR TRANSISTORS

Semiconductor transistors can also be considered resistances. Schematic symbols for junction types of transistors are shown in Figure 4–16. Junction transistors have one of two polarities: NPN or PNP. Today, the NPN type is used more in electronic circuits. The difference between these two types is their polarity. The arrow observed on the emitter element indicates the polarity. The junction transistor is a current-controlled device. Electron current flow through the transistor is indicated by this arrow. Current flow in the NPN junction transistor is

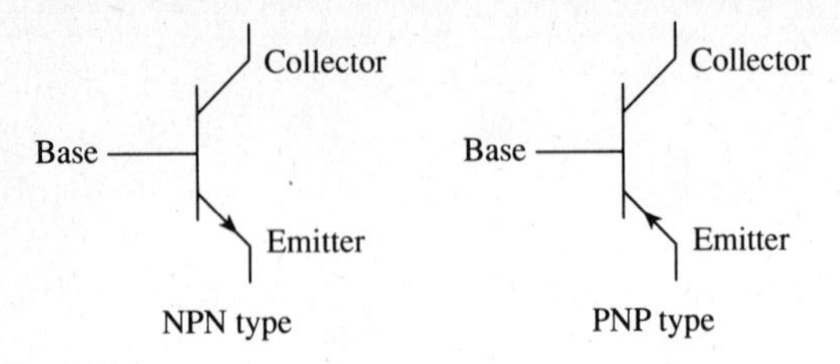

Figure 4–16 Schematic symbols for junction transistors. The arrow on the emitter lead indicates its polarity.

from the emitter to both the base and the collector. It is reversed in the PNP transistor, flowing from collector to base and emitter.

The internal resistance between the base and emitter elements is very low. This junction can be considered to be forward biased. The application of a small voltage to these elements will control the flow of electrons between its emitter and its collector elements. The total value of resistance in a resistive circuit determines the amount of current flow (if the applied voltage is held at a constant value).

Figure 4–17(a) is an example of a simple junction transistor amplifier circuit. Resistors R_1 and R_2 are used to establish a point of operation for the circuit. These are actually a series-connected voltage-dropping circuit. The voltage developed at their junction is used to set this operational point. Further study of amplifier circuits will explain the various types of amplifiers and their use.

The load resistor, R_{LOAD}, is series connected to the collector of the transistor. The emitter is connected to circuit common. Since the majority of current flow is from the emitter to the collector, this circuit will have the greatest amount of change. A small increase in the voltage applied between the base and the emitter elements will create an increase in the internal current flow between the emitter and the collector of this transistor. Since the load resistance is a fixed value, the only way this can occur is when the internal resistance of the transistor decreases.

The emitter-to-collector connections carry almost all of the current in the transistor. The junction transistor's action can be compared to that of a variable resistor in a circuit. This is shown in part (b) of Figure 4–17. A reduction in the voltage developed across the emitter-to-collector elements occurs as the internal resistance between these two elements drops. Since this is a series circuit, the voltage developed across the load will increase. This is based on Kirchhoff's voltage law.

The opposite action occurs when the voltage between the base and emitter elements is decreased. This will decrease the current flow in the transistor.

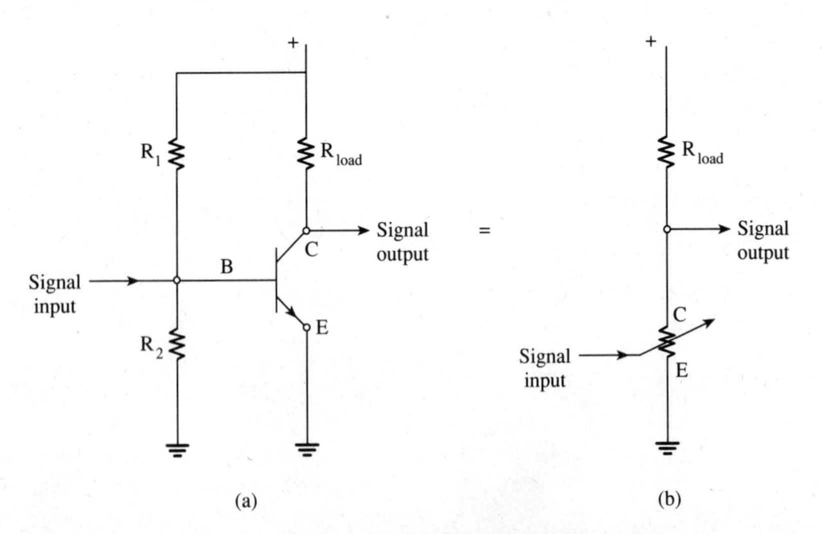

Figure 4–17 (a) The transistor amplifier circuit, and (b) how this circuit appears to act as a variable resistance.

Again, any decrease in current flow in a circuit can only happen when the resistance of that circuit is increased. Since the load resistor is a fixed-value device, the internal resistance of the transistor must increase. When the internal resistance of the transistor increases, the voltage drop that develops across its terminals also increases. Since this is still a series circuit, the voltage drop across the load will decrease.

In a sense, the junction transistor acts as a variable resistance in its emitter-to-collector circuit. This circuit is considered to be its output circuit in almost all of the configurations for transistors. The internal resistance, and thus the current flow, between emitter and collector elements is controlled by the difference in voltage applied between the base and emitter elements. In some applications the emitter is connected directly to circuit common, and its voltage remains at zero. A varying voltage applied between the base and emitter, but at the emitter, will control current flow in the emitter to collector path.

It is also possible to make the base element a zero-voltage device. This may be accomplished by connecting it directly to circuit common. Another method of doing this is to use a voltage-dividing bias network and then to add a decoupling, or bypass, capacitor between base and circuit common. When this type of circuit is used, the voltage at the emitter is varied. This will still create changes between the emitter and base elements and permit the transistor to act as a controlled variable resistance in the circuit.

Transistor Testing Transistor testing is performed with any one of several types of testing equipment. Some testers are capable of evaluating transistors while they are connected to the circuit. These will provide fairly accurate evaluations of the transistor. A voltmeter also can be used to measure the base-to-emitter voltage when the circuit is under power. Another method is to use the oscilloscope to compare input and output waveforms. Any of these can provide sufficient information for the service technician.

One test that is relatively easy to perform is the measurement of the base-to-emitter voltage. A silicon transistor's base-to-emitter voltage should be about 0.7 V. If this measurement is made, and a value of about 0.7 V is obtained, the transistor's input circuit is functional. A second test is then made from the collector to circuit common. This voltage should be much higher. If it measures the same as the source voltage, then there is an open in the circuit between the point of measurement and circuit common, and further testing is then required.

I do not support the concept of removing the transistor for testing purposes. Circuits adjacent to the transistor can be damaged while attempting to remove the suspected transistor. At times, the foil on the circuit board can be overheated, and the adhesive holding it in place will fail. When this happens, the foil has to be replaced or the board repaired before the circuit will work again. It is also possible to install the transistor incorrectly when returning it to the circuit. Expert service technicians have been known to accidently place the leads of the transistor in the wrong set of connections on the circuit board.

All of the above efforts are time consuming and do not make for efficient testing and diagnosis. I strongly suggest that the last thing the service technician should do when analyzing a circuit is to remove any component. Learn to make valid tests—ones that produce qualitative answers—before the suspected component is removed for further testing. This will lead to the development of a good set of test procedures and ultimately make you a more efficient service technician.

VACUUM TUBES

The reason for understanding vacuum tube theory was stated at the beginning of this chapter. Almost all video display terminals and television receivers use vacuum tube theory. The exception to this is found in many of the laptop and notebook portable computers currently available, which use liquid crystal displays. This section will only present theory related to basic vacuum tubes.

The vacuum diode is illustrated in Figure 4–18. While working on the electric light bulb, Thomas Edison discovered that he could create a current flow from the filament/cathode of the unit when the plate, or anode, was positive. This was the first form

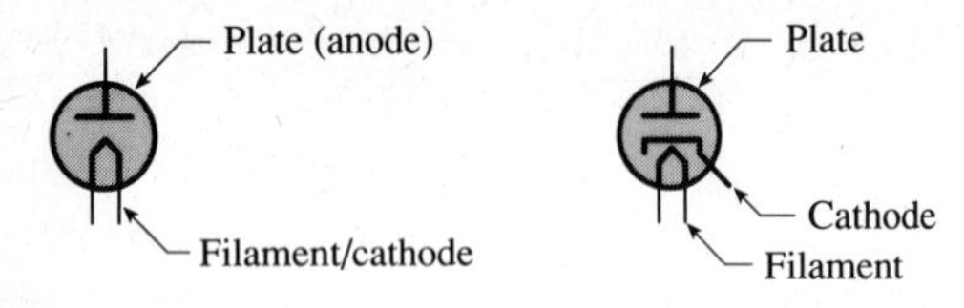

Figure 4–18 Schematic symbols for the vacuum tube diode. The cathode emits electrons, which are attracted to its plate.

of rectifier device. The filament, or heating element of the tube, was used as its cathode. Current could flow from the cathode to the plate of the tube. Today, the vacuum tube diode often has a separate cathode. This makes the tube more efficient. The vacuum tube diode is illustrated using schematic symbols.

The filament, or heater, element is used to boil electrons off of the cathode element. These electrons are attracted to the positive plate element of the tube. Current flow in the vacuum tube occurs in this manner. Figure 4–19 illustrates the vacuum tube rectifier system. Two individual voltages are required to operate the tube, one of these is the filament supply. Many vacuum tubes operate with filament voltages ranging between 2.0 and 12.0 V. This voltage can be either AC or DC, depending upon the specific application. A second voltage source is also required—the "high voltage" for the load.

The high voltage is applied to the plate and cathode of the tube. Vacuum tube operation is very similar to that of the solid-state semiconductor diode. When the voltage at the plate is positive compared to that at the cathode, the vacuum diode will conduct. Its internal resistance reduces to a relatively low value, and the majority of the applied voltage appears across the load. When the polarity of the applied voltage is reversed, the diode does not conduct. Almost all of the applied voltage develops across its elements, and the load voltage drops to close to zero volts.

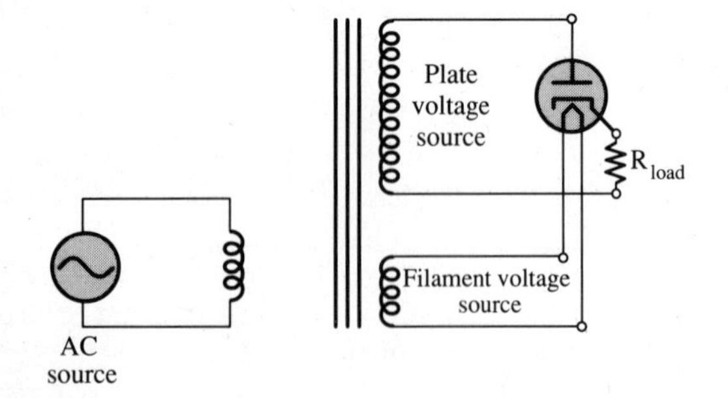

Figure 4–19 Schematic diagram for a vacuum tube half-wave rectifier circuit.

This action of the vacuum tube diode is either "on" or "off," and there is little control for circuit current. This is remedied by the addition of a third element, called a control grid. The schematic diagram for the vacuum tube triode is shown in Figure 4–20. (The filament, when separate, is not recognized as an element in the tube.) The control grid is used to vary the current flow inside the tube. It is placed inside of the tube and surrounds the cathode. A voltage placed on the control grid will cause a variation in the internal current flow of the tube. This is similar to the action of the semiconductor junction transistor.

The cathode-to-plate resistance of the vacuum tube is varied by the use of the voltage applied to the control grid. This changes the voltage developed across the series-connected load and thus controls the operation of the specific circuit.

Vacuum tubes often have more than three elements. The additional elements are used to enhance the operation of the tube in a circuit. Tubes can be tested using a dedicated vacuum tube tester. The experienced service technician can often use a meter or an oscilloscope to evaluate its operation.

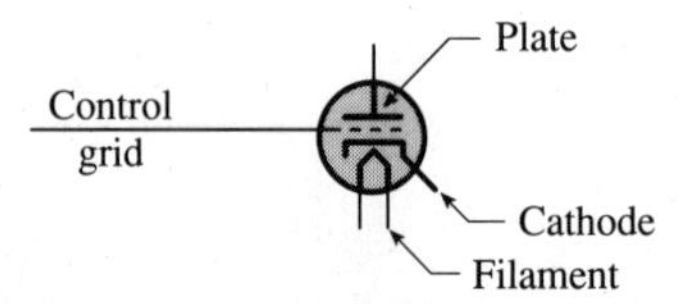

Figure 4–20 A third element, called a control grid, is used to vary the cathode-to-plate current in the vacuum tube.

REVIEW

Electronic components generally fit into one of two categories—static devices and active devices. Static devices are those whose values do not change when the component is operational in a circuit. Resistors, capacitors, and inductors are included in this group.

Active devices are those whose internal values change during their operation. The most commonly known active devices are semiconductor diodes, semiconductor transistors, and vacuum tubes.

The primary functions of resistors are to limit current flow in a circuit and to create a voltage drop necessary for proper circuit component operation. Values for resistors used for either purpose can be determined by using Ohm's law, Kirchhoff's law, or Watt's law. One must be aware of power ratings when determining the values of resistors used in any circuit. Selection of replacement components must also take this into consideration.

Resistor failures are often observable. You will notice a charring, or burning, of the body of the resistor, and in extreme cases, the body of the resistor can actually break into two pieces.

Capacitors have several uses—as bypass, coupling, and timing devices. All capacitors are constructed of two metalized surfaces and an insulating, or dielectric, material. The capacitor's plates accept a voltage charge and then attempt to maintain that value of voltage.

Capacitor failures fall into three categories: a short circuit, an open circuit, and a condition known as leakage. Opens and shorts often can be identified by use of either an oscilloscope or a multimeter. Leakage testing requires the use of a capacitor testing instrument.

Inductors attempt to maintain circuit current. A magnetic field is created by current flowing through the inductor. Once established, the properties of this magnetic field try to maintain the value of circuit current.

Failures of inductors include shorts, opens, and partial short circuits. These may be tested with an ohmmeter or with an inductor tester.

Active devices include semiconductor diodes, transistors, and vacuum tubes. The semiconductor diode operates in either one of two conditions. When it is forward biased, its internal resistance is low. Basic theory reminds us that low-resistance components develop small voltage drops and permit relatively high levels of current flow. When the diode is reverse biased, the opposite set of conditions occurs. Its internal resistance is very high, a high voltage develops across its terminals, and little, if any, current flow is permitted.

Diode failure can be a short, an open, or a leakage. These can be tested with a multimeter or with semiconductor-testing equipment.

Transistors are three-element semiconductors. They can be classified as being controlled variable resistances in a circuit. The major current path through the transistor is from its emitter to its collector elements. The amount of current flowing is controlled by the emitter-to-base voltage.

Transistor failures are similar to those of the semiconductor diode. Transistors can be tested with an ohmmeter or a semiconductor tester.

Vacuum tubes are still in use. The most common use is the cathode-ray tube used in video display terminals and television receivers. This tube operates when a heated filament gives off electrons. Some vacuum tube diodes use a cathode element that surrounds the filament element. The electrons are boiled off the surface and attracted toward another element, called a plate. The plate has a positive charge. Current flow is controlled by the voltage on the plate of the tube.

A more sophisticated vacuum tube uses one additional element to control current flow—the control grid. A small voltage charge on the control grid is used to vary the amount of current flowing between the tube's cathode and plate.

Vacuum tube failures often are limited to their filaments. The tube can be tested by use of a tube tester. Often you can look for the lighted filament, which glows as a dark orange/red color.

REVIEW QUESTIONS

1. Why is knowledge of vacuum tube operation important to the electronic service technician?
2. State the two basic purposes of resistors in electronic circuits.
3. Which basic electronic law is used to provide the value of a voltage drop?
4. Why is the power rating of a resistor important?
5. When a resistor appears to be overheated, what should the electronic servicer look for?
6. What is the most common failure mode for carbon film resistors?
7. How does a capacitor react to changes in circuit current and voltage?
8. What is the basic principle of operation of an inductor?
9. Describe the construction of a capacitor.
10. What is the purpose of the dielectric material in a capacitor?
11. Describe the function of a bypass capacitor.
12. Describe the operation of a coupling capacitor.
13. What types of failures can occur in a capacitor?
14. How does an inductor react to changes in circuit current and voltage?
15. How does a transformer differ from an inductor?
16. What is an active device?
17. How does the semiconductor diode act as a resistor?
18. How does the transistor act as a resistor?
19. What is the importance of the base to emitter voltage measurement of a transistor?
20. Give a brief description of the operation of the triode vacuum tube.

SECTION 2

The Troubleshooting Process

CHAPTER 5

Testing Equipment and Its Use

INTRODUCTION

One portion of the process of locating the area of trouble in any device requires selecting the correct type of testing equipment. The service technician is performing analysis to be able to localize the fault area in the device. Success depends on many factors. These include knowing what to test, where to make a test, what to expect to find when the test is made, and knowing which piece of equipment will provide the most information from the test results.

In the early days of electrical and electronic servicing, the test bench consisted of a meter, a vacuum tube tester, a signal generator, and possibly an oscilloscope. If the oscilloscope was present, it most likely had a thick layer of dust on it. Very few, if any, service technicians really understood the oscilloscope. It was purchased to meet the manufacturer's approved service center requirements. In those days there was little need for this device.

Today's service test bench has much more equipment on it. Most of it is necessary if the technician is to diagnose successfully and repair the sophisticated electrical and electronic equipment being produced. Those who succeed in the field of electrical or electronic servicing need to know several things, including:

1. What signals the equipment produces.
2. The function of each piece of testing equipment.
3. The limits of each piece of equipment.
4. How to use each piece of equipment.
5. When to use the specific equipment.
6. What to expect to find when the test is performed.

The purpose of this chapter is to provide information about each of the more common types of test equipment. The material will describe typical types of test equipment including meters, multimeters, and oscilloscopes and will introduce the topic of specialized test equipment. The complexity of the electronic equipment used today by business, consumers, and industry makes it imperative that those who maintain, service, and repair it know what testing equipment to use and how to use it. The days of observing the conditions on the face of the television receiver and knowing how to repair it by replacing a vacuum tube are long gone. Today's receivers are controlled by microprocessors; they provide on-screen channel selection information, offer split-screen video displays, inform us of the correct time, and maintain the quality of both the colors and the video displayed for our comfort and enjoyment. The ability to diagnose and service such a sophisticated piece of equipment requires knowledge of not only how it functions, but which type of testing equipment should be used and how it should be used.

In this chapter you will learn the basics of testing equipment. In addition, the material covers the various types of testing equipment and how and when to use it. All of this information will help the student/learner on the road to becoming the world's best electronic repairperson.

OBJECTIVES

Upon completion of this chapter, the reader/student should:

1. understand the function of basic testing equipment;
2. recognize the need to use basic testing equipment;
3. learn how to use this equipment;
4. recognize the need for safety precautions when making tests;
5. understand the limitations of specific testing equipment; and
6. have a better understanding of which piece of test equipment to use for a specific test procedure.

KEY WORDS AND PHRASES

analog meter: A test instrument using a dial and moveable needle to display the value being measured.

audio frequency spectrum: A range of frequencies that can be heard by the human ear; these range from 20 Hz to 20 kHz.

automatic polarity (autopolarity): Circuitry within the meter used to indicate the polarity of the voltage or current being measured.

automatic ranging (autoranging): Circuitry inside the meter used to automatically change the range of the value being measured.

circuit common: The point of reference for voltage measurements. Normally the negative lead of the power source is connected to this point.

deflection: The movement of the meter needle from its "at rest" position to a position indicating a flow of current through its armature winding.

digital readout meter: Use of a digital, or numeric, display as the output indicator for a test instrument.

function generator: Test instrument capable of producing a multitude of electronic signals, such as sine, square, and triangular waves.

multiplier resistor: A resistor connected in series with the basic meter. Used to extend the voltage-measuring range of the meter movement.

oscilloscope: Test equipment designed to display on a cathode ray tube both the amplitude and the time rate of an electronic voltage signal.

radio frequencies: A range of frequencies that are not heard by the human ear. These range above the 20 kHz upper audio frequency limit.

schematic diagram: The electrical or electronic "blueprint" that displays the path for current and the specific components used in the device.

shunt resistor: A resistor connected in parallel with the basic meter. Used to extend the current measuring range capability of the meter movement.

signal: A varying voltage used as a means of carrying information.

signal generator: An electronic device capable of creating test signals.

tolerance: Acceptable deviation from the marked value of a component.

V-O-M: A multipurpose test instrument designed to measure voltage, resistance, and current (volts-ohms-milliamperes).

BASIC TESTING EQUIPMENT

What do you measure when testing a circuit? The most common measurements are of voltage and resistance. Another important measurement is that of current flow in the circuit. The measurement and evaluation of electrical and electronic signals in the circuit can be accomplished by using the **oscilloscope**. The processes of signal tracing and signal injection are used for the same purposes. Which one to use and when to use it are also important to the service technician.

The influence of small, electronic integrated circuits has drastically changed the type of testing equipment currently in use. For many years the basic multimeter, as illustrated in Figure 5–1 was in use in the industry. This multimeter was known as a **V-O-M** (volt-ohm-milliamp meter) since it measured voltage, resistance, and current. The V-O-M consisted of an **analog meter** movement and several internal components. The internal components were selected by a switch, which permitted the basic meter to be used to measure a variety of values of voltage, current, and resistance.

BASIC ANALOG METER MOVEMENT

It is important to understand how this meter functions because there are many V-O-Ms using the analog meter movement still in use in the service industry. The diagram shown in Figure 5–2 is typical of the manner in which the basic meter movement is constructed. This meter consists of an armature, a permanent magnet, a needle or dial indicator, a dial, and some springs. The armature consists of a coil of wire wound on a form that is able to be temporarily magnetized. It develops electromagnetic properties when a current flows through its windings. A horseshoe-shaped permanent magnet is also included in the basic meter. The poles of this magnet are on either side of the armature.

An interaction occurs between the permanent magnet and the armature's electromagnet when an electric current flows through the armature. This is in the form of magnetic field interaction. A basic rule for magnetic devices is that opposite polar fields will attract and similar polar fields will repel each other. Current flow through the armature creates an electromagnetic field. The interaction of the two fields forces the armature to rotate on its axis.

Figure 5–1 The analog meter movement with a multimeter dial scale. (*Courtesy Triplett Corporation*)

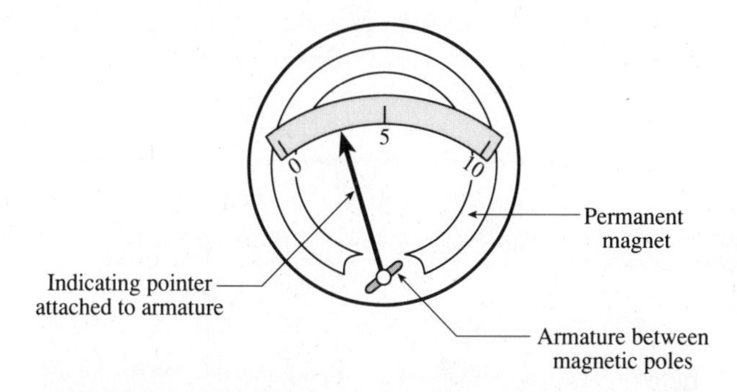

Figure 5–2 The basic analog meter movement uses magnetism for the movement of the armature and indicating needle.

This process is similar to that used to create the rotating motion in an electric motor. The basic difference here is the need for partial rotation instead of a continuous rotating device, like the motor.

Inside the meter movement, a needle indicator is attached to the armature. This needle indicator will swing, or deflect, as the motor action of the meter's armature occurs. The direction of this deflection depends upon the direction of current flow through the armature. Most analog meters are designed with clockwise rotation. Thus, the needle rests on the left side of the dial scale. The scale is calibrated, or marked, to indicate a maximum value of current flow. Current flow creates the rotational movement of the armature. The quantity of current flow and the design of the meter movement determine how far it will deflect. A pair of springs connect the rotating armature and the test lead connections of the meter. These springs serve two purposes. One is the electrical connection between the armature and the wiring of the meter. The second purpose is to return the rotating armature to its at-rest position at the left-hand side of the calibrated dial scale after a reading is made.

Each basic meter movement has electrical qualities. These include a value of voltage, a current value, and its ohmic value. The armature is wound with a wire having some value of resistance. The meter movement is classified by the maximum amount of current flow necessary to make the indicator needle deflect from its normal left-hand, at-rest, position to the right-hand, or full-scale, **deflection** position. The only manner in which a current can flow in a circuit is when a voltage is present and a load is connected to the circuit. This presence of voltage creates a difference in electrical pressure between two points in the circuit, and a current will then flow through the load. The quantity of current depends upon the applied voltage and the amount of load resistance to the flow of current. Thus, all meter movements are rated with a full-scale deflection value of current, the amount of voltage required to develop a full-scale deflection, and the value of armature resistance.

When a technician uses an analog type of meter movement, the deflection of the armature and its indicating needle depends solely upon the amount of current flowing through the armature.

This statement is true for all types of measurement with the analog type of meter. The illustrations of Figures 5–3 through 5–6 illustrate this concept.

The circuit shown in Figure 5–3 consists of a source voltage, a load, and an indicating meter. The meter is connected in series between the source and the load. Technically, the armature of the meter is connected in series with the circuit. All current in the circuit flows through the armature of the meter as well as the load. This is the way current is measured in a circuit with an analog meter movement. Current flow is limited to the quantity of current that can successfully flow through the wire windings of the armature.

It is possible to measure current when its flow is greater than the design capabilities of the meter movement. This is accomplished as shown in Figure 5–4. Bypass devices are used to provide an alternate path for some of the current. The bypass device is a resistor with a very low ohmic value. When used, the bypass, or **shunt resistor** as it is called, will direct a percentage of the current flow around the meter's armature winding. The specific value of this shunt resistance will determine what percentage of current flow bypasses the armature. The dial scale of the meter is calibrated to show the sum of the current flow in the shunt as well as the current flow in the armature of the meter movement. Shunt resistances are seldom built into the meter movement. Usually they are added to the terminals of the meter as an "outboard" device.

You must remember that the rules for parallel circuits apply to the application of the current extension meter shunt circuit. Kirchhoff's rules for current entering the junction and current leaving the junction also apply. Keeping these rules in mind will make it easier to understand the application of the meter shunt circuit.

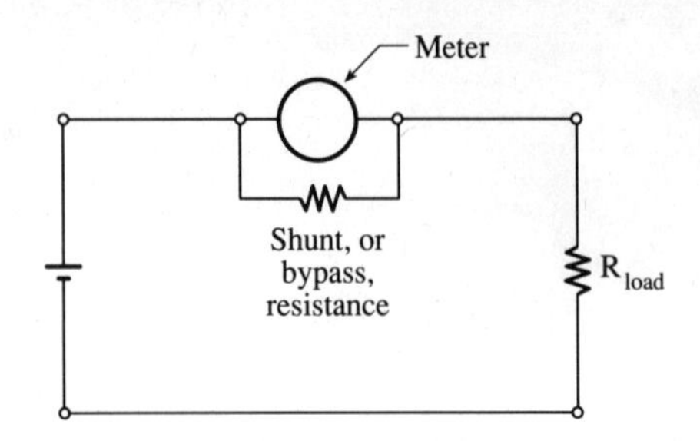

Figure 5–4 Use of a shunt, or bypass, resistance to increase the range of the basic meter movement.

Figure 5–5 shows a switch to illustrate that the amount of shunt current can be selected by use of multiple shunt resistors and a selector switch. This procedure is used in multirange current-measuring devices. The normal range for current meters is limited to rather low values by the size and current capacity of the wire used to wind the armature winding. Typically, it is less than 0.100 A, or 100 mA, and is often in the range of 1 mA or less. When currents in excess of the 100 mA maximum value for this example meter are to be measured, the shunt system is used for the balance of the current in the circuit. Shunt resistances usually have less resistance than the armature of the meter movement. Since current divides in reverse proportion to the value of the resistance in the parallel circuit, the

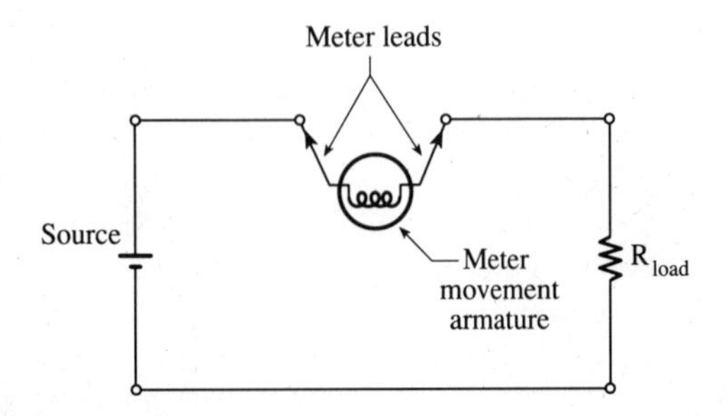

Figure 5–3 Total circuit current flows through the meter's armature.

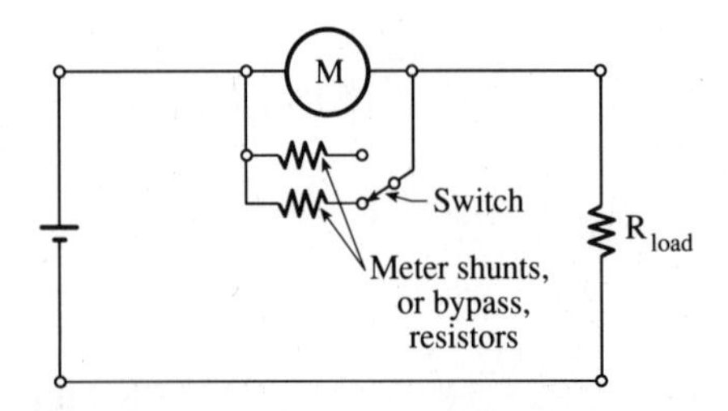

Figure 5–5 Additional shunt values may be selected by using a switch when more than one range is required.

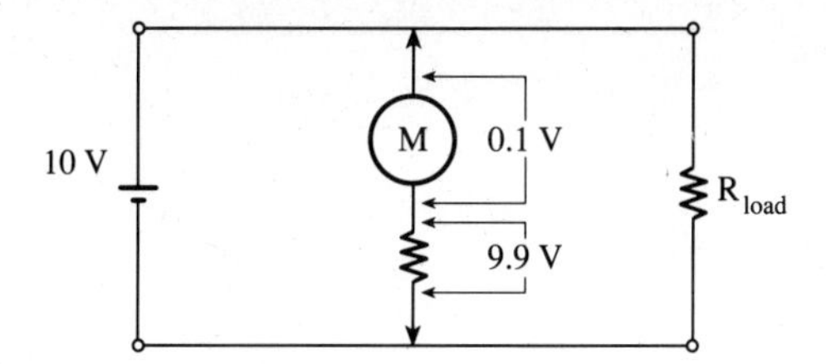

Figure 5–6 **The voltmeter uses a series-connected multiplier resistance to extend the range of the basic meter.**

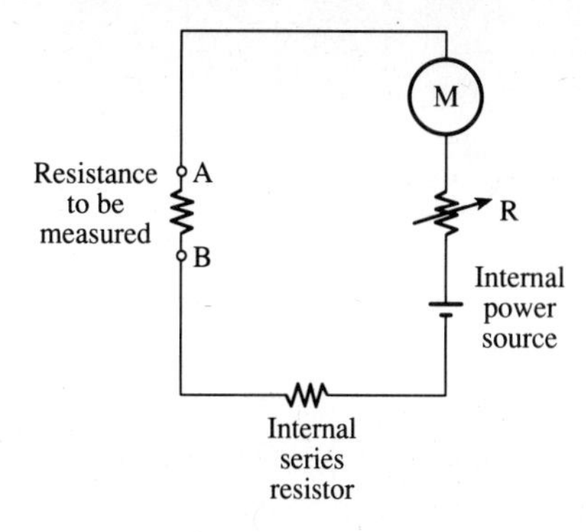

Figure 5–7 **The ohmmeter has an internal power source and a current-adjusting control.**

majority of current flow often occurs through the shunt resistance.

As illustrated in Figure 5–6, the current-measuring analog meter is also used to measure voltage. The addition of a voltage-dropping resistor in series with the meter movement converts it into a voltage-measuring device. The series-dropping resistor is called a **multiplier resistor** in this configuration. Its value is normally much greater than the resistance of the armature winding of the basic meter movement. In this illustration, the quantity of voltage necessary to create a full-scale deflection is 0.1 V. When a value of 10.0 V is to be measured, the resistance value of the multiplier resistor will limit the voltage drop across the meter's terminals to 0.1 V, The balance of the applied voltage, or 9.9 V, will develop as a voltage drop across the multiplier resistor. Changing the ohmic value of this resistor also will change the total voltage that can be measured.

To return to basics, voltage drop in the series circuit (this meter and multiplier resistor make up a series circuit) is directly proportional to the individual ohmic values of the components in the circuit. The ohmic value of the multiplier resistor is much greater than that of the meter movement. This permits the extension of the voltage value being measured. In the practical multirange V-O-M, a switching circuit changes the value of the multiplier resistor to accommodate a variety of voltage ranges. This is another application of the current flow through the meter movement. In essence, you are still using the current flow through the meter movement. The rules for series circuits, voltage drops across resistances, and the relationship of the voltage drop value to the ratio of resistances in the series circuit are all applied when considering how the meter multiplier circuit functions.

Figure 5–7 shows that current flow through a meter's movement is also used to measure resistance values. The ohmmeter has an internal power source in addition to an adjustable resistance. A basic electrical circuit is created when leads A and B are directly connected to each other. This circuit creates a current flow through the armature to the meter. The resistance, R, is adjusted to indicate a full-scale deflection on the meter's face. The leads are then separated and current ceases to flow. An additional, or unknown, value of resistance is added to the circuit. This is the value of resistance to be measured between points A and B. Current flows in the circuit since it has again been completed. The addition of this "extra" resistance will increase the total amount of circuit resistance. When additional resistance is added to a series circuit and the voltage is held at a constant value, the amount of current will decrease. This is an application of Ohm's law. The needle will deflect to less than its full-scale value. For example, if the original circuit requirements were for 1.0 V to create a current flow of 1.0 mA and the internal resistance of the meter's armature was 1000 Ω, then the addition of another resistance in series with the armature will increase the total resistance in the circuit. Let us assume that a value of 1500 Ω was to be measured. The total circuit resistance will now be 2500 Ω with this addition. The original 1.0 V pressure creating a current flow through the total resistance of 2500 Ω will reduce the current to a value of 0.0004 A. This is only 40 percent of the maximum current flow possible for this meter. The indicating needle will move to a position that is 40 percent of the full-scale value of the meter.

The dial scale of the meter is calibrated in ohmic units, which are determined by calculating the value of voltage and the amount of current flow in the circuit. Someone in the design area of the meter manufacturing company has developed a dial scale that indicates a resistance value of 1500 at the 40 percent point on the dial scale. This dial scale is nonlinear in its values. The ohmic values on the left-hand side of the scale are expanded, and the values on the right-hand side of the scale are compressed. A general rule to follow when making resistance measurements is to attempt to use the left-hand side of the scale whenever possible. This will provide a more accurate measurement.

Changing the value of the series resistor also will change the range of the measurement. Use of a switch will enable the meter to measure several different ranges of ohmic values. This is the method of changing ranges for resistance measurements used by many multipurpose meters.

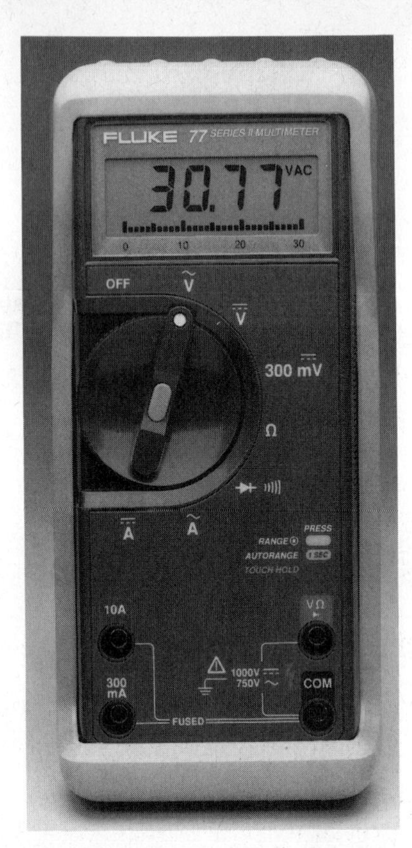

Figure 5–8 **The digital voltmeter, or digital V-O-M, has a numeric readout.** ***(Copyright 1991 John Fluke Mfg. Co., Inc. Reproduced with permission from the John Fluke Mfg. Co., Inc.)***

BASIC DIGITAL METER MOVEMENT

The introduction and development of integrated circuits played an important role in the field of electronic measurement. The older, analog meter system is being replaced by the **digital readout meter**. A typical digital readout V-O-M is shown in Figure 5–8. Notice that the basic difference is that the analog meter movement has been replaced by a digital readout. The functions of the two systems are similar, but the manner in which the digital meter creates its display value is entirely different. Several electronic circuits are used for this type of measuring device. Each of them measures a voltage instead of a current. The principles developed by Ohm, Kirchhoff, and Watt are still used. A block diagram for a basic digital electronic meter is shown in Figure 5–9. In this system, the voltage measured at the input to the integrated circuit is processed and converted into a signal that will operate, or drive, a digital display. Use of range-extending resistance values permits the system to measure many circuit values.

Each of the two basic types of V-O-M has some advantages over its counterpart. One of the principle advantages of the analog type is that the time response for its needle to indicate a change in circuit values is much faster than that of the digital readout. This is very important when the technician is attempting to adjust a circuit component for a maximum (or minimum) value of voltage or current. The faster response time permits a much more accurate adjustment. Often, when you rely on the digital readout for an indication of either a maximum or minimum value, the display indicates circuit values *after* the actual adjustment reaches the desired point. The technician, using the digital meter, does

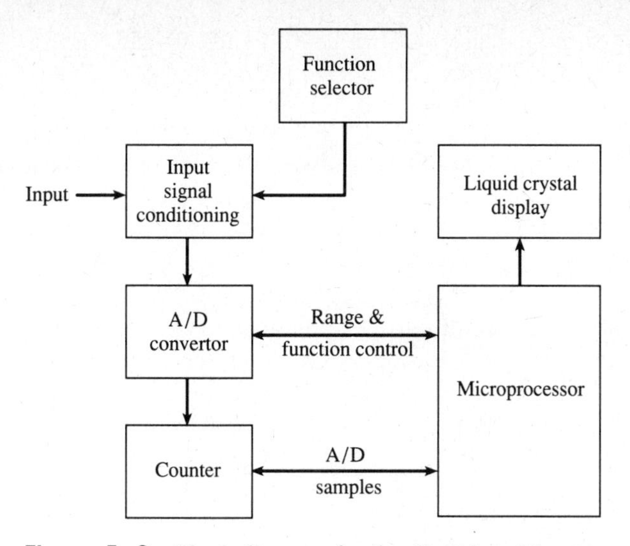

Figure 5–9 Block diagram for the digital multimeter.

not realize that this has happened and will continue to try to adjust the circuit. The end result may be a circuit whose adjustment is not at the desired level.

The digital meter's electronic circuitry permits the inclusion of circuits capable of automatic range selection instead of the manual switch rotation found on the analog type of meter. In addition, the digital electronic circuitry may include an **automatic polarity** display. This, too, is not normally available on the analog meter.

MEASURING VOLTAGES

Voltage measurement is one of the more common types of tests conducted during a service procedure. When making this type of measurement, keep in mind that voltage is the *difference* in electrical energy between two points in a circuit or system. Two procedures are commonly used to measure voltage. One of these is to use **circuit common** as a point of reference for the measurement. The other is to measure the potential, or voltage, difference between any two points in the circuit.

Two types of test equipment are used to measure voltage values. One of these is the often-used volt-ohm-milliammeter, or V-O-M. The other one is the oscilloscope. Few service technicians consider the oscilloscope a voltmeter. In reality, this is one of its functions.

Consider what the display on the face of the oscilloscope offers in the way of information. One of the more common values displayed is related to the frequency of the waveform; this appears on the horizontal axis of the display. The other one is related to the vertical information observed on the display. This is the voltage value of the signal, and it is measured using the vertical deflection circuit adjustments on the scope. These are seen as "volts per division" or "volts per centimeter" values on one of the switches on the front panel of the oscilloscope. Adjustment of this switch is made to insure that the full height of the waveform is observed on the faceplate, or graticule, of the oscilloscope. The values of these waveforms are measured using a process of counting the number of divisions displayed and multiplying this number by the value established for each division. When a display of 4.6 divisions is observed and the volts/division switch is in its 5 V position, the total voltage value is 23 V ($4.6 \times 5 = 23$).

Voltage measurements using the oscilloscope can be made in any circuit. They do not necessarily have to be signals. Many oscilloscopes are capable of measuring both AC and DC voltage values. The AC voltages can either be signals or operating voltage values. The same is true for the DC voltage measurements. Learning to use the oscilloscope in this manner is just one more technique practiced by the expert service technician.

Circuit Common Circuit common measurements are used more often than any other type of voltage measurement. In Figure 5–10 one of the test leads is connected to circuit common. The fact that the lead is connected to circuit common normally is not mentioned when discussing the value of measured voltage. In other words, you would most likely say, "The voltage at the junction of resistors R_1 and R_2 is 100 volts." What you are actually saying is that the difference in electrical potential, or voltage, between circuit common and that specific point in the circuit is 100 V. With this type of measurement, circuit common as the point of reference is assumed. In the majority of circuits, the negative lead of the

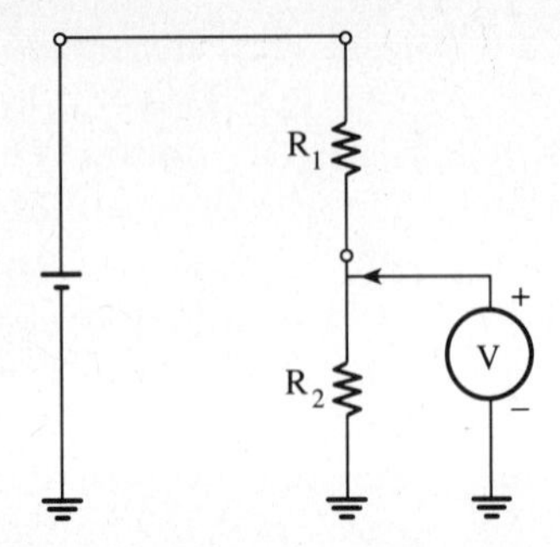

Figure 5–10 Voltages are usually measured using circuit common as a point of reference.

power source is connected to circuit common. Thus, the negative lead of the voltage-measuring device, or meter, is also connected to circuit common. This is illustrated in Figure 5–10. In this illustration the negative lead of the meter V (−) is connected to circuit common. The positive lead, indicated by the (+) sign, is connected to a different point in the circuit. The voltage measurement is the difference in electrical potential between the two points.

Component Reference Measurements Component reference measurements are made by placing one of the test leads of the voltmeter on one side of the component and the other test lead on the other side. Measurements in this manner are described as being "across the component." A measurement as shown in Figure 5–11(a) would be described as measuring 40 V across resistor R_1. Measurements of this type can be made across any component in a circuit. The difference is the description of the measurement. Figure 5–11(b) offers an alternate method of measuring the voltage drop across a specific component. This method takes a little longer, but may be more efficient at times. The process involves two measurements. One of these is the measurement of the voltage drop from point A to circuit common. The second measurement is the voltage drop from point B to circuit common. The difference in the two measurements is the voltage drop across the specific resistance in this circuit. The mathematical equation for this is:

$$V_A - V_B = V_{A \text{ to } B}$$

In each of the above examples, the manner of measurement is the same. Each example shows that the meter leads must be connected to different points in the circuit. The key to this type of measurement is to remember that voltage is actually the difference in electrical potential between two points in any circuit. Placing both of the leads of the voltmeter at the same point in the circuit will not indicate any potential difference, thus you would not obtain any type of measurement, or reading, on the meter.

One additional point to remember is that it is not possible to make a voltage measurement across two points in any circuit when both the test equipment and the device being tested have a connection from

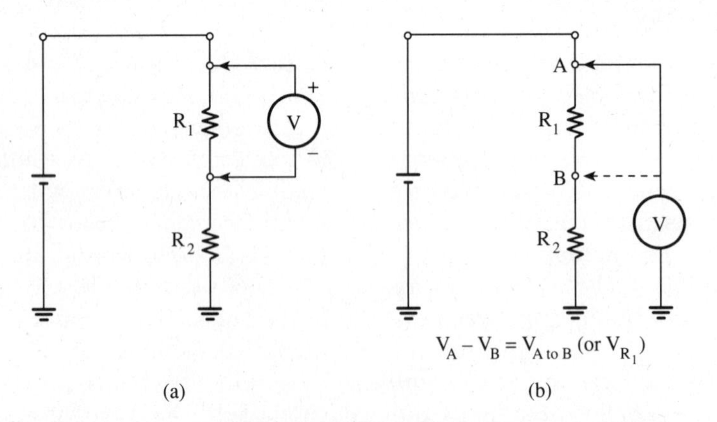

Figure 5–11 Voltage may be measured across a specific device or component using either of the illustrated methods.

circuit common to one side of their power source. Attempting to place the common lead of the test instrument on a component in the circuit will effectively short out the balance of that circuit and provide a false reading of the circuit value.

Polarity Polarity is one more factor to keep in mind when making a voltage measurement. Almost all analog meters and many digital meters are only able to read positive voltage values. The negative lead of the meter must be connected to the more negative point in the circuit. When this is done, the dial, or needle, will deflect to the right and provide a valid reading. If the test leads are reversed, the needle will attempt to deflect to the left. Since it is resting on the left side of the meter, it has no place to move and a reading cannot be made.

Many of the more sophisticated digital meters currently in use include a polarity-sensing circuit. The readout of the meter will indicate either a (+) or a (−) as a part of the digital value of the voltage being measured. With this type of instrument, all you have to do is to note the placement or arrangement of the test leads to determine whether the measurement is positive or negative. (If the test leads in Figure 5–11 were reversed, the display would indicate a negative voltage value instead of its actual positive value.)

Tolerance **Tolerance** of the component being measured or of the values provided by the manufacturer also must be taken into consideration as you evaluate the circuit. The values shown on service literature may not be exactly the same as those found in the equipment. Most manufacturers of components indicate their tolerance from their marked values. These can vary by as much as 10 to 20 percent from their marked values! The service technician must remember that measured values that are close to those provided in the service literature are often acceptable and do not identify major areas of trouble or problems. The technician is looking for a much different measured value from the ones provided by the manufacturer in the manual. This statement is accurate for almost all measurements. Deviations from published values of 10 percent or more are often acceptable as initial test results. Exact deviation values will vary on different pieces of equipment. Experience will assist you in differentiating the acceptable ones from those that indicate trouble in a specific circuit. The exception to this rule also may occur. Some very critical circuits are used today in electronic equipment, and when they are used, the service literature will identify that an exact value must be present for the device to operate successfully.

Test Evaluation Test evaluation determines which of the brackets described in the previous chapter are to be moved. The service technician is looking for a major deviation from the published values described in the service literature. Minor differences often are normal. Often the new technician fails to realize that small differences are not the initial reason for making the circuit measurement. Major differences indicate problems with components or circuits. The primary effort of the servicer is to locate the major area of trouble. Once this is accomplished, the specific component, board, or module can be repaired or replaced. If problems persist, then further testing is required. Ultimately you may have to look at the minor discrepancies in values to determine a specific component failure. This is done only when there are no major discrepancies and the circuits still fail to operate properly.

One word of caution is necessary at this point. Before spending much time looking for the reasons for the minor variations, stop to consider if this is really the proper place in the system to be investigated. There are times when we all get "locked in" to an area of the system and fail to realize that we may be on the wrong path for successful diagnosis and repair. This is one of those times when it may pay to stop and reconsider the diagnostic process and what the results you have found mean.

Safety Safety is of utmost importance in any test situation. Voltage is one of those things you cannot see. However, it is possible to "feel" it during any work on electrical or electronic equipment. Keep in mind that an electrical shock could be the last thing you ever feel. None of us really wants this to happen! The best rules to follow when testing or evaluating any equipment are:

1. Plan what you are attempting to measure and how you are going to make the measurement *before* you connect any equipment to the system.
2. Using the service literature, determine the approximate values you expect to measure at the point of the test if the circuit is functional.
3. If you are working with high voltages, turn off the source power before connecting any test equipment.
4. With the power off, make the test connections. Be sure to discharge all electrolytic capacitors before attempting to do this. The discharge of the circuits's electrolytic capacitors will help reduce the possibility of an electrical shock when making these connections. Then, and only then, turn on the power to observe the values being measured.
5. Be sure the test leads are connected to the proper places in the circuit. Also be certain that the test leads are not accidentally touching any other components or test points.
6. Always connect the negative, or common, lead of the test instrument first. Since most voltage measurements are made using circuit common as a reference, this will minimize the possibility of electrical shock.
7. When working under "power on" conditions, use only one hand to hold the test lead. Keep your other hand away from any metallic surfaces. One of the better places for the other hand is in a pocket or behind your back.
8. Always remember that it is not the voltage that can harm you; it is the application of voltage that can create a current flow through your body that actually may cause severe electrical shock or even death. This is the reason for the process described in #7, above.

THE AMMETER

The measurement of electrical current is often a necessary activity for the service technician. A typical method of making this measurement is illustrated in Figure 5–12. Two ammeters are shown, both indicated by the circuit containing the letter "A". The circuit has two places where the letter "x" is shown. Both have dots indicating a connection to the circuit.

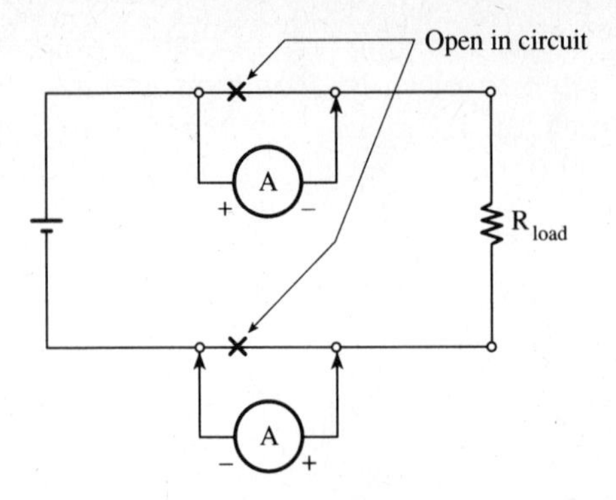

Figure 5–12 **Ammeters are normally inserted into the circuit to measure current.**

The "x" indicates a break, or disconnection, in the circuit. The circuit must be opened at these points when the ammeter is used in this arrangement. All the circuit current will then flow through the meter and through the circuit. This provides an accurate reading of the current flow in the circuit. Each ammeter has polarity indications alongside its graphic symbol. Current flow must conform to the polarity markings on the meter. Reversal of these leads or connections will make the armature of the meter attempt to move in a counterclockwise direction. Because this movement is not possible for an analog meter movement, either an invalid reading or no reading will be displayed. If the current meter is a digital unit with automatic polarity circuitry, then the current flow value will be negative when the leads are reversed. When this type of meter is used, the value of the reading is correct and only the polarity is wrong.

METHODS OF MEASURING CURRENT

DIRECT METHOD

The direct method of measurement is just as indicated in the previous section. The circuit is "broken," or disconnected, at some point, and the leads of the ammeter are inserted at these points. This will complete the circuit and permit a flow of current. Most current-indicating meters, whether they are analog or digital, have a very low internal resistance value. The addition of one of these meters has little effect on the total current flow in the circuit. The measured values are accurate within the tolerance limits of the components in the circuit.

INDIRECT METHOD

The indirect method of measurement is becoming much more popular for servicing. This method uses one of the basic laws described earlier in this book—Ohm's law for determining circuit current. The formula for this version is:

$$I \text{ (current)} = \frac{E \text{ (volts)}}{R \text{ (ohms)}}$$

Manufacturers of electronic devices are starting to use this formula and method to determine the value of current flow in a circuit. The direct method of current measurement requires that the circuit be opened, but opening the circuit is not always convenient. Several things often happen when the process of removing one end of a component for testing is performed. It is possible to destroy a component while attempting to remove one of its leads, to reinsert it incorrectly, or to electrically damage the component by placing the meter into its circuit. It is also not an efficient way to make this measurement. The indirect method makes current measurement simple and effective. All that is required is a voltmeter and a calculator, as is shown in Figure 5–13.

This circuit contains two resistances and a 15 V source. The values of the two resistances are 5 kΩ and 10 kΩ. The 10 kΩ resistance is the load in this circuit. The two resistances have a 1:2 ratio of resistance values. The expected voltage drop across the load resistance should be two-thirds of the total voltage or 10 V (2/3 × 15 = 10). Measurement of the voltage drop across the load resistance actually measures 10 V. Using the formula for current flow:

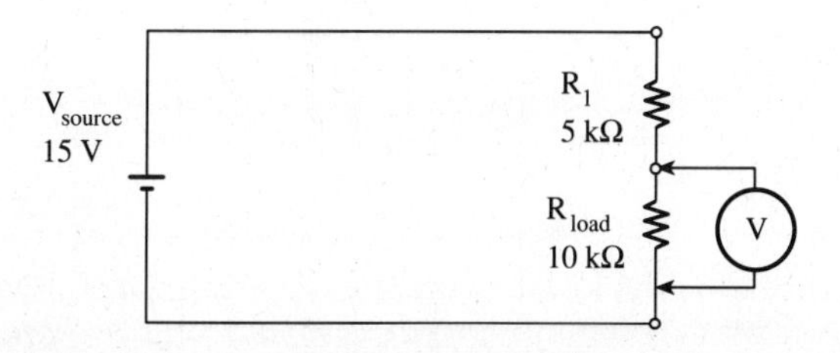

Figure 5–13 Current may also be measured by using Ohm's law. The voltage measured across a known resistance is used to calculate current flow values.

$$I = \frac{E}{R} = \frac{10}{10{,}000} = 0.001 \text{ A}$$

This method is often used in circuits containing transistors. It provides an alternate and comparatively easy method of determining current flow. Figure 5–14 shows a typical transistor amplifier circuit. The manufacturer has provided the value of the emitter resistor and the voltage drop that is expected to exist across this resistor. If you were to calculate the value of current flow, it would be 3.4 mA for this circuit. Let us examine what occurs when the voltage measurement is greater than the 0.34 V indicated on the diagram. If you measure a value of 1.0 V at the test point, then the current flow in this circuit is:

$$I = \frac{V}{R} = \frac{1.0}{100} = 0.01 \text{ A}$$

When current flow increases, a problem exists in the circuit; either the bias on the transistor has changed, the transistor is defective, or the load resistor has changed its value. Any of these conditions indicates a problem that requires further investigation. Any increase in the total amount of current flowing in a circuit has to be due to a decrease in the total amount of resistance in that same circuit. This relates to Ohm's basic laws for current flow values based on total resistance and total applied voltage. The same is true when the voltage measurement is less than the value provided by the manufacturer. A decrease in measured voltage indicates an increase in a resistive value in the circuit. Further testing and evaluation of the circuit is required. This is also illustrated in Figure 5–14. This chart shows the various values of current flow as determined by measurement of the voltage drop across the emitter resistor.

Another method of performing a current measurement without opening the circuit is similar to the previously described method. This system uses a

$V_{measured}$	R	A
1.0	100	0.01
0.34	100	0.0034
0.015	100	0.00015

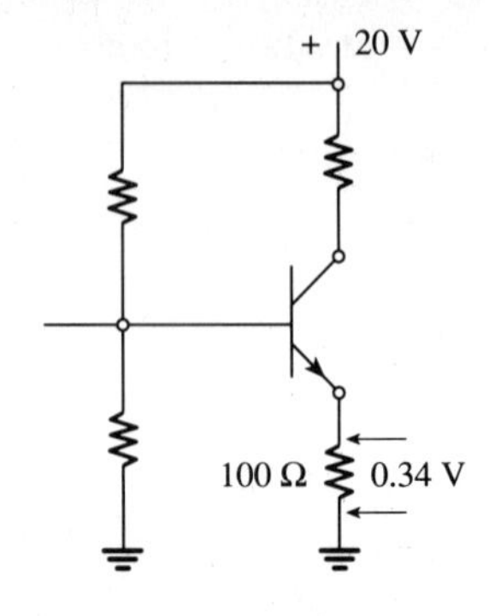

Figure 5–14 The use of a range voltage value and Ohm's law to establish current readings, as described in the text.

meter that does not have an **automatic ranging** circuit. The nonautomatic ranging meters usually are designed with voltage ranges in multiples of 1, 10, 100, and 1000. These ranges can be used with an application of Ohm's law. For example, if a 100 Ω resistance is in the circuit and the range switch on the meter is set on the 100 V range, the maximum meter reading will be:

$$I = \frac{E}{R} = \frac{100}{100} = 1.0 \text{ A}$$

Actual readings will range from 0 A to 1.0 A depending upon the percentage of deflection of the dial needle as the voltage across the 100 Ω resistance is being measured. Switching to another voltage range provides a different set of current ranges. (This is illustrated in Figure 5–15.) For example, if the voltage range is changed from the 100 V range to the 10 V range, the current values will increase. The maximum current flow under these new conditions will be 10.0 A, due to the change in the range of readings on the voltmeter.

Tolerance values for current measurement also depend upon the specific component values. Tolerance deviations for most circuits can vary as much as ±10 percent and still not affect circuit operation. Keep in mind that evaluation of any circuit initially looks for major discrepancies. Those measurements showing values that are close to those provided by the manufacturer do not normally indicate major problem areas in the system. At the expense of sounding repetitive, the major problems in almost

Resistance value	Range switch position on voltmeter				
	0 – 0.1 V	0 – 1.0 V	0 – 10 V	0 – 100 V	0 – 1000 V
1 Ω	0 – 0.1 A	0 – 1 A	0 – 10 A	0 – 100 A	0 – 1000 A
10 Ω	0 – 0.01 A	0 – 0.1 A	0 – 1.0 A	0 – 10 A	0 – 100 A
100 Ω	0 – 0.001 A	0 – 0.01 A	0 – 0.1 A	0 – 1.0 A	0 – 10 A
1000 Ω	0 – 0.0001 A	0 – 0.001 A	0 – 0.01 A	0 – 0.1 A	0 – 1 A

Figure 5–15 The voltage range switch and how to use it for current measurement. Each range is able to provide current flow values, as shown on the chart.

any electronic device seldom occur due to small changes in component values. There are some very critical timing- and frequency-sensitive circuits where this statement is not valid. In the vast majority of circuits, small deviations from the values marked on the components have little, if any, negative effect.

SAFETY

Safety is an important factor in the measurement of current. Consider the effect of current measurements as they relate to safety. Look at Figure 5–16. This circuit shows a source voltage of 1.5 kV; a load resistance, R_1; a device that could be any type of load; and a second resistance value, R_2. Voltage measured across R_1 is 750 V. Voltage drop across R_2 is 1.0 V. The current in this circuit is to be measured. Since this is a series circuit, the current flow is equal throughout the circuit. In terms of safety, the place with the lowest value of voltage is across R_2. Point A to circuit common is the place to make a voltage measurement to determine the quantity of circuit current, because it has the lowest voltage value in the circuit. Remember that the test leads of the meter expose you to the voltage values of the circuit. Testing at low voltage points in the circuit is the safe way to make this measurement.

If you wanted to permanently insert an ammeter in this circuit, point A is still the place to use. Again, the voltage values are very low and safety is of primary importance. Meters can expose you to the voltage values in the circuit. Use of the lowest voltage values is the safest way to prevent this. If the meter was inserted at point B, the voltage at the meter would be approximately 750 V, not a very safe operating procedure to follow, and it can be hazardous. This is true for a measurement using a test meter as well as for the installation of a permanent meter in the circuit.

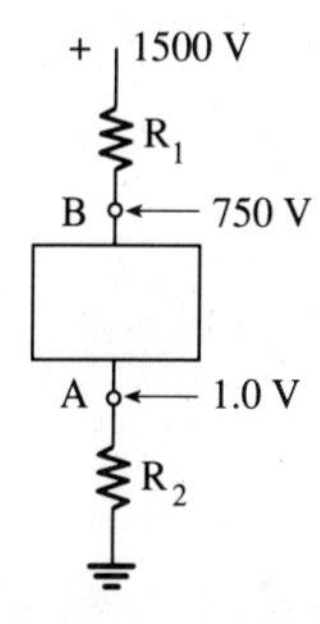

Figure 5–16 Placement of the current meter at the lowest voltage point in the circuit is a safe practice. Point A in this circuit has the lowest voltage and will still provide total current information for the circuit.

RESISTANCE MEASUREMENTS

POWER OFF CONSIDERATIONS

The need to measure circuit resistance values is one additional process in the area of servicing and troubleshooting. One of the most important rules to remember when attempting to measure resistance values is that this procedure must be done with circuit power turned off. The ohmmeter uses an internal power supply to create a current flow. The use of an external power source can either damage the test meter, provide a false reading, or both. The internal power source also might include charged capacitors in the system being tested. Be certain to discharge any major power supply capacitors before making resistance measurements.

SCHEMATIC ANALYSIS

You should review the **schematic diagram** before making any resistance measurements to determine the approximate amount of resistance in the circuit. This will assist in the evaluation process. Knowing what you expect to find simplifies evaluation of the circuit. A second reason for reviewing the schematic diagram is that it may indicate a resistance value that is connected in parallel with the one being measured. If you do not first determine what the circuit contains, the measurement could produce false readings. This can provide information that will send the technician on the wrong trail for diagnosis. If nothing else, it will provide confusing information and delay the ultimate successful diagnosis and repair.

This possibility is illustrated by the circuit shown in Figure 5–17. The specific resistance to be meas-

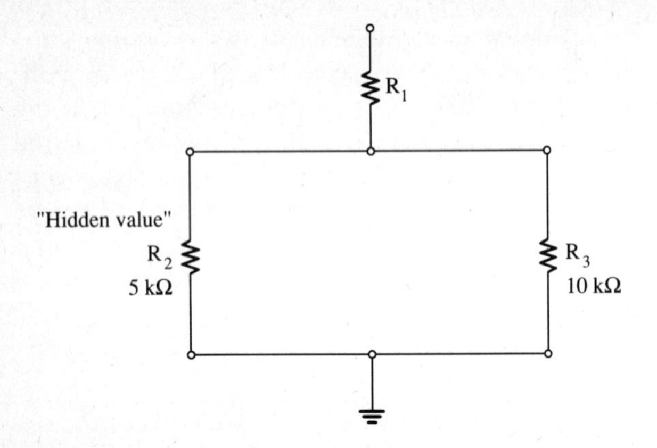

Figure 5–17 "Hidden" resistors, such as R_2, will provide false resistance readings if you do not first review the schematic diagram of the unit.

ured is R_3. This resistor has an ohmic value of 10 kΩ. A review of the circuit shows that a second resistor, R_2, with an ohmic value of 5 kΩ is connected in parallel with R_3. The total resistance of this parallel branch is about 3.3 kΩ. This resistance value, measured with an ohmmeter placed across resistor R_3, could lead you to believe that something was wrong with this circuit. This, of course, would not be a valid analysis for this circuit. A review of the schematic diagram for the circuit prior to making the measurement would provide proper information about the total circuit. An ohmic measurement of slightly more than 3.3 kΩ is the correct value for this circuit.

THE MULTIMETER

The multipurpose meter is used today instead of individual volt, amp, or resistance meters. The principal advantage of these meters is their ability to switch between various functions. The use of the multimeter, as it is called, offers a small, compact unit capable of measuring these basic functions. Multimeters are available as portable, battery-operated units and as AC-powered units in both analog and digital form. Selection of a specific type or brand of instrument is up to the individual. All have approximately the same ranges and functions. Some offer automatic range selection, others offer automatic polarity selection, and some instruments have both capabilities. Individual preferences and the amount of money you are willing to spend will determine the specific unit selected.

When operating any multimeter certain cautions must be observed. These include, but are not limited to:

1. Plan what you are going to measure and where you are going to make the measurement before connecting any test leads to the circuit.
2. Always connect the negative lead of the instrument first.
3. Do not switch between functions when the multimeter is connected to the circuit. If you inadvertently have the function switch on the wrong function, remove the instrument from the circuit before changing the position of this switch. You may damage the internal components of the meter when switching is done while the instrument is connected to a powered circuit. One type of meter has a built in fuse to protect it from improper use.
4. If the measurement is very different from what you expected, double-check to be certain you measured at the proper place in the circuit.
5. If you are using a nonautoranging multimeter and are not certain of the value you are going to measure under power on conditions, select the highest range on the instrument. If no reading is obtained, then move the range selector switch to a lower value and repeat the measurement. Continue in this manner until you have obtained a valid reading.

THE OSCILLOSCOPE

Signal-processing circuits found in much electrical and electronic equipment cannot be evaluated by use of the multimeter. The reason for this is that the multimeter's ability to evaluate signals is limited to those having either a frequency of around 60 Hz or a total lack of any frequency, as found in DC voltages. Individual circuits process electrical and electronic signals. The signals are used to control individual stages inside the unit. They can also be used to control the entire device. An example of this is the conversion of a switch, or key, on a computer keyboard into an electronic signal. The signal is processed by the computer and is ultimately used to print a letter on a paper in a printer. The signal will probably change in shape, magnitude, and possibly frequency as it is processed by the computer system. A failure in the chain of events used to create the printed letter on a piece of paper cannot be identified by use of a multimeter. Another measuring device is required to measure electrical or electronic signals. This device is called the oscilloscope. Figure 5–18 shows a typical commercial oscilloscope. Its function is to provide a visual display of the signal. The oscilloscope's display provides both amplitude and timing information to the experienced service technician. Amplitude information relates to the level of voltage present at the test point; timing information relates to a frequency determination for the display of the vertical portion of the waveform.

Definition of the sig
The electronic signal a
inside the system. This
component of the ope
Its amplitude indicate
ular module, block, c
The timing, or frequen
rect for proper operat
cilloscope will provid

VOLTAGE MEASUREMENT

Voltage may be measured with an oscilloscope. Figure 5–19 shows one part of the front panel of a typical oscilloscope. Displayed is the amplitude determining switch. Note that it is rated in volts/cm. This informs the technician that each centimeter (cm) on the faceplate of the oscilloscope has assigned to it a voltage value, which is indicated by the position of the switch on the face of the instrument. The amplitude of the display, when multiplied by the value of each centimeter, will describe the voltage value of the signal being measured.

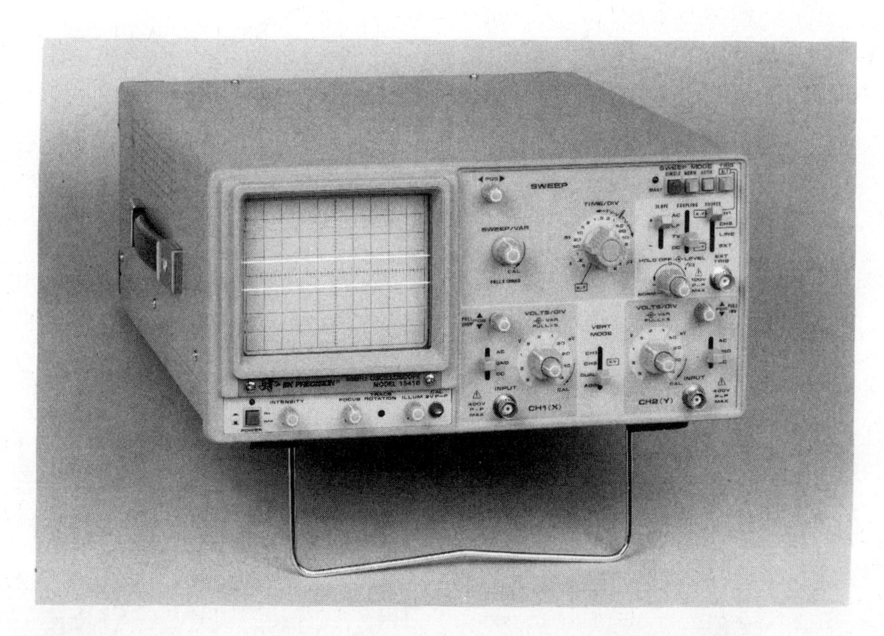

Figure 5–18 **A typical oscilloscope used in the electronic service field.** ***(Courtesy B&K Precision Instruments Division, Maxtec Corporation)***

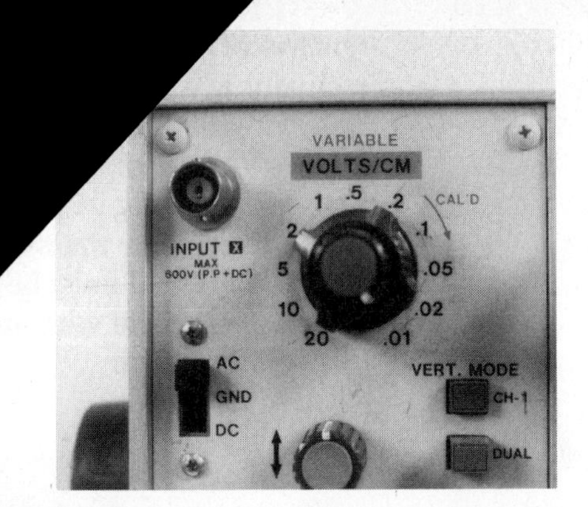

Figure 5–19 **The volts/cm switch assigns a voltage value for each centimeter on the oscilloscope display. (*Photo by J. Goldberg*)**

FREQUENCY MEASUREMENT

Frequencies may also be measured with the oscilloscope. Frequency is a function of time; actually, it is inversely proportional to time. The formula for frequency is:

$$f\ (\text{frequency} \in \text{Hertz}) = \frac{1}{\text{T}\ (\text{time} \in \text{seconds})}$$

Another portion of the face of the oscilloscope is shown in Figure 5–20. The faceplate, or graticule, of this oscilloscope is calibrated in units of the centimeter. Other oscilloscopes may use a different faceplate calibration system. This is the time/base section of the instrument. The dial will provide a time value for each horizontal centimeter on the face of the oscilloscope. If another faceplate calibration system is used, then the time/base switch will indicate those values instead of centimeters. The horizontal displacement of the display determines its time frame. Multiplying the total number of centimeters (or whatever value is used) for one complete cycle of the display by the values provided on the time/base switch will determine the total time for the complete cycle. This value is divided into one (for its reciprocal). The reciprocal of the time is the frequency of the display. Some recent-model oscilloscopes can make this conversion and display the value on the screen.

Figure 5–20 **The time base generator inside the oscilloscope determines the rate of beam movement and the frequency of the displayed signal. (*Photo by J. Goldberg*)**

SIGNALS

Signals are traced through any electrical or electronic device by use of the oscilloscope. Many manufacturers include signal processing information in their service literature. A typical bit of information is shown in Figure 5–21. Here, the manufacturer offers a DC voltage value and oscilloscope information. The oscilloscope information contains both amplitude and frequency values. The service technician is able to follow, or trace, the signal through the system by using the oscilloscope. The service technician compares the signal information provided by the manufacturer of the equipment to that observed on the face of the oscilloscope and by the time values provided on the oscilloscope's time/base switch. If these agree, then signal processing is correct at the point of measurement. The service technician now selects one of the rules presented earlier for signal flow systems. This will determine the direction and placement of the next test. Selection of the proper rule will enable this next test to be both efficient and valid.

THE FUNCTION GENERATOR

Function generator is another name for a **signal generator**. Originally, the signal generator's capa-

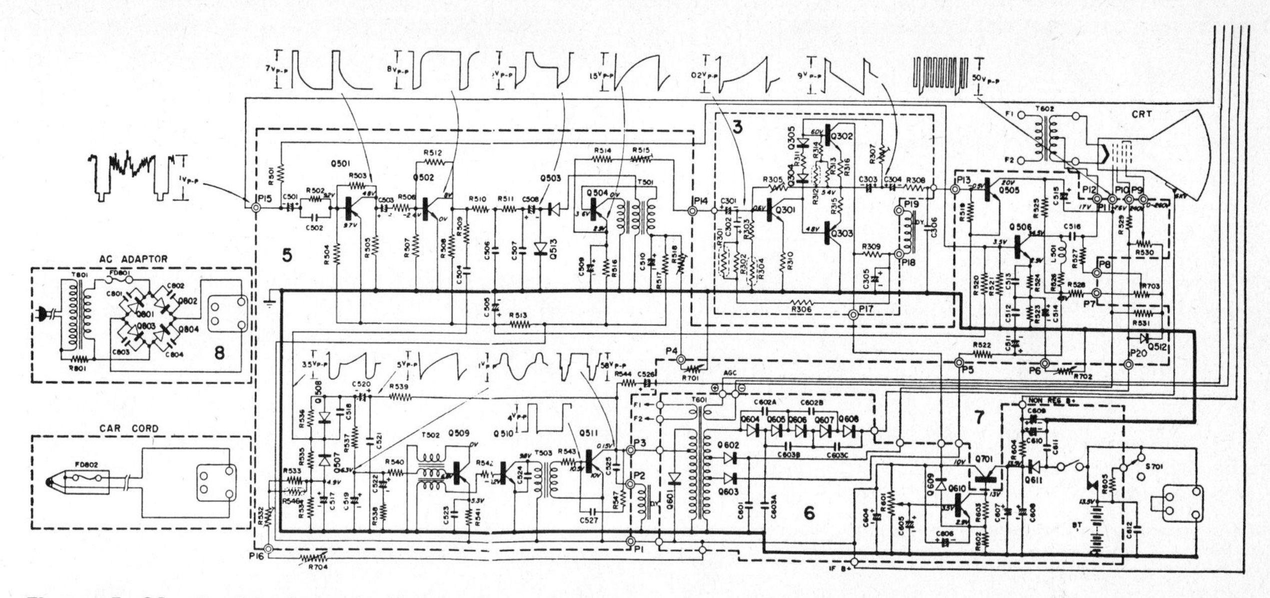

Figure 5–21 **Partial schematic diagram for a television receiver with both voltage and waveform information.**

bility was limited either to a signal in the audio range of 20 Hz to 20 kHz or to signals in the **radio frequency** range (higher than 20 kHz). Many of the function generators available today are capable of providing a range of frequencies from the low end of the audio frequency spectrum to well above typical radio frequencies. In addition, function generators can create nonsine-wave-shaped signals such as square waves and triangular waves. Some of these units can be modulated to create radio signals containing audio information.

The principal advantage of the function generator is its ability to create a wide variety of signals with a single instrument. The function generator can be connected to the input of a signal processing device. The oscilloscope is then used to trace the signal through the unit under test. Another method of processing is to use the function generator as a signal injector. The output of the generator is injected at various test points in the unit, during which the output of the unit is monitored. The injection probe is moved through the unit until the output signal is no longer distorted. The block or section between the last point of distortion and the first point of no distortion is the block that has the problem.

SPECIALIZED TEST EQUIPMENT

Often it is advantageous to use some very specific test equipment when servicing certain types of electrical or electronic equipment. These test units are designed to provide all of the necessary signals to service the unit. Examples of these types of test units include the test sets provided by manufacturers of commercial two-way radios, test units for sophisticated audio-processing units, television and radio analysts, and FM stereo test generator sets.

One model of tester that is often used in servicing two-way radio transmitting and receiving equipment is provided by the equipment's manufacturer. This tester is designed to connect directly into an existing test socket on the radio. Switching circuits in the tester selects the specific test the technician desires to perform. This type of equipment often is very expensive, due to its limited application. When a large portion of the service business at a specific location is related to just a few types of equipment, it is advantageous to purchase this type of test set. If the service organization does not specialize, then the more general types of test units described in this chapter are recommended.

REVIEW

Three basic types of testing equipment are discussed in this chapter. They are the volt-ohm-milliammeter (V-O-M), the oscilloscope, and the function generator. V-O-Ms are available in two basic styles. The major difference is the type of output display. One style uses an analog type of meter movement, and the other has a digital display.

The analog-style meter operates on the principle of current flow. A magnetic field is formed when current flows through the armature of this meter. The armature's magnetic field reacts with a fixed magnetic field in the meter. This interaction forces the armature to rotate. A dial needle is mechanically connected to the armature. The rotation of the armature on its axis is seen as a deflecting, or movement, of the dial needle. A faceplate on the meter is calibrated and displays the numeric value of the quantity being measured.

The fundamental value of the meter can be extended using either parallel or series-connected resistances. Parallel resistances, called shunt resistors, are used to increase the total current capability of the meter. Shunt resistors usually have less ohmic value than the resistance of the armature of the meter. Series resistances are used to increase the range of voltage readings. These are larger in their ohmic value than the armature's resistance value.

The addition of an internal power source and an adjusting resistor change the basic meter into one capable of measuring resistance values. The dial of the meter is calibrated into units of ohms, based on the interaction of voltage, current, and resistance.

The second meter is the digital type. This uses electronic circuitry to convert measured values into a form that can be displayed on a numeric readout. Many digital meters are also capable of displaying polarity as well as the numeric value. In addition, some of the digital meters have circuits that permit automatic range adjusting.

Tolerance must be understood by all service technicians. All components have some tolerance, or deviation, factor. This deviation is often as high as 10 percent of the marked value. Tolerance must be considered when evaluating voltage, current, and resistance measurements in a circuit. Measured values that are close to those provided in the service literature often do not indicate areas of trouble. Initially the service technician should look for major differences between the service literature values and those being measured.

Another factor to be considered when using any meter is safety. Use of safe habits will reduce the possibility of electrical shock and possible death.

The oscilloscope provides both voltage and frequency readings. It has a cathode-ray tube capable of displaying these factors. This is one of the few test instruments with these capabilities. Meters are only able to properly evaluate either 60 Hz frequencies or DC (0 Hz) frequency. The oscilloscope, by use of a time/base generator, can display a wide range of frequencies.

Function generators create sine, square, or triangular wave signals. These devices are able to create these signals over a wide range of frequencies. The function generator is used as a stable test signal while servicing.

Specialized test equipment is available for use on specific equipment. An example of this is a test unit for two way radio servicing. These units will create all of the necessary signals for analysis and alignment of the radio unit. Test units are often available from the manufacturer of the equipment.

REVIEW QUESTIONS

1. What basic principle is used in the analog meter?
2. How do you extend the range of an ammeter?
3. Explain why one measures ohmic values under power off conditions.
4. What is meant by the tolerance of a component?
5. What rotational direction is used in the analog meter?
6. How is the ammeter connected in the circuit?
7. How is the voltmeter connected in the circuit?
8. Explain how a function generator can be used as a signal tracing device.
9. What two values are displayed on the oscilloscope?
10. Explain the concept of measuring current at a point of lowest voltage in the circuit.
11. An oscilloscope's voltage/cm switch is at its 1.5 V/cm position. The display occupies 3 cm on its vertical axis. What is the total voltage displayed?
12. Explain why you should evaluate the service literature schematic before making a resistance measurement.
13. Why should you form the habit of working only with one hand when making voltage measurements?
14. What is meant by the term autoranging?
15. What is the point of reference for most voltage measurements?
16. Explain why you should refer to the service literature before measuring voltage.
17. How is it possible to measure circuit current without actually inserting the current meter into the circuit?
18. What are radio frequencies?
19. Explain the term "signal".
20. Explain, in electrical terms, how the meter shunt is able to increase the range reading of the basic meter movement.

CHAPTER 6

Methodology of Testing

INTRODUCTION

The ability to troubleshoot is a learned skill. Anyone having this skill will be highly employable. The skill may be applied to almost any field of endeavor. It is not limited only to those who work with electrical or electronic devices. The successful troubleshooter is able to use a very logical approach to almost any problem. The ability to locate almost all problem areas in many types of equipment is the goal of the troubleshooter. This ability can be developed by following a basic set of rules. These rules are not difficult. They do present a logical, systematic approach to almost all troubleshooting problems.

OBJECTIVES

Upon completion of this chapter, the reader/student should understand how to:

1. use a logical, systematic approach for troubleshooting;
2. perform a physical analysis of a nonoperable device;
3. use manufacturer's service literature as a tool for diagnosis;
4. analyze block diagrams to locate areas of trouble; and
5. develop a plan to successfully diagnose and repair a nonoperating, but repairable, electrical or electronic device.

KEY WORDS AND PHRASES 

bench test: Describes the operation of a device under test conditions in the workroom.

current flow path: This is a schematic diagram providing paths for operating current flow in the device.

mental barrier: Term describing a predetermined answer to a service problem.

service literature: Printed materials provided by the manufacturer of the equipment. Often contains schematics, parts lists, adjustment procedures, and troubleshooting information.

signal processing path: Information describing how signals are processed in a device using block diagrams as the reference for this information.

troubleshooting flow chart: A plan, often prepared by the manufacturer of the equipment, that describes a process of testing and where various test steps should be performed.

THE FUNDAMENTAL RULES OF TROUBLESHOOTING

The rules of troubleshooting have a very definite order, which must be followed when you are trying to identify the problem area. Review the following steps and try to keep them in order as you start on your path to learning how to become one of the greatest troubleshooters of all time. Anyone can do it; all that is required is the ability to approach the problem logically and use a logical method to analyze what you have found.

One of the more serious handicaps we have created for ourselves as we attempt to problem solve is the problem of attitude. We have been conditioned over the years to be "productive." This often means that we must seem busy and active. Recognition of the need to stop, think, and analyze what we have accomplished did not seem to be important. Often this type of "busy work" behavior was learned early in our schooling. This attitude has limited many excellent service technicians in their ability to survive as technology has changed. You must be able to break away from this prelearned behavior. Recognition of thinking time as a tool for analysis and repair is critical. Use of the basic concepts of electricity and electronics is critical for continued success in the repair field.

I remember a series of magazine articles that appeared in one of the more popular magazines many years ago. The scene of the articles was an automobile repair facility. The key characters were a more experienced, older mechanic and a young mechanic. Each article was designed to show the need to stop and analyze the information obtained from both the customer and the preliminary examination of the vehicle. Usually the younger mechanic rushed into the job and made some repairs to the car, but these repairs often did not solve the customer's problem. Ultimately the younger mechanic had to request assistance from the more experienced mechanic.

The first thing the more experienced mechanic did was to ask questions about the problem. Once he had this information, he sat down to analyze what he had learned. This was followed by some further questioning and suggestions for additional testing. Ultimately the problem was located and the reason for the analysis was explained.

The message in this series of articles was very strong. Basically it said that analysis and thinking about how to resolve any problem is of extreme importance. This applies to all types of service problems, not just those in the automotive or electronic fields. Once you learn the ability to problem solve using analysis techniques, you are on the way to becoming one of the world's best troubleshooters. Follow the steps described in this chapter and the balance of this book. You, too, can join the ranks of these troubleshooters.

THE PATH TO BECOMING THE WORLD'S BEST TROUBLESHOOTER

The first step in this process is to attempt to determine what the problem really is. This is accomplished through several important steps. The first of these steps is to talk to the operator of the machine or system. It really does not matter who the operator is or what the type of machine happens to be. Use your most tactful approach and question the operator. Given sufficient time, the operator will identify the problem for you. Ask for a description of the problem and ask how long it has been going on. Ask for a comparison of how the machine is operating at the present time and how it is supposed to operate. Also inquire whether the machine or system has done this in the past. Perhaps there is a history of past repairs. If so, this may simplify your work in restoring the equipment to its normal operational mode.

Keep in mind that the operator most likely is very frustrated due to the failure of the equipment. You will probably have to overcome this frustration in order to really hear what the problem is. This venting of frustration is a normal part of describing the problem. It is best to listen to it even though you really do not want to hear all of the operator's frustrations.

Second, be sure you understand how the equipment is supposed to operate. If you have worked on it before, check it out to see if it still operates in the same manner. Try to determine if anyone has modified it since you were last involved with it. You may have to review the operating manual. This is particularly true if you are not overly familiar with the unit, machine, or system. The **service literature** for all equipment should be reviewed prior to any testing, regardless of your previous knowledge about the equipment.

Attempt to operate the equipment. This is necessary in order to determine whether its specifications are still within those provided by the manufacturer in the technical literature. There are times when the operator has worked on a similar piece of equipment and expects this particular unit to operate in a similar manner. The other unit may be newer or may have been modified while the one you are to service is the older, unmodified type. Unusual operator expectations may be all that is wrong with the unit. Your ability to successfully convey this to the operator is important. It is hoped that the operator is not expecting more from the equipment than it is capable of delivering.

Next *stop* and *think* about your plan of approach to repairing the equipment. Exactly what do you have in mind about the method of analysis and repair? This is the area where you give the appearance of doing little, if anything, toward actually repairing the unit. Actually, the time spent in planning the approach and determining what should be found if the test area is operating properly is almost as important as your arrival to start the repair. Consider all of the symptoms involved in the malfunctioning equipment. This must be done before you pick up a single tool or piece of testing equipment. What you are now doing is similar to a doctor asking the patient what hurts and how it hurts. This requires the use of a set of equipment you brought with you on the job. This built-in set of tools are your senses of smell, sight, hearing, touch, and even taste.

It may be in your best interest to write down all of the symptoms of the unit at this point in your learning cycle. This will assist in identifying the problem area. List the likely cause of the problem after each symptom on the list. Whatever you do, do not assume anything as you approach the problem. Often one of the simplest problems for you to service is the cause for the service call. Examples of this include a disconnected cable, an unplugged power cord, no paper in a printer, and many other seemingly stupid things. If you should find that one of these is the problem, approach the solution in a very tactful manner. Remember that the operator does not want to appear to have done a foolish thing in calling for service. Treat the repair as a very professional activity. Help the operator understand the problem and show how it can be prevented in the future.

Review your list and attempt to place the causes of the problem in priority order. In many ways the approach to any service-related problem is a matter of playing the odds. Often the hardest-working section or components in a system are the ones that fail most of the time. Play the odds and consider this area as one of prime suspicion. After you have identified and listed the causes, then develop a troubleshooting procedure for each of them. Use the procedure you have established; do not attempt to take shortcuts around it. Experience has shown that these shortcuts often lead to additional and unnecessary work. You may jump over the problem area by skipping around and deviating from the procedure you have established. This approach often leads to extra time on the repair resulting in few, if any, productive results. If the service literature with the equipment has a troubleshooting section, use it to your advantage. After all, someone else has taken the time to develop an organized and systematic approach to problem solving for you. Use that person's knowledge and experience with the equipment to assist your repair procedure.

Avoid being one of those people who say, "Don't confuse me with the facts—my mind is made up," when servicing. Often we establish some **mental barriers** for our work or our lives. One example of this is illustrated in Figure 6–1. The challenge shown in part (a) is to connect all nine dots using four straight lines without lifting the pencil from the surface of the paper. People tend to place barriers around the perimeter of the blocks when starting to analyze the challenge, as seen in part (b). When you reread the challenge, there is no mention of these barriers. We tend to do this type of limiting based on some previous experience. This is a classic example of the mental barrier syndrome just described. The solution to this challenge is shown in part (c). The mental barriers have been ignored for this solution. Lines are drawn outside of the perimeter of the nine-dot box, and in this manner the problem is solved.

This type of thinking is critical when you are trying to locate a problem in a system or section of a unit. Another example of the need for critical thinking when attempting to identify a specific problem in a unit uses the partial schematic depicted in Figure 6–2 as its illustration. This is a partial schematic diagram of a citizens band radio. The problem in this unit was described as a "no signal" condition. Using the typical tools of the trade, the service technician found that the signal at transistor Q_2 was incorrect. Instead of the proper 0.455 MHz frequency at point C, a frequency of 10.455 MHz was measured at this point. The service technician determined that the stage to the left of the transistor contained the problem. A check of the waveforms and frequencies at points A and B verified the correctness of the frequencies at those points. The component of immediate suspicion was the diode D_1. This was replaced, but the unit still did not work. The technician continued to check around the area of the diode, even suspecting that the replacement diode was bad and again replacing it in the circuit. This did not solve the problem. The technician was baffled at this point and did not know what else to do.

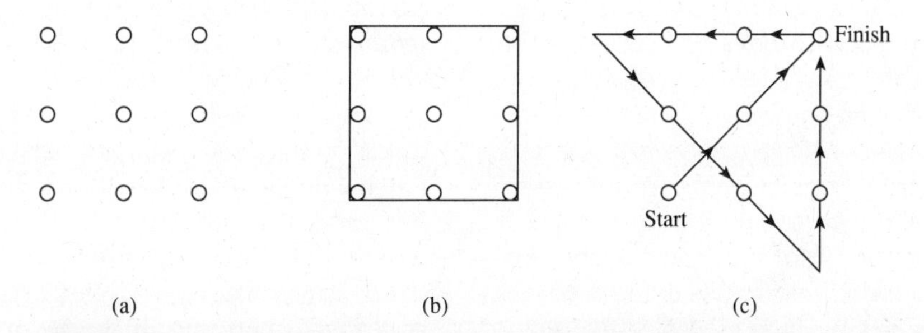

Figure 6–1 Do not be afraid to move outside of imaginary barriers in order to solve a problem. The box illustrated in (b) is one such barrier. The resolution to the problem is shown in (c).

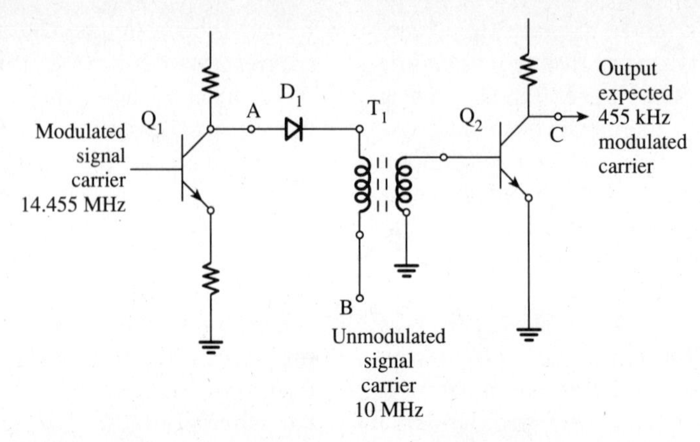

Figure 6–2 **Mixer circuit from a citizens band radio. A failure of transistor Q_1 provided a lower than normal signal to the diode.**

After all, these conditions existed:

1. proper frequency signals at both points A and B
2. proper operating voltages throughout
3. diode D_1 tested OK
4. diode D_1 replaced
5. no visible physical damage
6. transformer T_1 tested OK

The technician failed to consider other alternatives for solving this problem. The area of concentration was the diode and the transformer in the circuit. Since the frequencies of the respective signals were proper, the technician failed to remember two significant things: how the diode is supposed to operate in any circuit and how it should function as a frequency converter in this circuit. Actually, the failure on the part of the technician was directly related to not remembering how diodes are supposed to work. The anode of the diode requires a higher-value voltage level than the one on the cathode if the diode is to be forward biased and permit current to flow in the circuit. Proper signals were present. What was not present was the proper amplitude of signal voltage at the anode of the diode. A check of the system, using an oscilloscope, provided this information. Further testing indicated that transistor Q_1 had a partial failure. It did not amplify its input signal sufficiently, the result of a loss of the transistor's ability to act as an amplifier. The solution to this problem was replacing this transistor. Once the amplitude of the signal at the anode of the diode was correct, the radio operated properly.

This is a typical case of failure to consider all of the facts. The service technician's mind was made up to believe that the problem was in either the diode or the transformer of the radio. The facts were not fully analyzed. Signals of the proper frequency were present. Signals of the proper amplitude for diode action were not present. In addition, these signals were not checked out in order to prove whether the diode was functioning properly. The technician failed to consider all of the facts in this case. If the technician had considered the basic rule of diode action, the solution to the problem would have been more apparent. Use of a chart to list the problems and consideration of all solutions to these problems would certainly have resulted in a more efficient and rapid repair.

This example highlights still another suggestion to use in the process of troubleshooting: When you reach a dead end, or roadblock, in the troubleshooting process, take a break. Walk away from the equipment. Sit down with a cup of coffee or a soft drink and consider what you have done and what still has to be done. It is obvious that the equipment's problems still have not been resolved. A fresh start is required. A break will often help you decide on a fresh approach.

In another example, an automobile, an older foreign sports car, would stall while being driven. The operator of the vehicle had to be able to determine what was causing the problem even though replacement parts were not readily available. The problem was resolved by thinking about the normal operation of the vehicle. This vehicle had an electric fuel pump located on its fuel tank. The operator listened to the start-up noises and failed to hear the sound associated with the pumping action of the fuel pump. This led to an inspection of the wiring of the pump, and a corroded common connection was discovered. The solution was to remove the pump, clean up the connection terminal, and replace the pump. This solved the problem. This is a classic example of using all of the senses to determine where the problem was. The alternative to using the senses was for the operator to locate and replace typical components in the hope that doing so would solve the problem. This approach is both expensive and time-consuming. And there is no guarantee that the problem would be resolved in this manner.

Consider also that while you may be the world's greatest troubleshooting expert, we all need some additional help or information to solve problems. Don't be embarrassed to call for assistance when you need it. Remember that the equipment really does not have a mind of its own. It is not "out to get you." It was designed and built by human beings. Often someone else is experienced in and has great ideas for servicing. Another opinion often will help you repair the equipment rapidly and efficiently. Your ego should not be so big that you cannot recognize the need for additional assistance when you are attempting to solve a problem.

Also remember that your job is to repair the machine as efficiently and quickly as possible. Your responsibility is not in the field of engineering. Your work is not in the area of design. You must try to repair the unit, not redesign it. There may be times when your suggestions for circuit modifications are valid, but these must be cleared through the company's engineering department before any equipment modifications are performed in the field. Your efforts must be directed toward rapid return of the nonoperational equipment to its operational stage. The operator of the equipment is more concerned about its return to operation and its ability to be productive than anything else.

Be sure you find the real cause of the problem. Identify the "culprit" that created the problem. Figure 6–3 shows a transistor amplifier circuit. A physical inspection of the area shows that resistor R_3 has taken on an unnatural appearance. It may look charred, or burned. At other times the resistor may not look like a piece of burned material; it may only be that its sides have bulged out beyond their normal shape. The immediate solution is to replace this component. While this certainly is one of the procedures that will have to be done, it may not totally solve the problem. Of prime importance here is determining what caused this resistor to char. Obviously the charring was due to excess current and heat in the resistor. The service technician must consider why this happened. It seems obvious that some other component in the circuit created this excess current problem. This must be addressed in order to resolve the total problem. The solution is to test the other components in the circuit before completing the repair. This will find the true cause of the problem. It is very unproductive to have to spend time disassembling a unit for the second time if the primary repair cause was not identified.

A problem similar to the one described earlier is shown in Figure 6–4. This figure shows a typical auto radio audio output circuit using discrete components. The audio output power transistor had a history of failing in this circuit. No other components had any evidence of failure. There were no

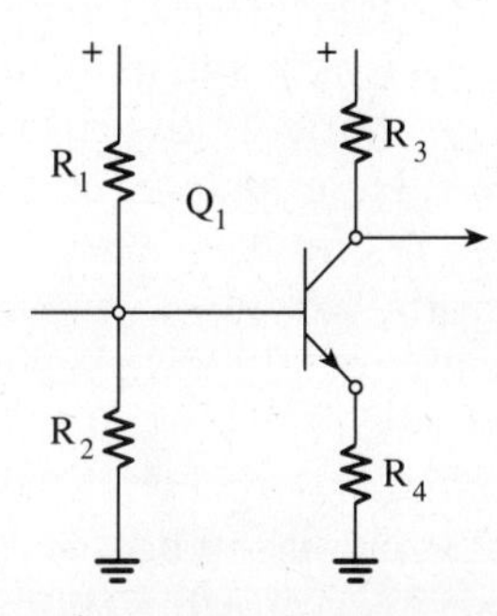

Figure 6–3 **Even though resistor R_3 was charred, the actual problem was found to be another component in this circuit.**

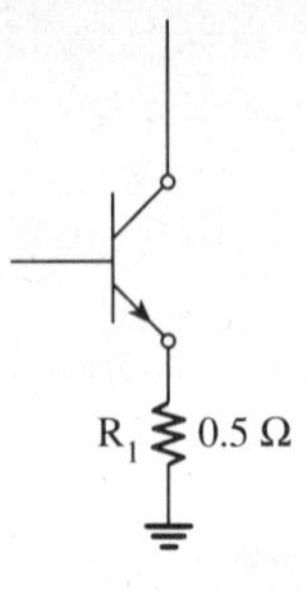

Figure 6–4 The 0.5 Ω resistor often fails in this auto audio output circuit.

visible indicators of charring or broken leads. The service technician replaced the power transistor, and the system was restored to its working condition. The time involved in installing the radio in the automobile was over a half hour. The service technician did **bench test** the unit for a short period of time and found that it seemed to be functioning properly. The next logical step was to return the radio to the automobile. Unfortunately for the technician, the bench test was too brief. One of the characteristics of this type of circuit is that the 0.5 Ω emitter resistor had a tendency to fail after operating for just a short time. The service work on this radio should have included replacing the 0.5 Ω emitter resistor in addition to the transistor. A longer bench test of at least a half hour would have indicated that the emitter resistor had also failed. In this case, the customer was on a vacation, and the radio failed shortly after the car left the repair station. The vacation schedule did not allow for time to return to the repair facility to correct the problem. The net result was a very unhappy customer and a recall repair at no charge to the customer after the vacation was over. Here, an extra half hour of bench test, or "burning in," of the repaired radio would have shown that the repair was incomplete. In addition, the customer would have been satisfied that the repair was correct.

Once the repair is completed, the unit must be tested. The test period has to be long enough to ensure that the repair was effective. If this is not done, the service technician may have to return to perform additional service work on the unit after it has been reinstalled or returned. Customers do not appreciate additional repair costs or their inability to use the machine. Employers also do not appreciate this type of work; recalls are not income-producing work. They are done at the employer's expense and often at the expense of the service technician. The above example is an excellent indicator of poor customer relations and bad servicing techniques.

The final step in this process is to keep a log, or notebook, of what has been done. Include in this data the cause of and solution to the problems faced in repairing the specific unit or system. Problems have a way of repeating themselves. Referencing these problems and how you solved them will assist you in future service calls.

BASIC METHODS OF APPROACHING THE PROBLEM

Now that you have had the opportunity to learn how to be the world's greatest service technician/troubleshooter, you need to look at the process again. Return to the procedures identified earlier in this chapter.

IDENTIFY THE PROBLEM FIRST

Preliminary Information Obtain preliminary information from the initial call for help. With a little luck, the person who takes the service call from the customer is trained to try to find out as much as possible about the problem at hand. Often the service department will develop a question sheet. The operator will ask key questions in an effort to provide preliminary information to the servicer. This will help to determine the specific kind of problem, the type of equipment involved, and other additional information required for a rapid and successful service call.

On-site Information On-site information is obtained during the initial service call step, when the servicer meets with the person seeking help. Time spent discussing the nature of the problem will pay off in time saved while working on the system or unit. Listen to what the customer is telling you, for he or she often will explain exactly what the problem is, where it is located, and even, in some cases, offer suggestions about how to repair it.

Several years ago, my new car had a problem. Being interested in how things operate, I had purchased the service manual for the car. When I identified the specific problem, I returned the car to the dealer for service. I described the problem in great detail and offered suggestions as to what was wrong. These suggestions were based on the information obtained from the service manual for the car.

Rather than appreciating my efforts to help, the service manager told me forcefully that it was none of my business as to what was wrong or how to fix it! The service manager also informed me that he was the only one qualified to make these diagnostic analyses. It so happened, in this case, that my diagnosis of the problem was correct. Even if I had not been correct, the service manager did little to make me feel confident of his ability. As a result of this confrontation, I decided not to return to this particular dealer in the future. This anecdote is just one more message about the importance of listening to the customer. Even if the customer is totally incorrect, just the knowledge that someone (in this case, the service manager) listened politely and considered what was said goes a long way toward maintaining good customer relations.

PRELIMINARY INSPECTION

A preliminary inspection of the unit helps to locate the problem area. If possible, attempt to operate the unit. This will provide information only you can obtain about how the system or unit is functioning. It will also help to reduce the possible area of trouble from the entire unit to perhaps one or two sections. While operating the unit, listen for unusual noises, look for smoke or other visual signs of trouble, and do not forget to use your sense of smell to assist in the diagnosis process. All of this should be done before you use any tools.

CLOSER INSPECTION

Inspect the suspected areas closely. This requires a careful physical inspection of the inside of the unit. Look for burned, broken, or charred components. Inspect for loose connections or broken wires. Locating any of these will assist in identifying the specific problem area. These inspections are accomplished without the use of any testing equipment. The only tools used up to this time are those required to remove the cabinet or back of the unit for inspection.

USE THE MANUFACTURER'S LITERATURE

Refer to the manufacturer's literature and review the functional block diagram of the unit. This will aid in locating specific blocks or sections requiring a more detailed inspection or testing. Block diagrams are a necessity if you are to understand the complex operation of most electrical or electronic devices. The ability to recognize the function as described in each of the blocks on the diagram will assist in the preliminary elimination of those blocks that seem to be functioning properly. At this point there is little, if any, need to test or check these blocks. Later on it may be necessary to return to one of them as you narrow down the area of suspected trouble.

In addition, the use of the manufacturer's literature can aid in the location of a specific section of the unit requiring further testing. Some of the more "enlightened" manufacturers will include troubleshooting information with their installation and service literature. Using this will certainly aid in the location of the problem area. Often the service literature will include photographs or line drawings indicating the location of boards and components.

Some manufacturers even include a **troubleshooting flow chart**, similar to the one shown in Figure 6–5. As you review this chart, notice that it shows a series of tests. Each of these tests is designed to eliminate a working part of the unit as a possible problem area. Many manufacturers even include the type of testing equipment to use. This is the basic function of any troubleshooting procedure. Use of these flow charts will often expedite the repair process.

IDENTIFY THE APPROPRIATE PATH

Identify the appropriate path for the process involved in your inspection. You are either looking

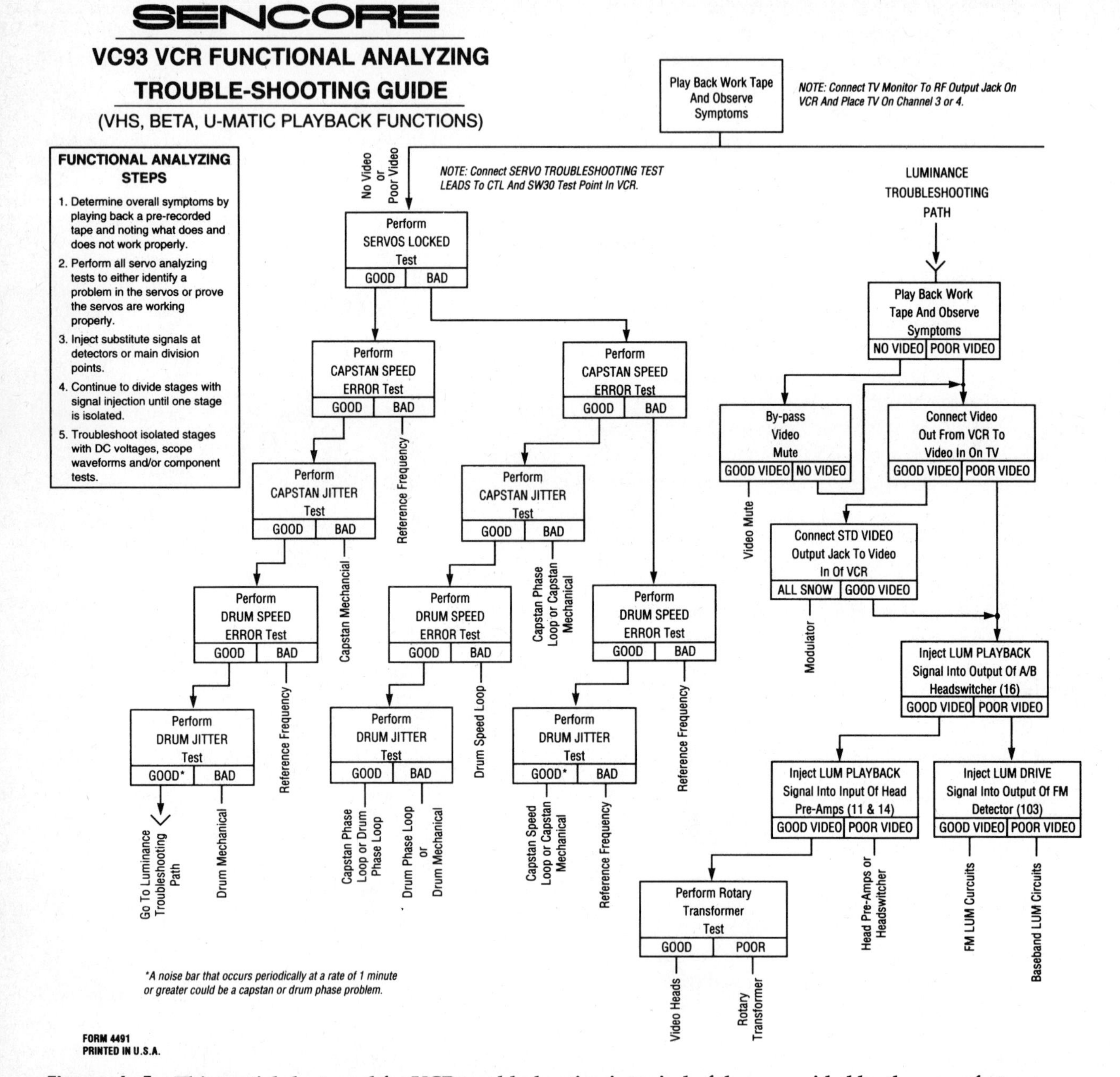

Figure 6–5 This partial chart used for VCR troubleshooting is typical of those provided by the manufacturer as an aid to diagnosis and repair. (*Courtesy Sencore Inc.*)

for a **signal processing path** or a **current flow path**. Initially you should be looking for the signal processing path. This will be indicated by the block diagram of the unit. If no block diagram is available, the schematic diagram will often provide similar information. For example, the schematic diagram shown in Figure 6–6 identifies the function of each semiconductor. Since almost all radios have similar block diagrams, it is possible to relate the function of the semiconductor to a specific block. Knowing the function of the block will aid in determining how it operates.

The second portion of the path identification system is related to the path for current flow. Most electronic technicians are taught to follow the electron current flow path. This path starts at circuit common, usually the power source negative, and flows through resistors and other conductive devices until it reaches the positive terminal of the power source. Most electronic devices are designed with one power source and several parallel current paths from that source. Current path analysis should only be performed after you have identified which specific block or section of the unit is not operating properly.

DEVELOP AN APPROPRIATE PLAN

Developing the appropriate plan to use approaching the service problem is of utmost importance. The world's greatest service technician will carefully plan the approach to be used in locating the problem. This planning includes identifying the initial test point, the type of equipment needed to make this test, and the expected test results. You should know what to find if the device is working properly before you conduct any tests. The planning phase of the repair should be a very thorough one.

Once the initial test point has been identified, the next step is to determine the type of test equipment to use. Selecting the proper type of test equipment will tend to reduce wasted efforts. Actually, the service technician should plan a complete set of tests. These should be planned using the initial planning worksheet described in an earlier chapter. Learning to plan on paper will ultimately lead you to planning the approach to the problem mentally. The use of a paper plan will help eliminate wasted steps during the testing process. Let us use a basic audio amplifier as the model for the identification of a plan to test a unit. The block diagram for one

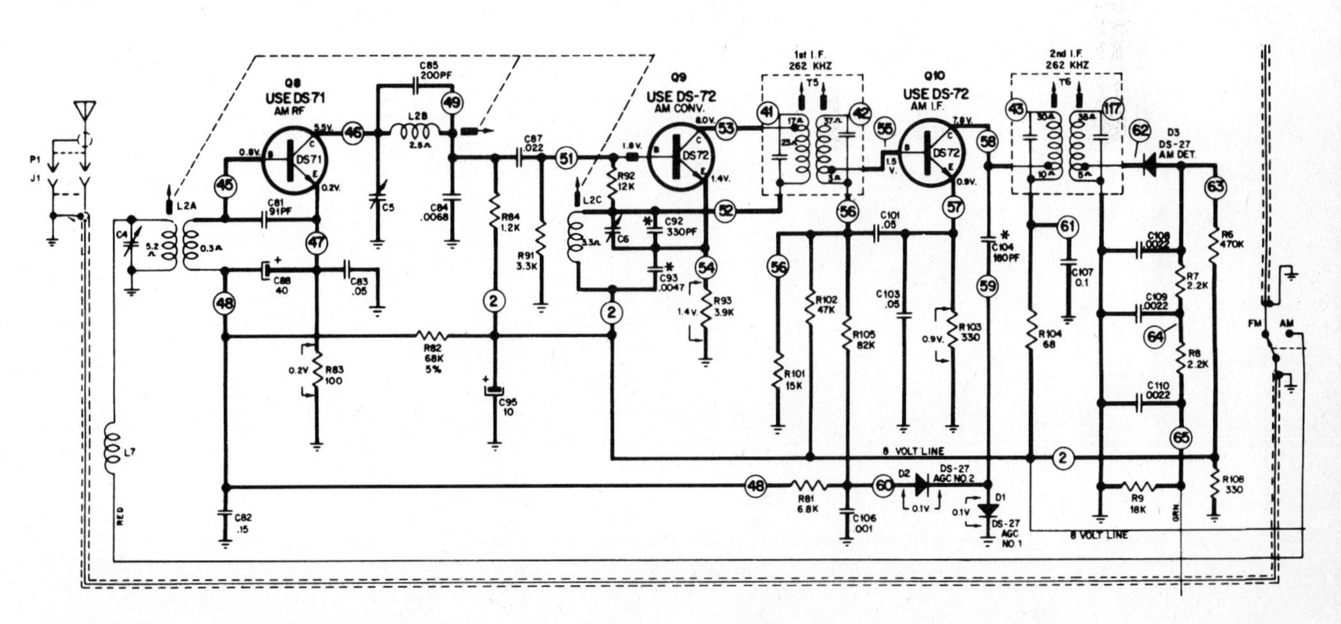

Figure 6–6 Component functions are often identified by the manufacturer on the schematic diagram. This aids in the location of functional blocks.

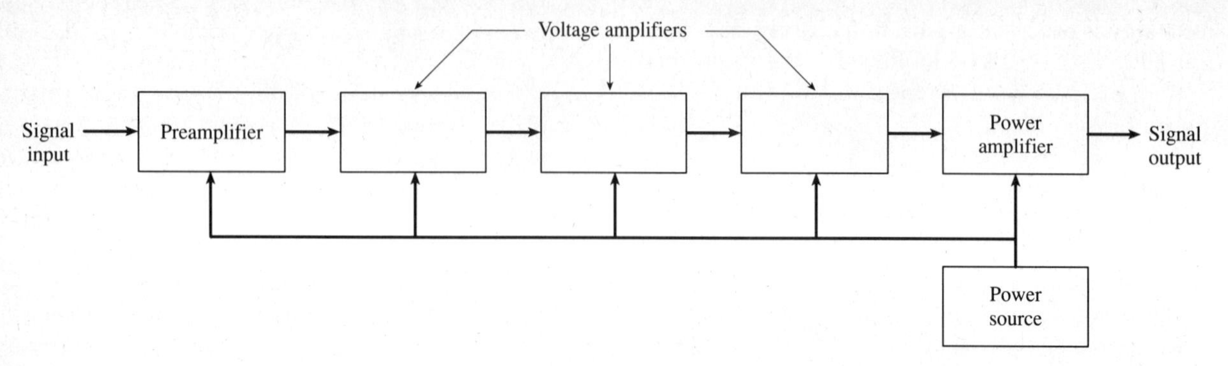

Figure 6–7 **Block diagram for an audio amplifier, showing both power supply connections and signal flow processing.**

such unit is shown in Figure 6–7. This audio amplifier has six active blocks. The input is a voltage amplifier identified as a preamplifier. Three blocks, or stages, of voltage amplification follow the input block. The audio output block is a power amplifier. Also included, and definitely not to be overlooked, is the power source for the system.

The problem is identified as a "no output" condition in this example. Initial brackets are placed around the entire system as shown in Figure 6–8. The paper approach must include answers to the following questions:

What is the initial type of system: current flow or signal processing?

What type of block system is used in this device?

Where are the initial brackets to be placed?

What type of test signal is to be used?

Where should the test signal be inserted into the system?

Where should the first test be made?

What should you expect to find if the initial test is OK?

Which bracket should be moved and where should it be moved to if the initial test is OK?

Which bracket should be moved and where should it be moved to if the initial test is not OK?

The answers to these questions are what you should try to find during the planning portion of the service call. After these questions are answered, then, and only then, should actual testing be initiated and conducted.

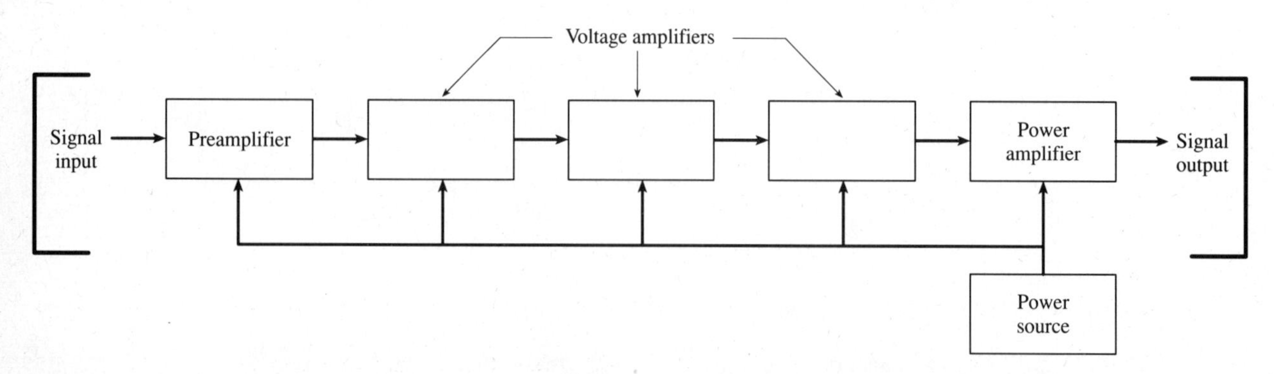

Figure 6–8 **The initial set of brackets is placed around the entire system as a starting point for diagnosis.**

IF IT ISN'T BROKEN, DON'T FIX IT

If the unit being serviced actually is working, there is no reason to attempt to fix it. While this statement may sound ridiculous, consider the possibility of an operator error. Consider also the possibility of a software problem if the unit is computer controlled. Keep in mind at all times that a human being is operating this equipment. Human beings are known to attempt shortcuts. These shortcuts may work some of the time; however, they may not work all of the time. It is these "other times" that are prone to creating problems.

YOU ARE NOT AN ENGINEER

This statement relates to making engineering design changes while performing a repair procedure. Several major repair companies call their field service personnel "field service engineers." This does not provide them with either the knowledge or the credentials of a true engineer. Regardless of your title, your responsibility is to repair the problem and return the equipment to operating condition as quickly as possible. Making changes in the design of the equipment is not one of your responsibilities. There may be times when, in consultation with the engineering staff of the manufacturer, a circuit is modified while the service technician is on the job. This is only to be done under the supervision of the engineering staff. Under no circumstances should you make design changes on your own initiative.

The previous paragraph is very explicit. This does not mean that you cannot be creative when attempting to solve a service-related problem. Use your brain to get the equipment up and running as quickly and as safely as possible. When a system requires a 10,000 μF capacitor for its power supply section and none is immediately available, the possibility of parallel combining two or more capacitors to obtain the proper quantity of capacitance is acceptable. You may have a problem with physical constraints when combining units, but who is to say that a temporary "fix" is wrong? The prime concern of the customer is to return the equipment to productivity as quickly as possible. How this is accomplished, assuming that it is safe to operate, is of little concern. A temporary "fix" may require returning for a permanent repair at a later date. Customers understand this and are willing to work with you. Your "hero" badge will not be tarnished when you do have the units operating again. Remember, this is why the original call was made. The main concern is to eliminate the problem and return the equipment to operating condition.

One additional thought on this process of repair: All good service technicians, at one time or another, will find a better way to perform some of their work. These thoughts should be shared with the company. If the company is willing to listen, and most are, these suggestions may pave the way for recognition of them. Some companies offer bonuses for time- and labor-saving suggestions. Others will reward the person with a salary increase. Regardless of what the individual company policy may be, the ability to recognize a better method and share this information with others is self-rewarding.

KNOW WHEN TO QUIT

There will be instances during the service process when the only solution to the problem is to replace the entire unit. This decision is to be made upon the recommendation of the servicer. The cost of the repair may exceed the value of the equipment. Newer equipment probably is able to do more than the unit requiring repair. The service technician has a responsibility to the customer to provide this information. There is nothing wrong with admitting that continued servicing of the unit is no longer cost-effective. Be honest with yourself and with the customer when making this decision. Both of you will benefit in the future.

REVIEW

The basic rules for troubleshooting must be followed if you are to be successful in this endeavor. These, applied with the fundamental rules for electrical and electronic circuits, aid in the rapid and successful repair of nonoperational equipment. One of the cardinal methods for success in this field is to use all of our senses when approaching any type of repair problem. The ability to review all of the information available and then to analyze it carefully before attempting any physical work is of extreme importance. This ability allows you to consider the basic rules and apply them to any type of device. These fundamental rules do not change as the applications of electrical and electronic technology change. You have to be able to apply them to all devices.

One of the best sources of service-related informational items is the customer. Listen carefully to what is said and attempt to apply all of the information to the problem at hand. Often customers will tell you exactly what is wrong and even how and where to fix the problem.

Be sure you understand how the equipment is supposed to work when it is functioning properly. This knowledge will help you to restore it to its original operating condition. If possible, operate the equipment to see how it functions for you. There have been times when operators think equipment is capable of doing more than it was designed to do.

Plan your approach to the repair problem. The novice repair technician should use paper and pencil for this initial planning stage. List the equipment to use, where to use it, and what you expect to find when you conduct the test. Stop during this process and review what you have accomplished. Modify your plans as you obtain additional information during the testing process.

Do not develop any mind-set about the solution to the problem or where, in the unit, the problem exists. This type of thinking limits your ability to resolve the problem. Expect that thinking in this manner will often increase the amount of time required to repair the unit.

Keep referring to the basic rules as you evaluate the test results to determine whether the section is functioning properly. Failure to observe the basic rules also often extends service time on the unit. If you fail to locate the problem in the unit, take a break from the service activity. Move away from the problem; relax and consider what you have done and where to go after your break. If necessary, call for assistance from someone more experienced than you. This willingness to accept help is a sign of the intelligent service technician, not a sign of weakness.

Use all of your senses while evaluating the repair. Look for damaged components. Inspect for broken wires or disconnected cables. You will develop a keen sense of smell, recognizing the unique odor of overheated components. Feel for "hot spots" on components or units in the device. All of these senses are important to the servicer. Read the service literature provided by the manufacturer of the unit. This will often provide information to make the job of repair more efficient.

Do not attempt to perform engineering-level changes on the equipment. Your role is to maintain the integrity of the system while restoring it to operating condition. If necessary, contact the engineering staff and discuss any modifications with them. Then, and only then, should circuit modifications be made.

REVIEW QUESTIONS

1. Explain the importance of questioning the customer as a start of the service process.
2. What information is provided in the manufacturer's service literature?
3. What is the function of the block diagram of the system?
4. Explain the difference between a signal flow path and a current flow path.
5. What information is provided when you see a burned or charred component?
6. Why is a visual inspection important to the service technician?
7. What is the result of a mind-set while servicing?
8. What is the difference between repairing and reengineering a unit?
9. What is the importance to the technician of a service record of the equipment?
10. Why is it important to plan the service procedure before actually starting to work?
11. Why should the service planning sheet be reviewed during the service process?
12. Why should you keep a record of the repairs you have made?
13. What should be included in the repair record file?
14. Why is it important to apply the basic electrical rules when servicing?
15. Why is it important of identify the type of path (linear, meeting, etc.) when servicing?
16. Explain the process of bracketing the system as a service technique.
17. Explain why it is important to know what expected values of voltage or signal are to be found during servicing.
18. List the five most important steps for servicing identified in this chapter.
19. Why is it important to know when to quit during a service procedure?
20. What is the importance of the bench test procedure?

CHAPTER 7

Test Result Analysis

INTRODUCTION

Owning an array of test equipment is very impressive. Knowing how to use this test equipment is doubly impressive. Who is being impressed by this array of electronic testing equipment is the major question. Is the owner trying to impress others? Does the owner have to have this equipment available to meet the qualifications for factory-authorized service? In other words, why is this array of equipment being used? If there is a true need for the equipment, then the most important thing is to see that it is used properly.

Observing the equipment being used successfully is very rewarding to its purchaser and to those overseeing successful repairs. Knowing how to analyze the results of the specific test is how one becomes the world's best troubleshooter. Chapter 4 discussed how signals and current flow paths should be analyzed. Chapter 5 presented information related to the operation and selection of testing equipment. This chapter will help you understand how to use the results of a variety of tests and measurements. The results of an individual test analysis determine where and how the next test is to be made. This analysis also aids in the identification of specific modules or components that have failed.

OBJECTIVES

Upon completion of this chapter, the reader/student should:

1. better understand how to analyze test results;
2. be able to relate signals to functions of blocks;

3. be able to analyze voltage measurements;
4. be able to analyze current measurements; and
5. be able to analyze resistance measurements.

KEY WORDS AND PHRASES 

open circuit: An incomplete electrical circuit. This circuit does not permit current flow.

oscillator: An electronic circuit capable of creating an electronic signal.

partial short circuit: An electrical circuit containing an alternate and undesired path for current. This describes a condition where part of the current flow is normal and part follows the abnormal path.

schematic values: The specific component descriptions and voltage levels that appear on electrical or electronic "blueprints."

short circuit: An undesired current flow path in an electronic device. This condition is often caused by component failure.

tolerance rating: Almost all electronic components are manufactured with some variance from their design and marked values. The tolerance rating is the permitted percentage of deviation.

ANALYSIS OF TEST RESULTS

A major tool required for the analysis of test results is the manufacturer's service literature and circuit, or schematic diagram. These schematic diagrams provide exact information about the device being analyzed for repair. A high-quality service literature package has much to offer the service technician. The better service manuals include schematic diagrams, physical parts layouts, and unit interconnecting cable connections. In addition to these, there often are circuit board and module parts placements, alignment procedures, parts lists for replacement parts, and unit disassembly instructions.

A typical schematic diagram is shown in Figure 7–1. There are two major considerations when you study any schematic diagram. One of these is the decision whether to look for signal paths or current paths. The second consideration is to understand how each of these is displayed on the diagram. Many schematic diagrams use an industry standard format for their layout. Both the signal flow path system and the current flow path system are explained in the following paragraphs.

SIGNAL FLOW PATHS

Signal flow paths are displayed in the same way you read a book or newspaper. The signal input to the system is at the upper left-hand corner of the diagram. It is typically processed in a row-by-row manner. This is illustrated in Figure 7–2, where the signal path is shown by the use of arrowheads on the diagram. Once the signal processing, or flow, reaches the upper right-hand side of the first row of the diagram, lines drawn on the diagram show it returning to the left-hand side of the schematic diagram. This progression continues until the output stage of the unit. Usually the output stage is shown on the right-hand side of the diagram. This is true for a linear, or in-line, type of circuit. Other circuits, such as meeting or separating types, require the use of a block diagram in conjunction with the schematic diagram to follow the signal processing paths. Typically the power source is displayed in the lower right-hand corner of the schematic diagram regardless of the type of signal flow path used.

CURRENT FLOW PATHS

Current flow paths differ from signal paths. Current flow paths are normally displayed in a vertical manner. These are identified in Figure 7–3 by the arrowheads on the schematic diagram. Normally there are multiple, or parallel, current paths in electronic equipment. Service technicians follow the electron current path as they analyze the specific circuit. Cur-

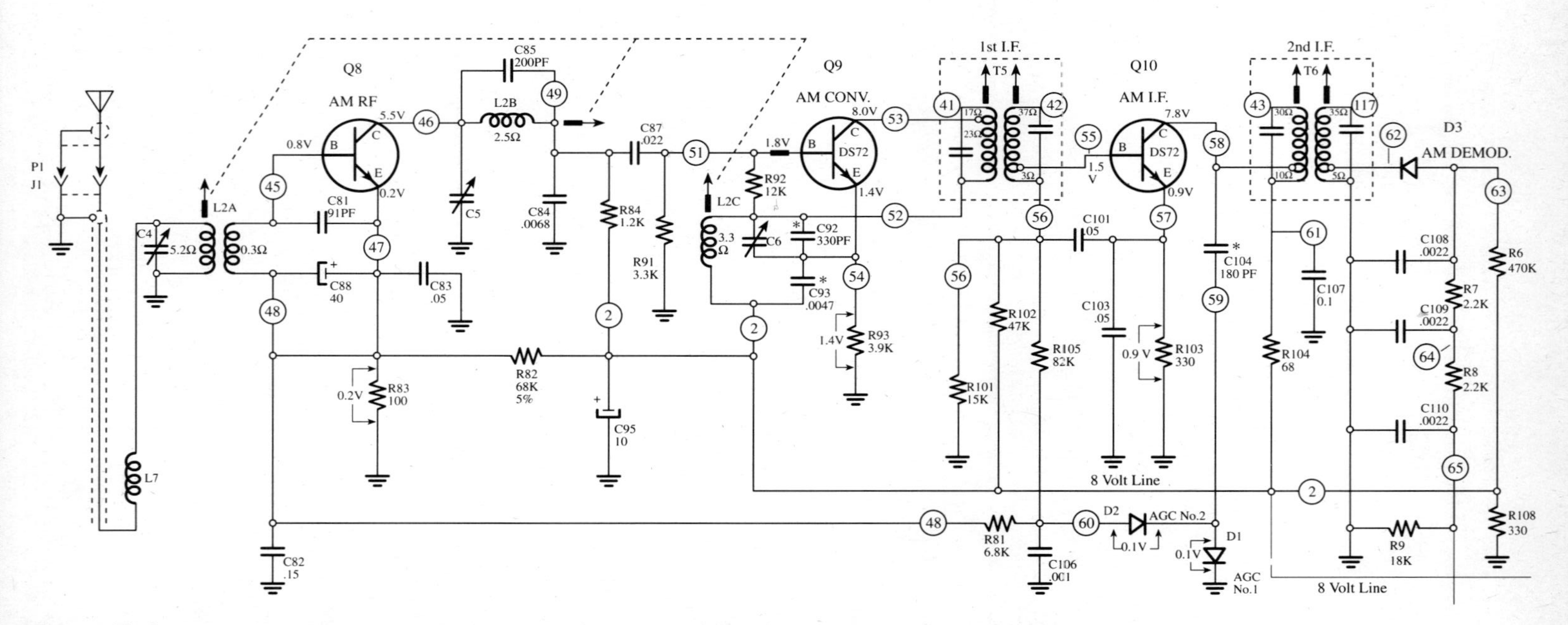

Figure 7–1 This schematic diagram is typical of those available for the electronic service technician.

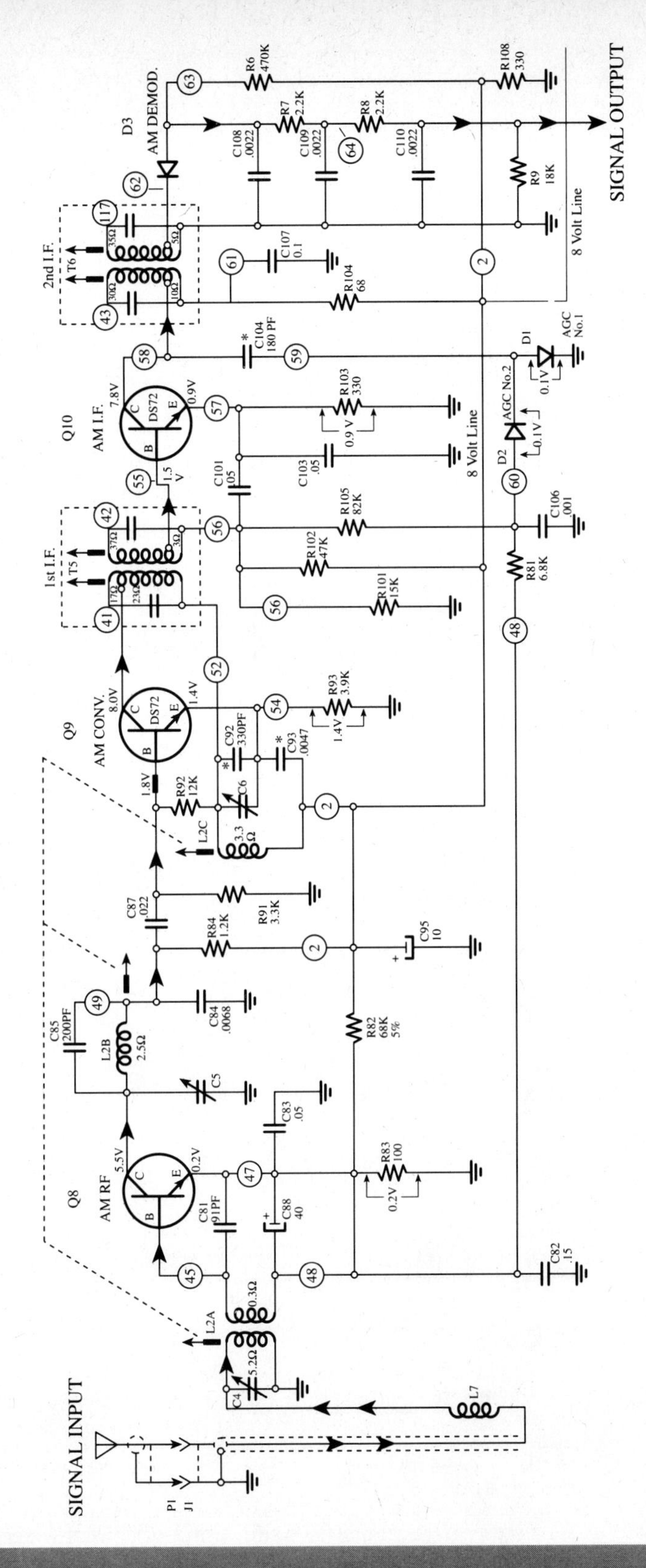

Figure 7–2 The signal flow path is highlighted with arrows on this diagram.

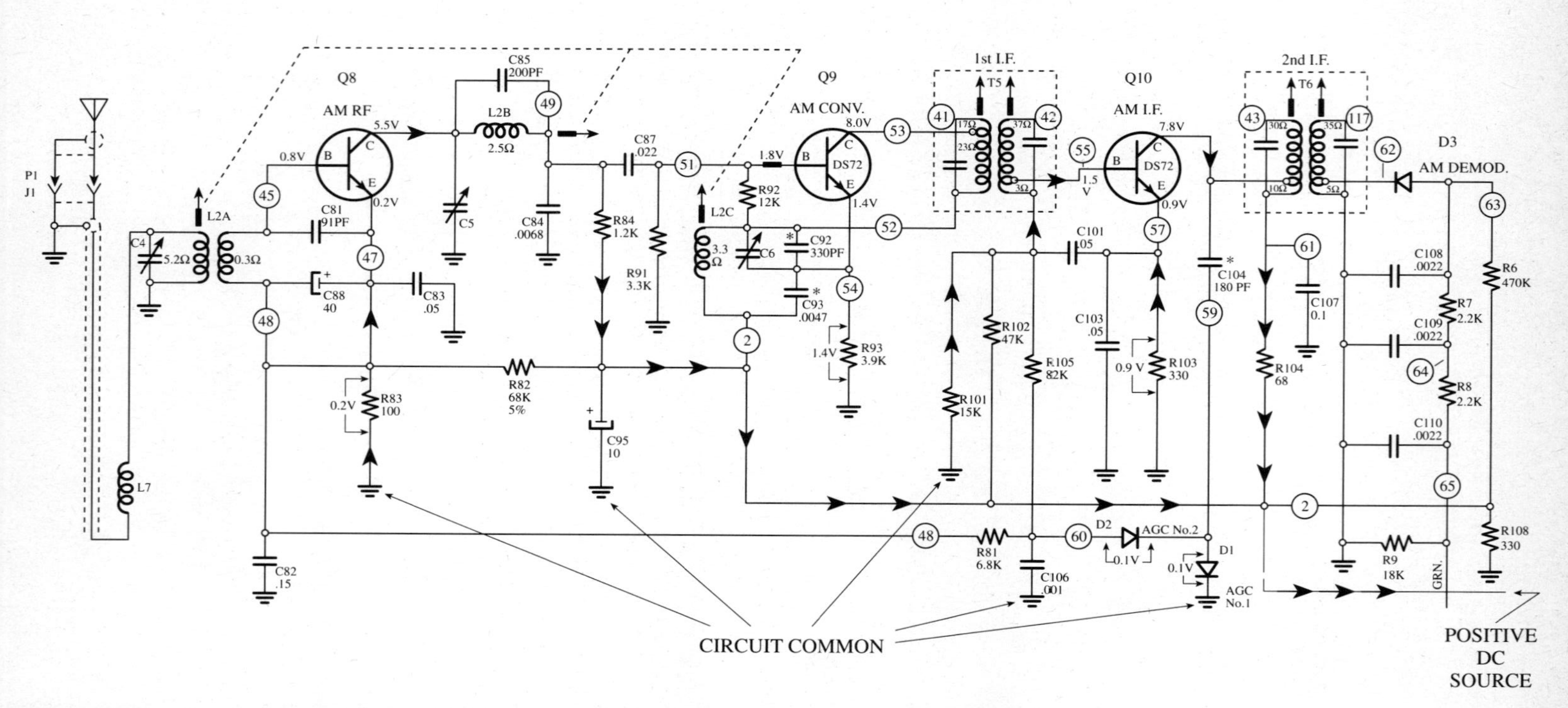

Figure 7–3 Current flow paths are highlighted with arrows on this version of the schematic diagram.

rent production electronic devices usually use the negative terminal of the power source as circuit common. Electron current flows from circuit negative common, through each circuit, and then to the circuit positive terminal of the power source. Each individual circuit uses this same system for current flow. Analysis of current flow paths requires the ability to trace the paths on the schematic diagram.

SCHEMATIC VALUES

Schematic values are those provided by the engineering staff of the company that produces the device. These values are those selected by the designing team. The same values are marked on the components in the circuit. Somewhere on the schematic diagram will be a statement indicating the type of instrument used to measure the values of the components and the acceptable tolerance values for them. This is normally on the lower portion of the printed page in a box marked "Notes."

Deviations from Schematic Values Deviations from **schematic values** are not only permitted, they are often normal occurrences in the process of servicing electronic products. The major question is, "How far can they deviate and still operate the circuit correctly?" There is no simple answer to this question. One major factor to consider is the **tolerance ratings** of the individual components in the circuit. The circuit and the chart shown in Figure 7–4 illustrates this point.

Each of these three resistors is actually within its rated tolerance value. If all three should be at the upper, or high, value of tolerance, then the measured value of the series string would be 17,300 Ω. Should all three be at the lower end of their individual tolerances, then the measured value of the string would be 14,700 Ω. Any measurement falling between these two limits is acceptable for this circuit. Even a slight variation from these limits is often permitted. The service technician is looking for *major* deviations. A major deviation indicates that something is wrong in that circuit. These areas are the ones requiring immediate attention. Anything

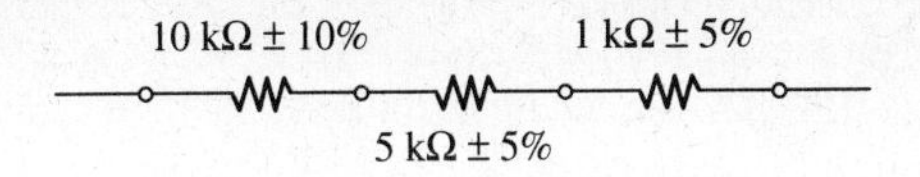

Marked value	Upper tolerance	Actual value	Lower tolerance
10 kΩ	11,000 Ω	10,000 Ω	9,000 Ω
5 kΩ	5,250 Ω	5,000 Ω	4,750 Ω
1 kΩ	1,050 Ω	1,000 Ω	950 Ω
Totals	17,300 Ω	16,000 Ω	14,700 Ω

Figure 7–4 The possibility of deviation from the marked values of components, such as resistors, often occurs. This chart shows the actual, marked values and the acceptable ranges for each resistor.

close to the tolerance value is acceptable for initial analysis.

Almost all electronic components have some tolerance rating value. The ability of the manufacturing industry to produce similar parts has improved over the years to the point where a standard of tolerance is often less than ±5 percent of the rated value. Some components have much greater tolerance factors. One of the classic examples of this extreme is the values of electrolytic capacitors. Some of these devices have been rated for tolerance with such extremes as +100 percent and −50 percent! Of course, not all capacitors have such a wide tolerance range. Many of the newer types of electrolytic capacitors have ratings of ±5 percent. These are equal to the tolerance ratings of other components. Look at the specifications of the components as identified on the parts list to identify the tolerance values for those components used in the device.

SIGNAL ANALYSIS

The skills you develop in the study of electronics are enhanced by the knowledge gained from looking at a block or a schematic diagram. Each block or section of this diagram is identified by its functional name. In a radio, one will usually find such blocks as IF amplifiers, demodulators, and audio voltage amplifiers. Electronic signals are processed by each of these blocks. These signals have some relation to the function of the block or the device. For example, you would expect to find analog types of signals in a radio. A review of the schematic diagram should identify many different amplitudes and, possibly, frequencies of signals. The ability to recognize the type of signal and its typical shape will assist you in identifying malfunctioning sections of the unit.

When you are servicing a computer or its associated equipment, the shape of the waves will be different. Typically the wave shape in computing equipment is in the form of a square wave. This wave shape is common to most of the signals. The service technician should be able to set up testing equipment to check for this type of signal, its frequency of operation, and its amplitude.

When repairing most audio amplifiers you should expect to see analog, or sine, waves instead of the square waves associated with the computer. The reason for this is that the audio amplifier, with only one slight exception, processes sine wave signals. This is part of the recognition of what you should expect to find when servicing a unit. Finding square wave-shaped signals in an audio amplifier would normally indicate an area of trouble. Further testing of the area in which the square wave is now observed is necessary to locate the defective component.

Area Signal Identification Area signal identification is one method the experienced service technician uses in the diagnostic process. The block diagram of a television receiver shown in Figure 7–5 will aid in illustrating this point. This receiver has several individual blocks, many of which can be grouped into larger sections. For example, the blocks in area 1 are processing the incoming signal. These blocks include those found in the tuner and the IF section of the receiver. The experienced service technician should expect to see signals with the classic shape of the one shown in the diagram.

A different type of signal should be present when you inspect the scanning circuits. The basic shape of this signal is that of the sawtooth waveform. The larger signals should use more than one frequency. The exact signal frequency and amplitude information for testing this section is available from the service literature for the unit.

When you are familiar with the typical signal processing blocks, it is possible to use the rules established for signal paths to maximize servicing efforts. One example of this is the expected output of the demodulator section of an AM radio. Proper signal processing in the radio up to this point will develop a demodulated audio signal. The demodulator block is approximately in the middle of the blocks making up the AM radio. The output of the demodulator block has the original audio signal created by the modulation process at the transmitter. One of the easily identified components at the output of the demodulator is the radio's volume control. Instead of checking each of the preceding stages, you could move to the demodulator's output and check for proper signal at that point. In many radios, the output of the demodulator is also the input to the audio amplifier sections. If the proper signal is present at that point, then all preceding stages are operating properly. This is one more example of using the half-split method to localize signal-related problems.

Block Signal Identification Block signal identification is similar to the area signal identification process. Each block has its own specific identification title. These titles often can be used as an initial method of locating signals. For example, a block identified as an **oscillator** should create some form of electronic signal. This will be in the form of a sine wave or a square wave, or it might be a sawtooth waveform, depending on the circuitry involved in creation of the wave. If the oscillator identification applies, the wave should be repetitive and have a constant amplitude.

Consider the type of signal when the block is labeled "amplifier." An amplifier, by definition, is a device using a small amount of power to control a larger amount of power. When the term "voltage" is used as part of the description for the amplifier block, you should have specific expectations. This block's output should have a larger value of amplitude when compared to the block's input signal amplitude. If the frequency of the signal is known, all you have to look for is an increase in the amplitude

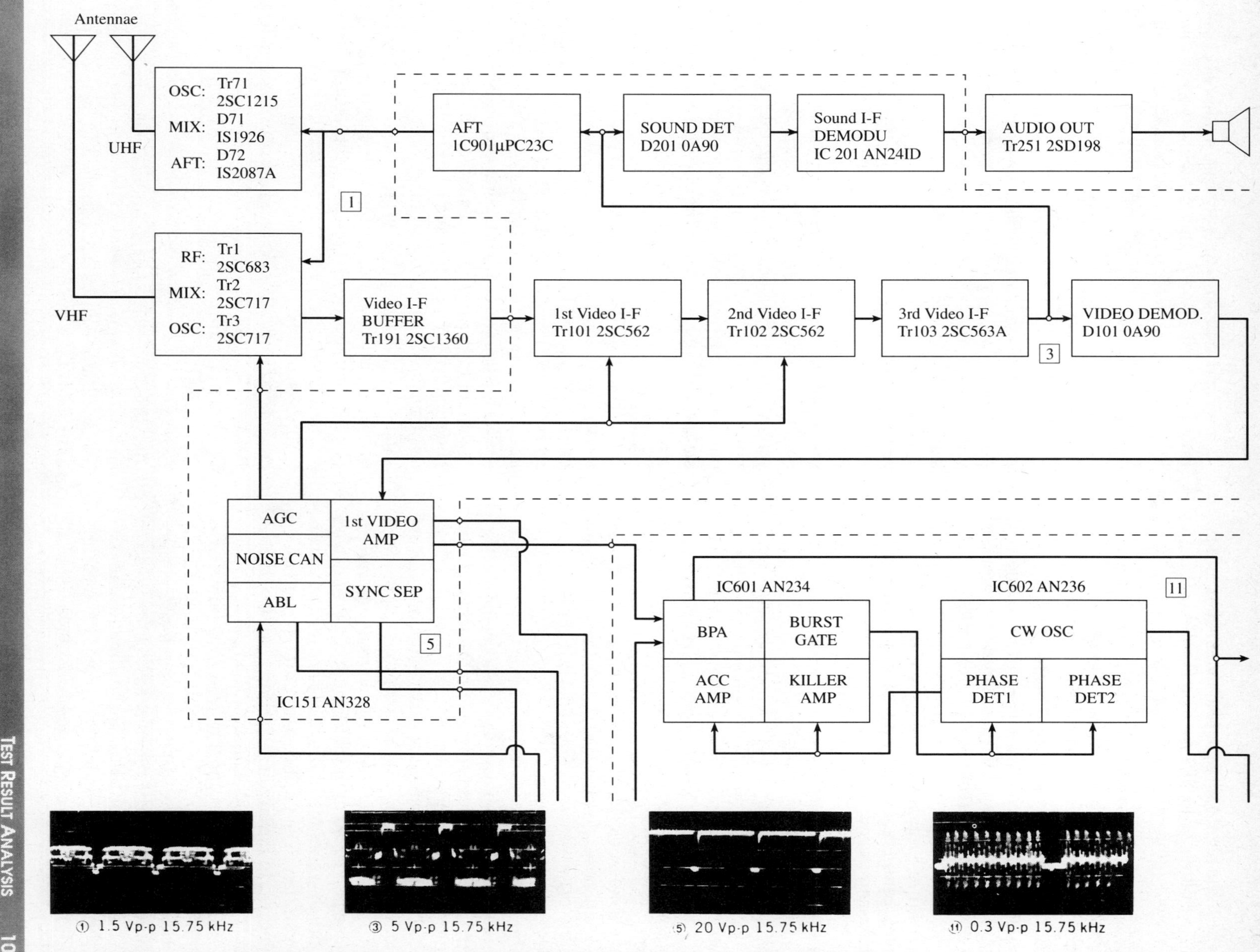

Figure 7–5 Partial block diagram for a television receiver and typical signals for the blocks.

of the input signal at the output terminal of the specific amplifier stage. Terms of this type will aid in predetermining the output from any block.

VOLTAGE MEASUREMENTS

This is one of the most often-made types of measurements. Traditionally the use of the voltmeter is one of the first things taught in the electronics classroom. This measurement is easy to perform. Three major points need to be made related to the measurement of voltages in the circuit. They relate to safety, equipment setup, and location of the points for measurement.

SAFETY

Safety must be a major consideration when measuring voltages in a unit. Voltage measurements are taken when the device's operating power is turned on. The possibility of an electrical shock is always present when working with equipment that is turned on. You must always think "safety" under these conditions. Work with one hand behind your back whenever possible. This will minimize the possibility of an electrical current flowing through your body and its resulting shock.

Carefully plan your moves when attempting to measure an operating voltage. Normally voltage measurement values provided on the schematic diagram or in the service literature are measured from circuit common. Often the service literature or the schematic diagram will include this information. This type of measurement is shown in Figure 7–6. The negative lead of the voltmeter is first connected to circuit common. This is necessary for the technician's safety. The positive lead of the meter is then connected to the circuit at the point of the measurement. Typically you will say, "There are 120 volts at that point." This cannot be true, since all voltages are measured *from a point of reference to another point*. What is really being stated in a shortened form is, "There is a potential difference of 120 volts when measuring from the test point to circuit common." The inference and assumption is that the measurement is being made from circuit common.

A second type of voltage measurement is illustrated in Figure 7–7. In this drawing the negative lead of the voltmeter is not connected to circuit common. A voltage is being measured *across* a specific component. Remember, voltage is a difference in electrical pressure between two different points in a circuit. When a current is flowing in the circuit, a potential difference, or voltage drop, develops across each of the components connected to the voltage source. This type of measurement is important when attempting to locate a component whose resistance value has changed during operation.

OPENS AND SHORTS

The terms "opens" and "shorts" are often confusing to the learner. The best way to explain these terms is to use electrical examples. Let's explore the open using Figure 7–8 on page 107 as the example. This is a typical electrical circuit. It contains one source, one load, and two lines, or wires, connecting the source and the load. When the circuit is functional, as shown in Figure 7–8(a), a current will flow from the negative terminal of the source using one of the lines as its path. Current flow will occur from the negative terminal of the source to the load. It then flows through the load and, using the second line, returns to the positive terminal of the source. As long as the current is flowing, the load will perform its designated work. For practical purposes all the resistance in this circuit is identified with the load. The conductive lines do have some resistance, but it is usually such a low percentage of the load's resistance that it is ignored. In this circuit the 100 V source creates a 1 A current through the 100 Ω load. The lines for this type of circuit will probably have a total of 0.010 Ω of resistance. Using the rule for total resistance in the series circuit, the total resistance is now 100.01 Ω. With an applied voltage of 100 V the current flow will be:

$$\text{I (amperes)} = \frac{\text{E (volts)}}{\text{R (ohms)}} = \frac{100}{100.01} = 0.9999\ \text{A}$$

The difference between the original value of 1.0 A and the new value of 0.9999 A is so insignificant that for practical purposes it is assumed to be equal to 1.0 A.

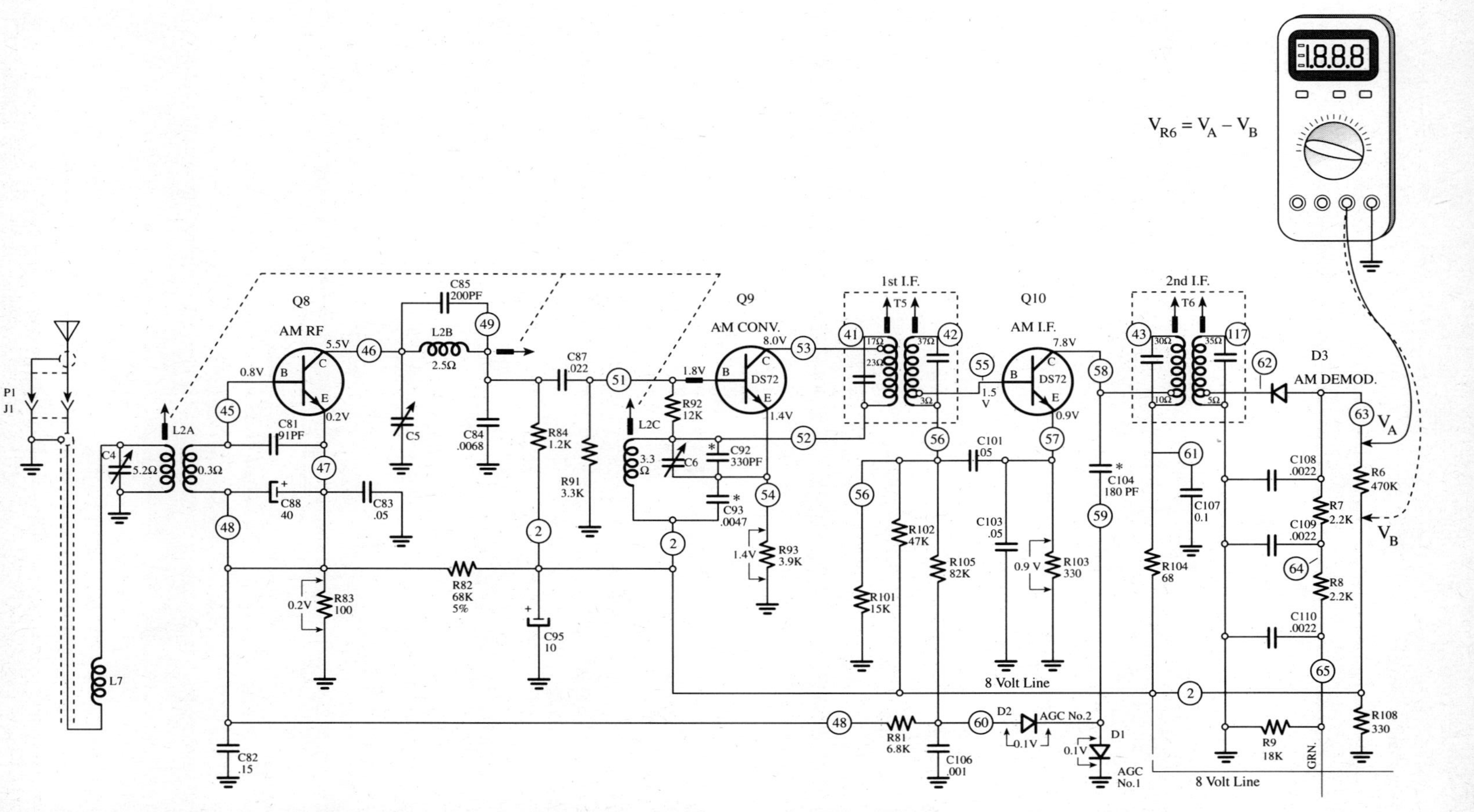

Figure 7–6 Voltage values are usually measured with one meter lead connected to circuit common, as shown here.

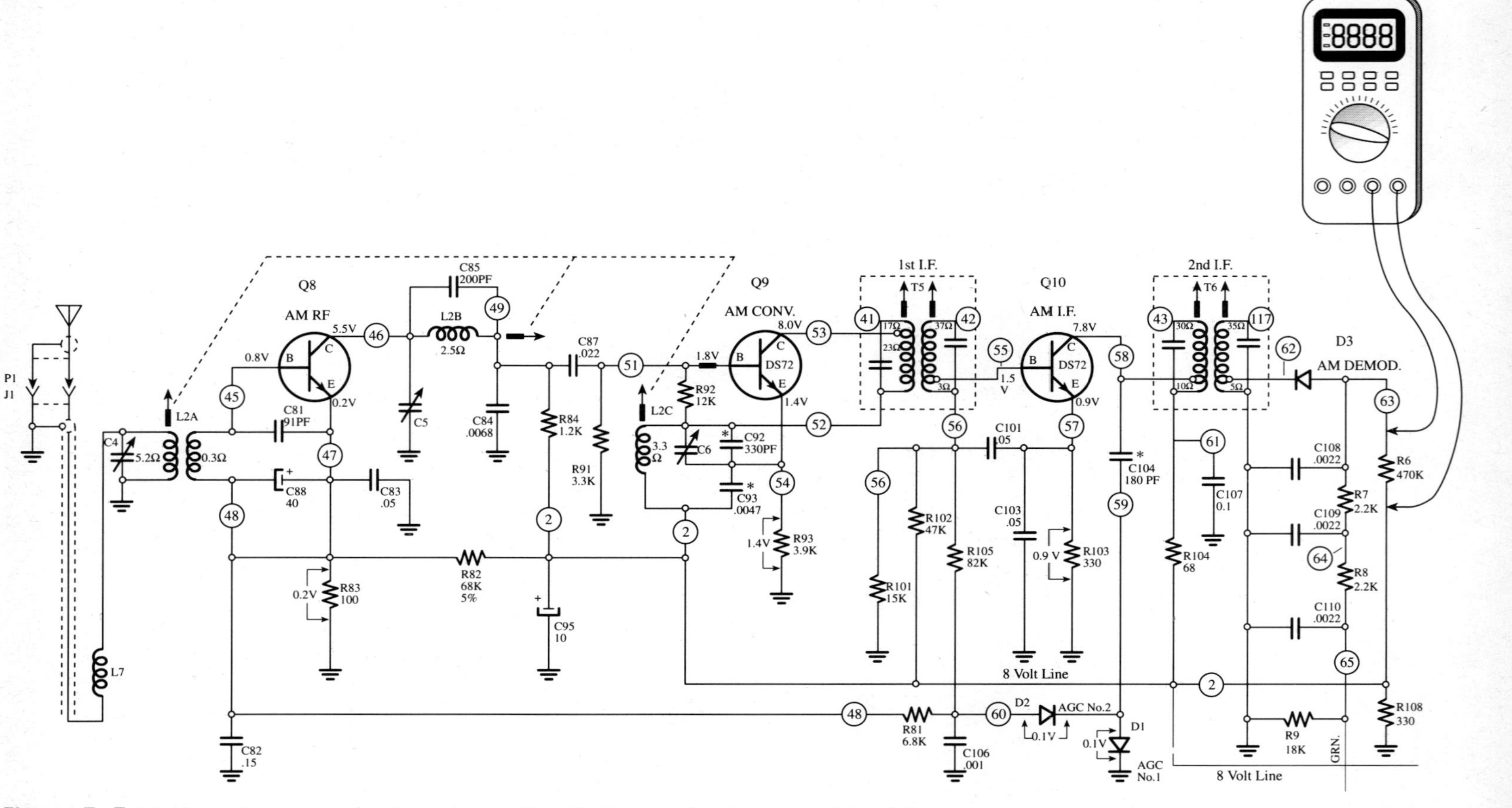

Figure 7–7 Voltage drops can also be measured by placing one lead on each side of the component.

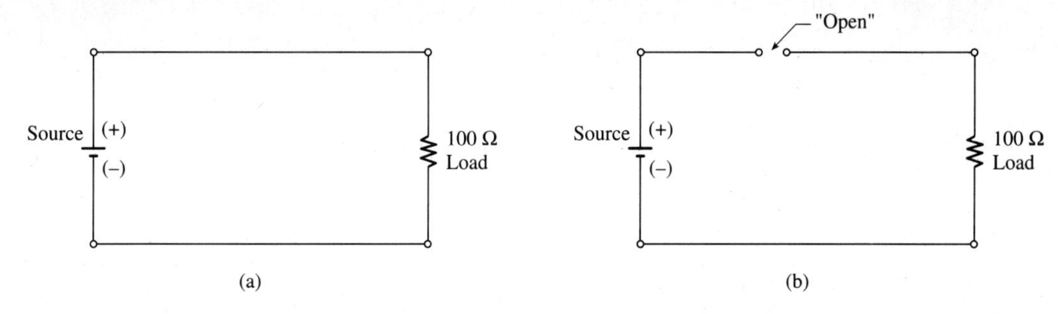

Figure 7–8 **Circuit (a) is a properly functioning one, while circuit (b) has an open. No current will flow when any part of the path is open.**

Next, look at the circuit shown in Figure 7–8(b). This is the same circuit except it has a break in one of its lines. This break is the "open" that is being described. When the **open circuit** condition occurs current cannot flow. The result is that the load is unable to perform its designated function. This condition can be described in electrical terms. Figure 7–9 uses the same values and the same components. The difference here is that the open is shown as a very high value of resistance. Instead of a single component series circuit, the circuit now has two resistances connected in series with the power source.

Line resistance value has now increased from its original value of 0.01 Ω to a value in excess of 100 MΩ. (The value of 100 MΩ is arbitrary and only used to illustrate the effect of the open acting as a very high value of resistance.) Current flow with the addition of such a high value of resistance will be greatly reduced. The current flow under these conditions is:

$$I = \frac{E}{R} = \frac{100}{100{,}000{,}100} = 0.000000999 \text{ A}$$

The amount of current flowing under these conditions is so small that it will be impossible for the load to operate in its normal manner. Actually, an open circuit will stop all current flow in the system. With a no-current-flow condition the load cannot operate at all. The final word in respect to the open condition of a circuit is that current cannot flow, and the load does not operate.

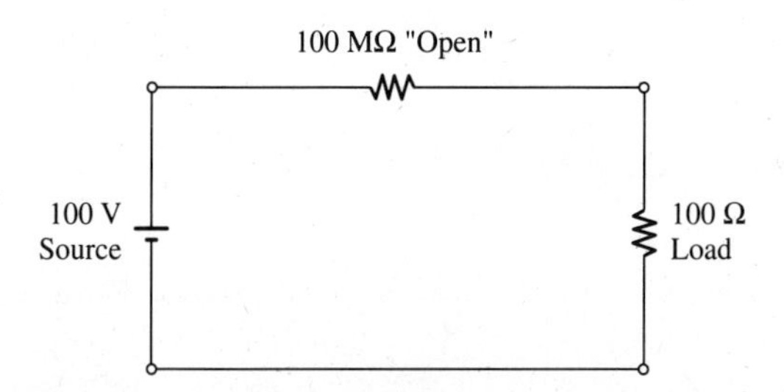

Figure 7–9 **The open in this circuit is represented by a very high value of resistance, 100 MΩ. This value is so high that little, or no, current can flow in the circuit.**

Another way to look at this type of circuit is, again, that the open is an extremely high value of resistance. In any series circuit, the voltage drops that develop when a current is flowing are directly proportional to the individual values of resistance. The largest voltage drop will develop across the highest resistance value in the circuit. If the open represents an ohmic value in excess of 100 MΩ, then the greatest percentage of voltage drop will develop across the open in the circuit. The original 100 Ω resistance now represents 0.001 percent of the total circuit resistance. Using a percentage of the applied voltage (100/100000000), the voltage drop across the 100 Ω load will measure 0.0001 V and the balance of the 100 V source, or 99.9999 V will develop across the 100 MΩ "open" resistance. For all practical purposes, the voltage drop across an open circuit component or wire will be equal to applied voltage. This is true for the series circuit. It is also true for any series circuit containing more than two resistances.

The **short circuit** is the opposite of the open circuit. A short circuit offers very little resistance to the flow of current. Usually when a unit has developed a short circuit, the excessive current flow will either trip a circuit protective device or cause a fuse to "blow." Look at Figure 7–10 and consider the effect of the short circuit based on electrical theory. The short circuit is actually an extremely low value of resistance. Again, for the sake of discussion and explanation, consider that this short circuit has a resistance of 0.1 Ω. Placing the 0.1 Ω value in parallel with the 100 Ω resistor will change the total circuit resistance. The basic rule here is that, in the parallel circuit, total resistance is always less than the lowest single resistance value. Thus, in the short circuit condition:

$$R_T = \frac{1}{\frac{1}{0.1} + \frac{1}{100}} = \frac{1}{10 + 0.01}$$

$$= \frac{1}{10.01} = 0.0999\ \Omega$$

Consider the value of current that will flow under these conditions:

$$I = \frac{E}{R} = \frac{100}{0.0999} = 1001\ \text{A}$$

This amount of current is a practical impossibility. Very few, if any, readily available power sources are able to produce this quantity of current. When an overcurrent such as this one occurs, a second condition is created. The demands on the power source are so excessive that its output voltage will drop to a very low value.

You may also use the typical voltage-drop values in the circuit for troubleshooting and location of components whose values may have changed during operation. Figure 7–11 is a typical three-resistance series circuit. The concept of voltage drops being directly related to the percentage of total resistance in the circuit may be used to roughly determine individual voltage drop values. In this circuit, resistance values are:

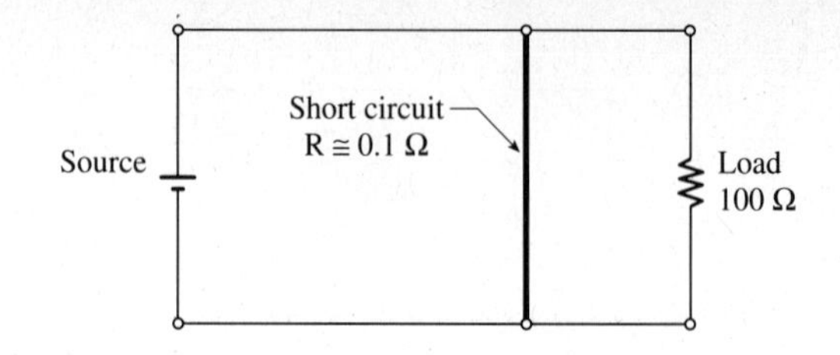

Figure 7–10 The short circuit places an extremely low value of resistance across the circuit.

$$R_{TOTAL} = 15.7\ \text{k}\Omega;\ R_1 = 4.7\ \text{k}\Omega$$
$$R_2 = 10\ \text{k}\Omega;\ R_3 = 1\ \text{k}\Omega$$

The individual percentages of the individual resistance values to the total resistance value are:

$$R_1 = 4.7\ \text{k}\Omega/15.7\ \text{k}\Omega = 30\%$$
$$R_2 = 10\ \text{k}\Omega/15.7\ \text{k}\Omega = 64\%$$
$$R_3 = 1\ \text{k}\Omega/15.7\ \text{k}\Omega = 6\%$$

Voltage drops created by current flow will be directly proportional to the percentages of individual resistances to the total resistance of the circuit:

$$V_{R1} = 100 \times 30\% = 30\ \text{V}$$
$$V_{R2} = 100 \times 64\% = 64\ \text{V}$$
$$V_{R3} = 100 \times 6\% = 6\ \text{V}$$

Even if the service literature failed to provide the individual voltage drops, the service technician could easily determine them using this process. In fact, you could also do some arbitrary "rounding" of values of the resistances and arrive at a total of 16 kΩ instead of the true value of 15.7 kΩ. For ex-

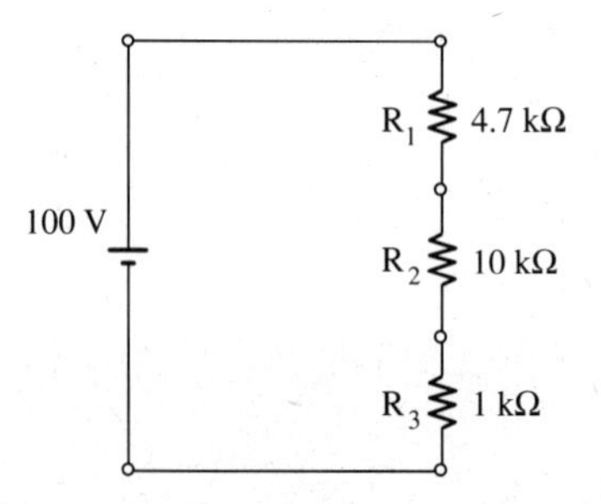

Figure 7–11 Voltage drops can be determined as a percentage of the total resistance, since they are directly proportional to each other.

ample, the 4.7 kΩ resistance could be rounded up to a value of 5 kΩ and the total circuit resistance could be rounded down to 15 kΩ. Then, this resistor represents 5/15, or one-third of the total circuit resistance. The expected voltage drop across this resistor would be approximately one-third of the applied voltage, or 33 V. The 10 kΩ resistor represents 10/15, or two-thirds of the total circuit resistance. Its voltage drop would be approximately 67 V. This process provides an approximation of the true circuit values and provides the service technician with some values that are close to the actual ones in the circuit. This, then, offers a simpler method of determining specific voltage-drop values when the information is not provided on the service literature.

CURRENT MEASUREMENTS

The procedure for measuring electrical current was briefly described in Chapter 6. The basic procedure for current measurement is to physically insert the current meter into the circuit. This procedure is illustrated in Figure 7–12. This circuit is opened at a point between resistors R_2 and R_3. Actually, any point in the circuit may be used to measure current flow. All of the circuit current will now flow through the ammeter as well as through the circuit. Since most ammeters have a very low value of internal resistance, the addition of the meter will not adversely affect the circuit values. In practice and given a choice, the meter's leads should be inserted as close to circuit common as possible. This is done for safety reasons.

An alternate method of measuring current flow is to measure the voltage drop across a fixed resistance value. Once this value is measured, Ohm's law is then applied. The answer is the current flow in the circuit. This method offers the advantage of not having to "dig in" and open the circuit up to insert the ammeter. When working with a series circuit, if the voltage drop is less than the value provided in the service literature, then another component's resistance has increased beyond its rated value. If the voltage measured across the component is higher than its reported schematic value, then another

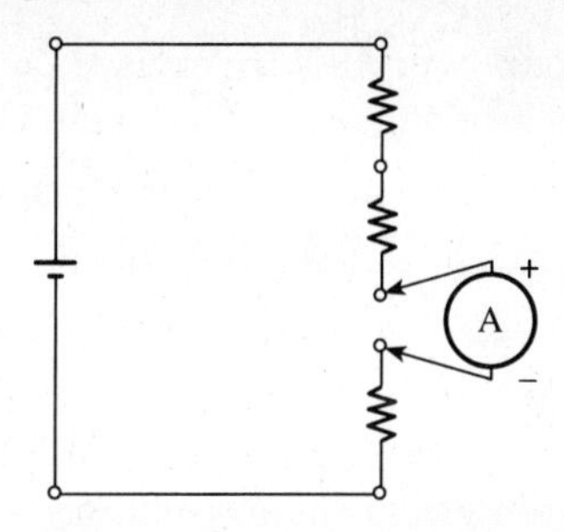

Figure 7–12 Current flow is normally measured by placing the meter in series with the current flow path.

component's resistance has decreased. These statements are based on the rules for voltage drops and resistances in the series circuit.

OPENS

Opens are relatively easy to identify in the circuit. The reason for this is that a true open stops all current flow in the specific circuit. The ammeter inserted into an open circuit will not show any flow of current. Once this type of condition is identified, the service technician is able to make some voltage-drop measurements. The voltage drop across the open component is equal to the value of the applied voltage. This is shown in Figure 7–13. The voltage drop across all other components in this circuit is zero volts. The reason for this zero volt drop is due to the basic rule for voltage drops in any circuit. The only time a voltage drop will occur is when a current is flowing in the circuit. When you encounter

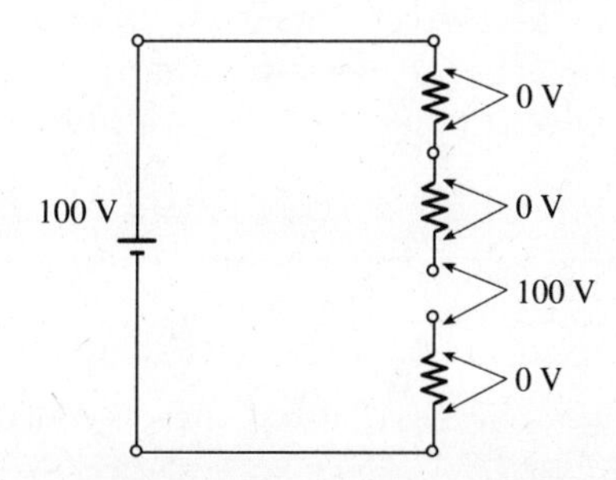

Figure 7–13 When a circuit has an open, the applied voltage can be measured across the point of the open, due to its extremely high resistance.

the open circuit condition, there is no current flow due to the lack of continuity in the circuit.

SHORTS AND PARTIAL SHORTS

Short circuits and **partial short circuits** create an entirely different set of circuit values. A reduced value of circuit resistance is caused by the short condition, which creates an abnormally high amount of current in the circuit. The method of locating this type of condition requires more effort than that for the open circuit. Figure 7–14 portrays a system that has a single power source and three load paths. This type of circuit is very common in almost all electronic devices. If the original values of load resistances were 3 kΩ each, then the total resistance in this circuit would be 1 kΩ. Normal current flow in this circuit would be 0.01 A, or 10 mA. If one of the load paths resistances changed to a value of 10 Ω, the total circuit resistance would be slightly under the lowest single resistance value of 10 Ω. The actual value works out to be 9.9 Ω. Current flow will increase from its original value of 0.01 A to 1.01 A.

The problem for the service technician is which approach to use to determine which of the three current paths has developed this lower value of resistance. Use the partial schematic in Figure 7–15 as a reference for this discussion. Many electronic circuits containing a single power source and a distribution system will incorporate some low values of resistance for circuit isolation purposes. These resistances are typically less than 100 Ω and do not affect the circuit negatively. Disconnecting the power source and measuring total resistance will not provide the correct answer. Neither will measuring across each of the individual paths. In this type of service problem the best solution is to separate each of the current paths. One of three procedures should be followed under these conditions. Procedure one is to remove one of the load path resistors and remeasure the circuit resistance value. If you are lucky, then the first resistance removed from the circuit will be the one connected to the defective circuit. If you are not so fortunate, the third resistance, or last, to be disconnected will be the one in the defective path. This process is time-consuming and may result in damage to the circuit board or the components being removed. It also can create a problem if, during the reinsertion process, the component is connected incorrectly or to the wrong place in the circuit. This is not unusual. Ask almost any service technician if this has occurred and the answer will be yes.

A second procedure requires the ability to analyze the current paths in the system. An experienced service technician should be able to determine the approximate resistance value for each branch from schematic information. Using an ohmmeter, resistance measurements are taken from the load side of each of the isolation resistors identified in Figure 7–15. The area of suspected trouble is identified when the path with the lowest resistance measurement is found. Further investigative work is required in that path. The interesting factor here is that, while current from the ohmmeter will flow through the other parallel circuits, this current must also flow through the isolation resistors. Working on the load side of these resistors will essentially eliminate their resistance when searching for the lowest resistance value path.

The third procedure requires cutting the circuit board paths. Although it is drastic, this does provide isolation to the circuits. The cut may be made on either side of the isolation resistors shown in Fig-

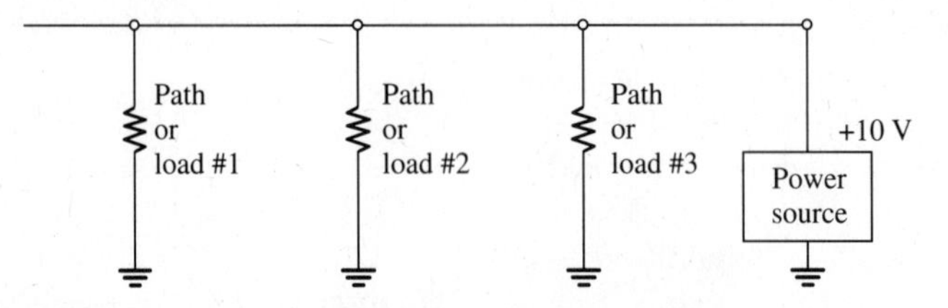

Figure 7–14 A short in one of the multipaths in this parallel circuit is more difficult to locate.

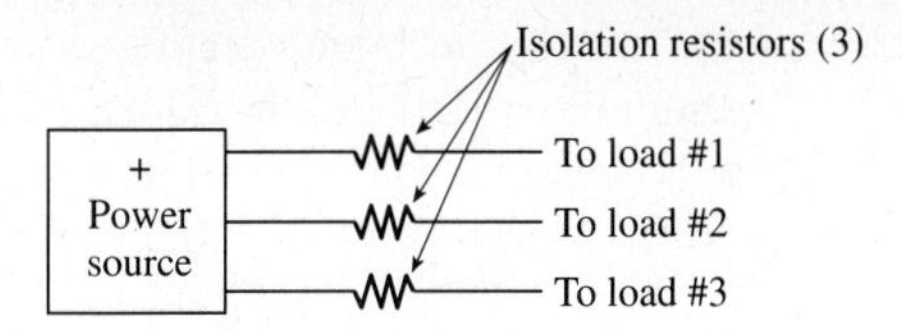

Figure 7–15 One method of locating a short in a parallel circuit is to cut the conductive paths for each branch until the one with the low resistance is identified.

ure 7–15. Resistance measurements may then be made on the individual circuits. These measurements will identify the circuit having a lower than normal resistance value. This, then, is the circuit requiring further inspection. The service technician will then repair the cut paths by soldering a jumper wire across each of the cuts.

A fourth procedure, and one that is not recommended, is even more drastic than number three. This procedure requires that the service technician turn on the circuit power and attempt to locate the problem area by the process of observation. Typically, an overcurrent condition will result in the components in the circuit overheating. Often these can be observed by close inspection of the unit. Overheated components may show up by the creation of dark heat spots on the circuit board containing the components. They may also heat and change the color of protective tapes or covers on the boards. This is a much better method of identifying the area of the problem than the "brute force" system used under power on conditions.

Service personnel have been known to say, in a humorous manner, that this process lets the smoke out of the components. They qualify their statement by remarking that all electronic components are constructed with some quantity of smoke inside them. When these components operate properly, the smoke stays inside. When the component fails during its operation, the smoke is let out of the component. While this statement is not true, the concept of using smoke to identify an overheated component and its circuit may be used, but it is not recommended.

This process involves turning on the power to the device and carefully observing where smoking occurs. The service technician immediately turns off the power and uses more traditional methods of identifying the specific components that have failed. One of the major disadvantages with this type of servicing is that the process often creates other problems in the unit. One of the more common additional problems is the failure of the power source. This type of servicing creates additional work and it may not locate the problem area in the unit. Remember when using this process that the overheated component may be the effect of the problem and not its cause. Changing a component because it has overheated may be necessary, but it does not always repair the problem. Further investigative work is required to determine the *cause* of the problem in the device.

RESISTANCE MEASUREMENTS

Many of the concepts involved in testing of resistors were discussed in the section covering current flow diagnosis. It seems almost impossible to separate these concepts, since one is dependent upon the other in all circuits. The concept of the need to isolate the individual resistance, or load, should be explored. Figure 7–16 includes an ohmmeter and a single resistance. The ohmmeter contains its own power source, an adjusting resistance, and a current-indicating meter, or readout. The initial adjustment of this type of meter requires that the two test leads be connected to each other. This completes the

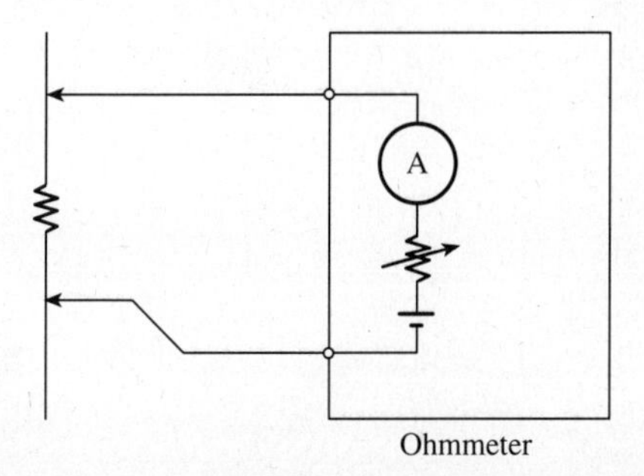

Figure 7–16 The ohmmeter's internal power source supplies a current to the circuit or component being evaluated.

meter circuit and permits a flow of current. The variable resistance inside the meter is then adjusted to indicate a full-scale deflection on the meter movement. The two test leads are then separated and connected to the resistance value to be measured. The circuit now has additional resistance, and less current will flow through the test resistor and the meter circuit. The scale on the meter's face is calibrated by means of Ohm's law to display a value of resistance. Additional resistance values are included in the meter's components when it is necessary to extend the range of this basic ohmmeter.

Now refer to Figure 7–17. This circuit has two resistances and the same ohmmeter as shown in the previous figure. Each of the resistance values is 1 kΩ. The service technician wishes to measure the value of R_2. This value should be 1 kΩ, but it measures 500 Ω, because when the ohmmeter's leads are connected across R_2, the resulting current also flows through R_1. Total circuit resistance for the two parallel connected resistances is 500 Ω and this is the measured value. If the service technician is not aware of the parallel connection, the indication is that the resistance value is incorrect. Replacing this component will not change the total resistance in this circuit.

Two things must be considered at this point. First, if the service literature had been consulted prior to taking the resistance reading, then the parallel-connected resistances would have been noted. The resulting reading of 500 Ω would have been expected and the servicer would have had to consider where to make an additional test. Second, for the most part, when resistances fail, an increase in their ohmic value will occur. Thus, the reading of 500 Ω should make you step back and reconsider what the results of this test actually indicated. This is not to say that resistor failures resulting in lower than manufactured values do not occur. I have several of these in my desk drawer. A failure curve for resistances is shown in Figure 7–18. Failures do result in a reduction of the initial value of the resistor. As the value decreases, the result is an increase in circuit current. This increase in current will create additional heat in the resistor and its value will ultimately increase to one much higher than its original resistance. Therefore, a resistor having a lower than normal value can be identified if you are able to catch the change before the resistance increases to a value that is several times the original manufactured value.

Each of the analysis procedures described in this chapter is required for successful servicing of electronic units. The service technician needs to know

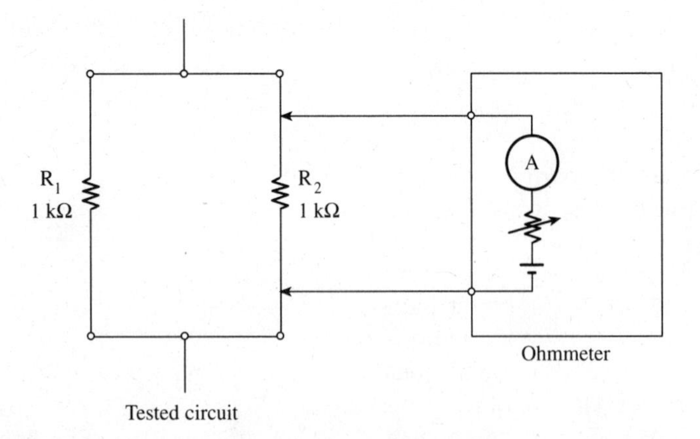

Figure 7–17 The ohmmeter cannot isolate a resistance. When two or more resistors are parallel connected, as shown here, current flow is through both and the measured resistance value does not agree with the schematic diagram's information.

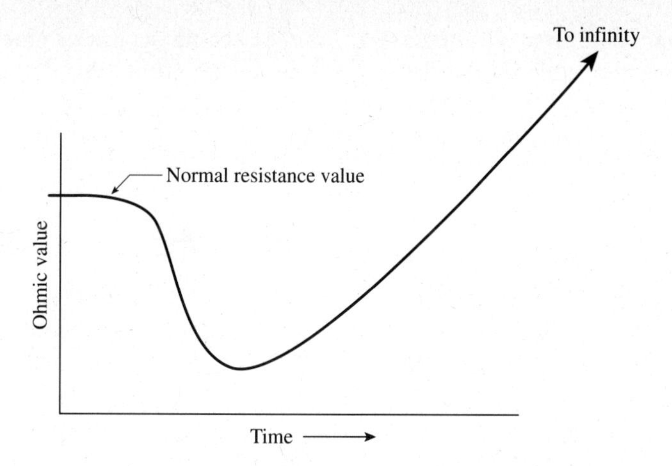

Figure 7–18 **The failure curve for a composition type of resistor. Often these are found to have a higher than normal value, as shown by the curve at the upper end of the chart.**

which one to use and what to expect when making the test. In addition, the technician must be able to connect the proper testing and measuring equipment so that its use does not further damage the device being tested. Using all of this information will enhance your entry into the world of successful electronic servicing.

REVIEW

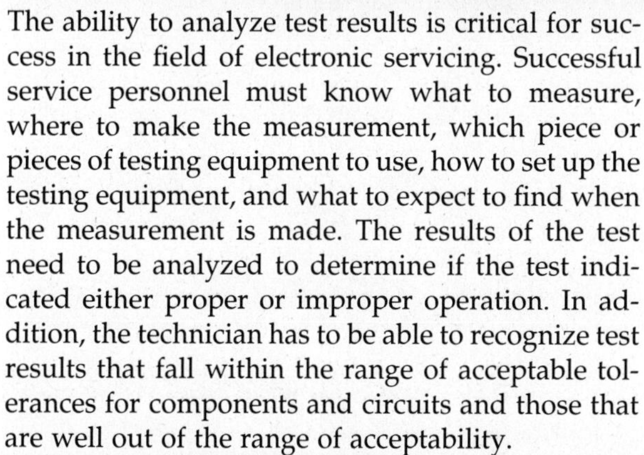

The ability to analyze test results is critical for success in the field of electronic servicing. Successful service personnel must know what to measure, where to make the measurement, which piece or pieces of testing equipment to use, how to set up the testing equipment, and what to expect to find when the measurement is made. The results of the test need to be analyzed to determine if the test indicated either proper or improper operation. In addition, the technician has to be able to recognize test results that fall within the range of acceptable tolerances for components and circuits and those that are well out of the range of acceptability.

Both signal flow paths and current flow paths must be recognized and analyzed. Each of these has its own set of characteristics. Signal paths are directly related to the block diagram of the unit. Current paths deal with the internal working of specific sections of the unit. Many of the better-quality service manuals provide both signal flow information and the "signature" waveforms found at specific locations within the unit. Current flow information is very seldom provided in the service literature. Voltage measurements are usually given. The successful service technician is able to use the values of voltage drop and resistance to calculate the current flow through the component. This information aids in determining whether the component or current flow path is acceptable or further investigation is required in that circuit.

Methods of measuring current flow in a parallel circuit include isolating the circuit from others in parallel with it, using schematic information to determine the ohmic value of the circuit, and the "watch for smoke" routine. The first two are recommended, and the "watch for smoke" routine is *not* one to normally use, since it often will introduce other failures into the system. The wise technician will use the senses of smell and sight to look for areas of the unit that have overheated. This inspection will aid in locating a specific area of trouble.

REVIEW QUESTIONS

1. In what direction do you "read" the signal flow path on an electronic schematic diagram?
2. Where, on the schematic diagram, is the normal position for the power source?
3. How does the path of the current flow differ from that of the signal on the schematic diagram?
4. What tolerance values, or percentages, are normally used for electronic components?
5. Explain why it is possible to use the rules for signal paths described in an earlier chapter for the process of troubleshooting.
6. What is the relationship between names of functional blocks and the signals they process?
7. Where are the meter test leads connected for most voltage measurements?
8. Why is it necessary to plan where and how to make all circuit measurements?
9. What is an open? Explain in electrical terms.
10. What is a short? Explain in electrical terms.
11. Why is the "turn it on and look for smoke" method of troubleshooting not recommended?
12. How do voltage drops across individual components in the series circuit compare to the percentage of total resistance in the same circuit?
13. Why is the total applied source voltage measured across an open in a circuit?
14. Why is there no measurable voltage drop across a short circuit?
15. Explain why circuit power is removed when attempting to make a resistance measurement.
16. Why should you remove at least one end of a component when measuring its resistance value?
17. Does a small variation from the manufacturer's tolerance rating indicate a major source of trouble in a circuit?
18. What is the process involved when a circuit board is cut in order to make a circuit measurement?
19. How do you insert an ammeter to make a current measurement in a circuit?
20. Explain how a voltage measurement can be used to determine circuit current.

CHAPTER 8

Basic Rules and Test Results

INTRODUCTION

The expert troubleshooter *continuously* applies basic electrical and electronic rules during each analysis and repair. Technicians often apply these without realizing that they are being used to locate and repair the inoperable unit. The troubleshooter/repair student must also be able to apply these rules. The major difference between an expert and a novice learner is that the latter requires more time to think about what to do, what the various test results indicate, and how to make the repair. Material in this chapter is presented to assist the novice in developing a set of skills that will provide the first steps in becoming an expert troubleshooter.

OBJECTIVES

Upon completion of this chapter, the student/reader should:

1. be able to relate the basic rules to troubleshooting procedures;
2. use Ohm's law to aid in locating malfunctioning circuits;

3. relate the effect of abnormal circuit values to malfunctioning circuits;
4. use Watt's law to locate malfunctioning circuits; and

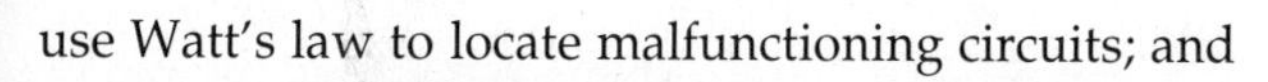

5. use Kirchhoff's rules to locate malfunctioning circuits.

BASIC RULES AND TEST RESULTS

The expert troubleshooter develops a procedure for efficient diagnosis. This procedure is based upon the application of a basic set of rules and laws. You need to know these rules before you plan which tests to make and how to make them. Tests that produce meaningless results are a waste of time. They tend to confuse the tester by providing data that does not offer the information required to rapidly locate the component or unit that has failed. Often too much information is just as bad as having insufficient information for the repair process. This is why one of the major themes of this book is that you first stop and think about the repair procedure before actually starting to work on the unit.

The first step in any repair procedure is developing a plan. This requires little, if any, hands-on activity with testing equipment. It does require thinking of how to approach the problem. This step will lead to identification of the malfunctioning area of the unit. The basic rules developed by Ohm, Watt, and Kirchhoff are required knowledge for anyone who diagnoses and repairs electrical and electronic equipment. For example, recognizing the concept of why current increases in a series circuit will aid the repair technician's diagnosis and location of a malfunctioning component.

Each of the tools available to the service technician requires knowledge of not only how to use them, but also where to use them. Information gained from their proper application is then used to determine where and how to make additional tests. Remember, the purpose of a plan is to *systematically* reduce the area of suspicion from a large one to the specific part. This can only be accomplished by using a systematic test-and-analyze approach.

A specific repair activity that highlights this procedure is presented below. Recently, a call was made to service a combination radio transmitter/receiver. The problem was that no signals were being processed by the receiver section of the unit. When the service technician arrived at the installation, he started by making some initial tests *before opening the unit and before using any test equipment.* The reasoning behind this activity was to attempt to localize the problem using knowledge of how the system functions. This unit had a digital frequency readout as part its front panel layout. Tuning, or turning the frequency selection knob showed that the digital frequency readout changed as the knob was turned. In addition, turning on the transmitter section showed that radio frequency energy was being created by the unit. The service technician also observed that all antenna cables were connected to their proper terminals on the unit. This information was used, along with a block diagram of the unit, as a preliminary step in the troubleshooting process. A block diagram for a portion of this transmitter/receiver is shown in Figure 8–1.

Studying the block diagram, the service technician was able to apply one of the basic rules of troubleshooting—the rule eliminating seemingly correctly working sections of the unit as an initial step in the diagnostic process. In this unit the digital readout is one output from the oscillator/mixer section (OSC/MIX). It is identified as the "counter" block on this diagram. Since the readout changed as the tuning knob was rotated, this block was eliminated as an area of trouble. Correct functioning of the transmitter section of the unit eliminated those blocks marked "FINAL" and "LL DR." Noting that the antenna was connected in a linear flow path to the SWR (standing wave ratio) section, an LPF (low pass filter), and then the signal was split into two paths—the T/R (transmit/receive) switch and the FINAL, the technician decided to make the initial test at the antenna, or input, circuit of the unit. The service literature revealed that when this unit is in the static, or at rest, condition, the signal flow path was through the T/R block, into the BPF (band pass filter), and then into the receiver signal processing sections of the unit.

At this point further information was obtained from the diagram shown in Figure 8–2. This diagram shows the wiring between the individual circuit boards of this unit. A study of it showed that there was an alternate path for the receive signal. This path bypassed the T/R switch and those blocks associated with the transmitter. The receiving antenna connection went directly to the RF mixer board. The service technician made the decision to make several tests in this order:

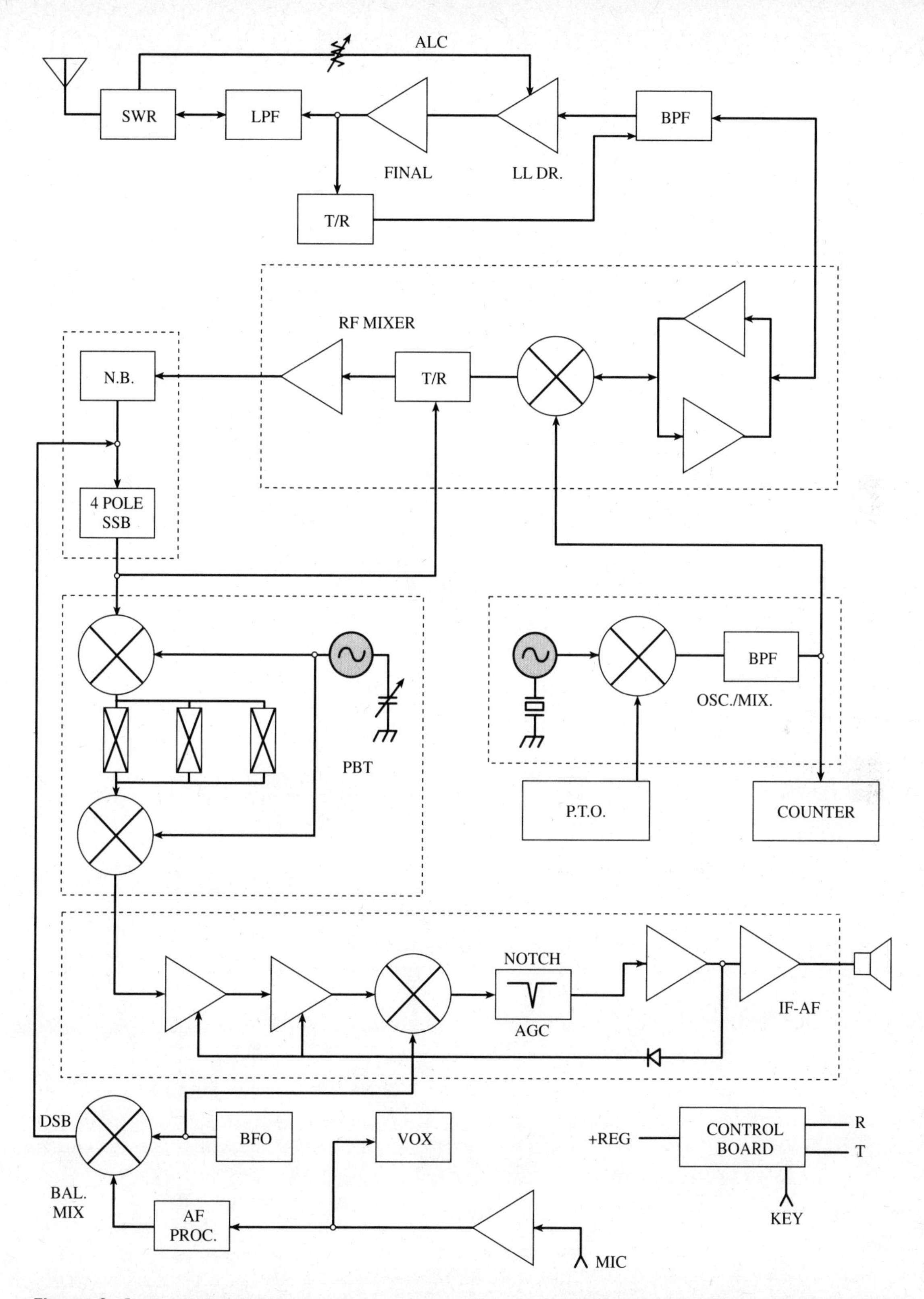

Figure 8–1 **Partial block diagram for a commercial amateur radio transmitter-receiver unit. (*Courtesy Ten-Tec, Inc.*)**

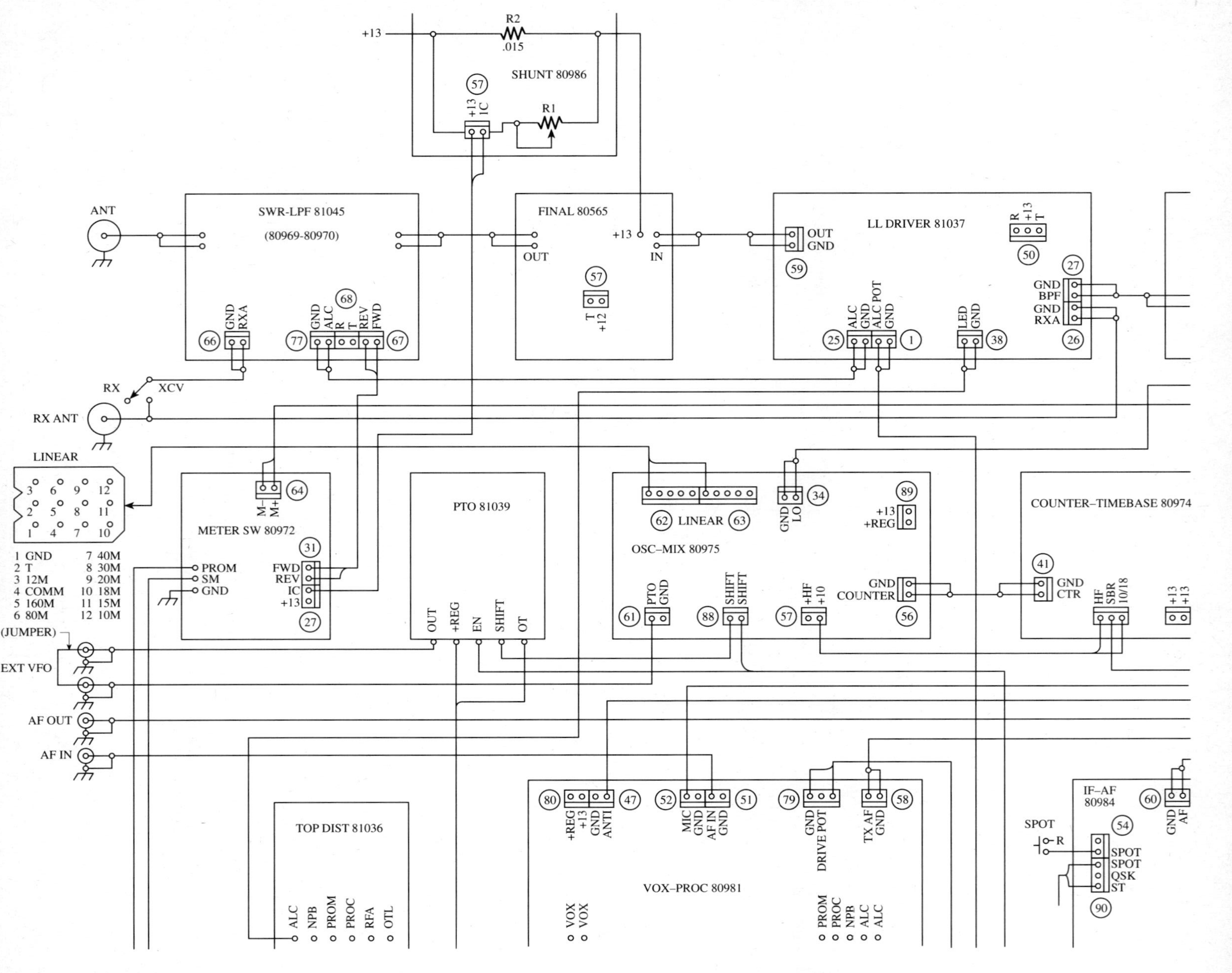

Figure 8–2 The block diagram also will show interconnecting wiring. (*Courtesy Ten-Tec, Inc.*)

1. Inject a signal at the ANT terminal of the unit.
   a. If a signal was present at the output of the unit, then it was functioning in a normal manner.
   b. If no signal was present, then further analysis was required.

Presuming that the injection of a test signal at the ANT terminal did not indicate a correctly functioning unit, the following step had to be taken:

2. Inject a test signal at the RX ANT (Receive Ant) terminal.
   a. If a signal was present, then the problem was between the RF MIXER block and the ANT input connector.
   b. If no signal was present, then the problem was in a block that was common to both transmit and receive after the RF MIXER block. (This was highly unlikely, due to the common components of the system.)

The service technician decided to make these tests before considering other additional tests on this unit. The first step involved the use of a test signal. This signal was created by an RF signal generator and injected into the ANT terminal of the transmitter/receiver. Tuning of the frequency selection control of the unit created an output signal at the speaker. This first step provided information indicating that the problem area was *outside* the unit. The service technician then had to stop and develop a second set of analysis techniques. Logic indicated that, since a test signal injected into the antenna terminal of the unit provided the proper output, the area of suspected trouble was elsewhere. The only components still to be considered included the outside antenna and the cable connecting this antenna to the antenna terminal of the transmitter/receiver unit. Since the problem seemed to be located between the actual antenna and the input terminal to the unit, this seemed like the next most logical place to look for a problem. An inspection of the coaxial cable and the antenna connections to the unit was performed. This included both a visual inspection and a physical inspection. The latter incorporated the process of "wiggling" or moving each individual section of the coaxial antenna wiring.

The visual inspection of the coaxial antenna wiring disclosed no observable problems, such as disconnected cables, cables twisted into forms that would not permit passage of signals, or cut or damaged wiring. The physical inspection further discovered that the internal wire conductor on one of the connectors had failed to make a connection to its mating component. This connector was removed and replaced and the repair was successfully concluded when the signals were again processed and reproduced by the receiver.

This process emphasizes the need for a plan and the willingness to follow the plan as well as knowledge of how the system functions. A service technician might start by disconnecting the unit from its power, antenna, and control cabling. The next step might be to remove those bolts or fasteners used to keep the covers on the unit. This would be followed by a visual inspection of the internal components, wires, and boards in an effort to locate signs of overheating or cable separation. The next steps would include using testing equipment to determine if the power supply was functional. Most likely the last step would be to inject a signal to determine whether the unit was functional. After all, didn't the operator call for service because the equipment was not working? In practice, one of the most often-heard remarks from people involved in "field service" activities is that, once the unit stops working properly, the operator immediately calls for assistance and does not attempt to make the repair. Operators seldom want to know why the unit is not working; all that seems to be critical is the restoration of normal operations.

The use of the "plan of attack" approach accomplished several things, including a successful repair, a rapid diagnosis procedure, and a relatively short time interval to return the unit to normal operating condition. All of these are requirements for successful servicing. This same type of procedure has to be used in all service applications. The previously described example is applicable to all levels of servicing. Often the problem area in the unit is localized to one board or section. When this is the case, the service technician should be able to apply a similar thought process to develop a plan for locating the specific components that have failed. This step approach occurs continuously during the analysis process of servicing. It is not unique to electronic devices, since the same style of servicing will occur for almost all products. Eventually the service technician will either have to remove and replace a module or board, or service the equipment at the component level. When servicing is required at the component level, the technician will require further information. This information is usually contained in the schematic diagram, which will provide actual circuit information. The service technician's plan will include:

1. *What to measure.* The plan must include the kind of measurement to be made, which can be obtained from the service literature for the specific unit.
2. *What equipment to use.* Will this require the use of an oscilloscope, a digital multimeter, or some specialized type of testing equipment?
3. *Where to measure.* The service manuals will usually identify specific points in the unit where measurements can be made. With some luck, photos of the units will show these points.
4. *What to expect to find when making the measurement.* Often this will include the values of voltage, current, or signals expected at those points when the unit is functioning normally. In addition, this information should tell the technician what scales or ranges are necessary for the initial setup of the testing equipment.
5. *What the actual measured values indicate.* Do these values indicate a normal condition? If not, what specific information is obtained from the measurement?
6. *The next step(s) after analysis of the measurement.* Once the specific information is obtained, the technician must consider the next steps in the service procedure. If the test indicates an abnormality, what does this actually mean in terms of the basic laws of Ohm, Kirchhoff, and Watt? Information in the rest of this chapter will help make these determinations.

SCHEMATIC INFORMATION ACCURACY

The average service technician has a tendency to depend on the accuracy of the information provided on the schematic diagram. Most of the time this is very true. Sometimes, however, the schematic diagram information is in error. The expert service technician is able to recognize these errors because he or she has the knowledge of how the circuit rules and laws are applied to working circuits. The partial schematic diagram shown in Figure 8–3 illustrates this type of misinformation. The circuit is one part of an audio amplifier system. Its functional name is audio frequency voltage amplifier. The analysis of the information provided on the schematic diagram should show that this circuit cannot possibly function properly.

The reason for the previous statement is based on an analysis of how typical voltage amplifiers are biased in functional circuits. Typically the voltage drops created by the emitter-to-collector circuit should rise from zero at the negative circuit connection until they reach the level of source voltage. Collector voltage in a circuit of this type is often close

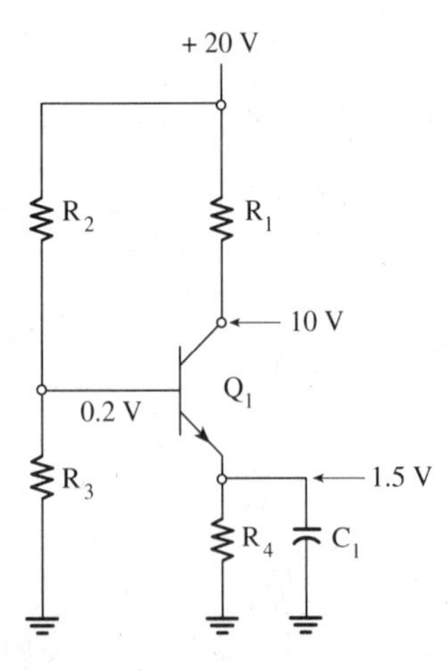

Figure 8–3 **It is possible to find incorrect values on the schematic diagram. While this seldom occurs, the technician has to be able to use his or her knowledge of circuit operation to determine if this has happened. In this circuit the voltage at the base is incorrect.**

to one half of the applied source voltage. The values of voltage rise in an orderly fashion from zero to the value of the source, in this case, 20 V. Using the previous rule, the collector voltage should be close to 20 ÷ 2, or 10 V. This is the voltage shown on the schematic. You can assume that the voltages for this section of the circuit are correct. Referring to the basic rules of voltage divider networks, there is a ranking of voltage values on almost all schematic diagrams. The bottom, or lower end of the circuit, is usually the lowest value of voltage. This connection is usually zero volts, or circuit common. Voltage values increase in a vertical manner, with the highest value of voltage being located at the upper end of the circuit. Often this is connected to the positive terminal of the power source. In other words, the voltage drops that develop in this series circuit have a ranking. Their values change as the notations on the schematic diagram rise from the lowest point on the specific circuit diagram to the highest point on the same circuit diagram.

The experienced service technician should question the voltage shown at the base of the transistor. Almost all schematic diagrams will provide voltage measurements. A note at the bottom of the page of the schematic diagram will tell the technician about the reference point for voltage measurements. This is, except in unusual situations, circuit common. Information on the schematic diagram indicates a value of 0.2 V at the base, when measured from circuit common. If this is true, the transistor is biased into cutoff and it would not be able to conduct any current there. The reason for this is that the emitter-to-base voltage for a silicon NPN type of transistor, as shown here, is typically a value of about 0.5 V. This measurement is taken by placing one lead of the voltmeter on the emitter element and the other lead on the base element of the transistor. Voltage values much less than 0.5 V will create an operational condition called "cutoff." When the transistor is in its cutoff stage, its internal resistance is extremely high and little, or no, current flow occurs through it. Signal tracing analysis would indicate an output signal level of zero. An amplifier of this type does not operate in its cutoff condition. Almost all audio frequency voltage amplifiers of this type operate under class A conditions. They are biased at the midpoint of their operational range. Signals at the input of this circuit create a shift of operational levels that range between cutoff and saturation, but never into either of these extremes. You would have to look at the voltage level indicated on the schematic diagram with suspicion. Voltages measured between the emitter and the base of this type of transistor should show an *increase* in the value of the base voltage when it is compared to the emitter voltage. If the transistor is a silicon type this value should increase by about 0.6 V to 0.8 V with no signal applied to the circuit.

The experienced service technician will recognize that one of two things is wrong with these circuit voltage values. The primary error is the value of voltage shown for the base of the transistor. This voltage should be higher in value than that at the transistor's emitter. Either the emitter and base voltages have been reversed on the schematic diagram or the value of the base voltage was printed incorrectly as 0.2 V instead of 2.2 V. A measurement of the actual value of the voltage would indicate which of the above is true.

The previous example is just one of the things that the experienced service technician needs to recognize as equipment is being analyzed for repair. I am not suggesting that this type of occurrence is typical, but these errors can occur because the people developing the service literature and schematics are capable of making an occasional mistake. The message is this: When you find a circuit value that appears to be unique or out of the ordinary, it is best to consider some of the basic circuit laws. Application of these laws will reveal whether the schematic is in error. Often a note to the manufacturer identifying the error is appreciated. Corrections of this type are often included in later editions of the service literature.

TEST RESULTS AND CIRCUIT LAWS

The relationships of voltage, current, and resistance are presented in Ohm's law. These relationships are fundamental tools used by all service personnel. Because you have learned these laws so well, you will not even realize you are applying them during the analysis process. The need to recognize and use these relationships is one of the basic requirements for successful servicing.

SERIES CIRCUIT ANALYSIS

There are two basic types of circuits in use in electronic equipment. One of these is the series circuit, the other is the parallel circuit. Each circuit uses its own approach for troubleshooting because the rules for series and parallel circuits differ. In the series circuit the current is constant while the voltage drops differ. The parallel circuit has a constant voltage and the current varies in each branch. Because of these differences the approach to diagnosis and repair is also different. Recognition of the type of circuit is of utmost importance. Also important is the application of the proper set of testing rules for the two different types of circuits.

VOLTAGE MEASUREMENTS

$$E = I \times R$$

This equation represents the format used to show Ohm's law. Let us examine it. The letter E represents electromotive force; this is measured in units of the volt (V). The letter I represents electrical current. It could be stated that the I represents the intensity of electrical current. Electrical current values are measured in units of the ampere (A). The third letter, R, represents resistance, in units of the ohm (Ω). These three values were presented in the formula shown above by Georg Ohm. They indicate that the amount of voltage is directly proportional to the quantity of current times the value of opposition, or resistance, to the flow of the current in a circuit. Another representation of this formula is: If the voltage is held at a constant value, a rise in current flow will occur only when there is a drop in the opposition, or resistance, to this flow. Conversely, when the resistance increases the current will decrease. Understanding these interactions is extremely important for the service technician's ability to diagnose problems in equipment.

KIRCHHOFF'S LAWS FOR VOLTAGE AND CURRENT

Another German scientist developed two sets of rules for electrical circuits. One of these relates to voltage drops and the other to current flow. These rules are applied during the service analysis process. Kirchhoff's voltage law is applied to the closed loop, or series circuit. Most electrical and electronic circuits can be considered as series circuits. Even those with parallel paths often can be simplified into multiple-series circuits. One simplification is shown in Figure 8–4. Part (a) of this illustration shows the typical method of drawing this type of circuit. Part (b) shows the individual current paths connected to a common power source. Both circuits are the same; their only difference is the method of drawing them.

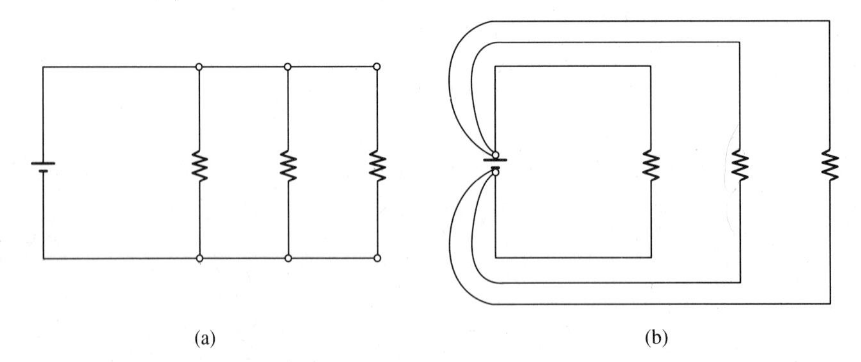

Figure 8–4 Two methods of illustrating current paths in the parallel circuit. The one shown at (a) is most often used.

Kirchhoff's voltage law states that the voltage drops developed in a closed loop, or series circuit, add to equal zero. This process is accomplished by inserting polarity values for each component in the loop. An example of this is shown in Figure 8–5. Each component, including the power source, has its polarity identified. You may start at any point of the circuit and add each of the voltage values. The sum of all of the drops, including the source, is equal to zero. Another way of stating this rule, and one that is more commonly remembered, is that the voltage drops developed in the circuit equal the applied source voltage.

Kirchhoff also presented a rule for current flow. This rule states that the current entering any junction is equal to the current leaving the same junction. This is shown in Figure 8–6. Point B is the junction of R_2 and R_3. The current flow occurs through each of these resistors. The arrows on the drawing show the individual currents. They meet and join at point B. The sum of the currents leaving point B is equal to the sum of individual currents flowing through each resistance.

The reverse is true at point D. Here, the current entering this junction is equal to the sum of the individual currents leaving the same junction. Therefore, if the current through R_2 was 2 A and the current through R_3 was 3 A, then the current entering this junction would be 5 A, or the sum of the two current path values.

In analyzing the circuit shown in Figure 8–7, you must first determine its operating values. Total circuit values in this circuit are determined by using

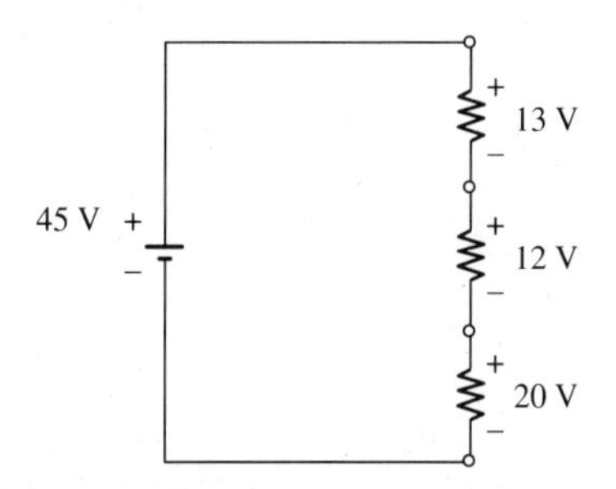

Figure 8–5 Kirchhoff's voltage law shows that the sum of the voltage drops across the components is equal to the applied voltage.

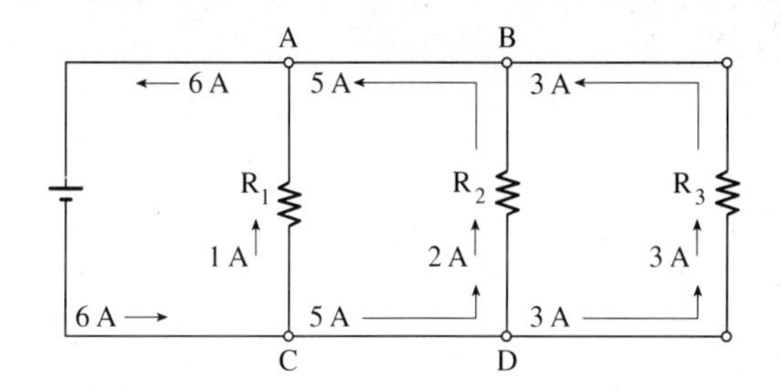

Figure 8–6 Kirchhoff's current law describes the fact that the sum of the current entering a junction is equal to the sum of the current leaving the junction.

the information at hand. For example, current flow is calculated by using the voltage drop developed across resistor R_1. The ohmic value for this resistor is 5 kΩ. The voltage drop, as provided on the schematic, is 10 V. An application of the formula for Ohm's law, $I = E \div R$, provides a value of 0.002 A, or 2 mA for current flow through this resistance. Next you need to identify the type of circuit used. In this case, the circuit is considered to be a series one. The rule for current flow in the series circuit is that it is equal throughout the circuit. If this rule is to be applied, then the current through the 100 Ω resistor, R_2, is also 0.002 A. The voltage drop developed across this resistance as current flows in the circuit is calculated using Ohm's law, as well:

$$E_{R2} = I_{R2} \times R_{R2}; \text{ or } E_{R2} = 0.002 \times 100 = 0.2 \text{ V}$$

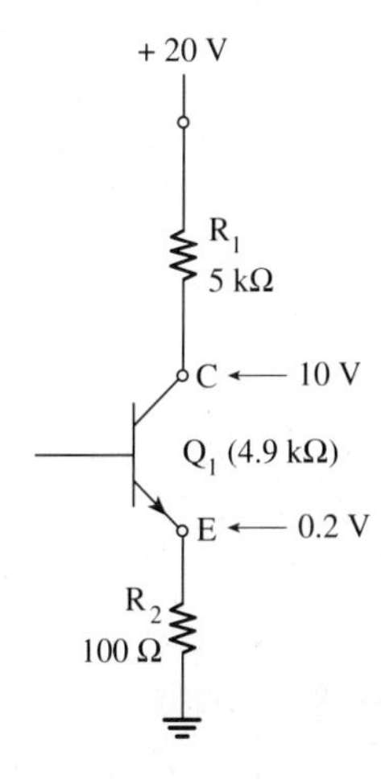

Figure 8–7 This circuit can be considered to be a series circuit for purposes of analysis.

The difference between the sum of the voltage drop across R_1 and R_2, when subtracted from the applied voltage, provides the voltage drop across the transistor's emitter-to-collector elements. In this circuit this value is 9.8 V. Therefore, the resistance of the transistor in this circuit is: $R_{TRANSISTOR} = 9.8 \div 0.002$ or 4.9 kΩ. This value is indicated in parentheses on the diagram. These are the values for correct operation of this amplifier circuit.

Using the basic Ohm's law rule for voltage drops is one method of analysis that can be applied in this circuit. Looking at the basic formula, the value of voltage is determined by the combined values of current and resistance. The relationship of voltage and current may be used since the two values are in inverse proportion to each other. A rise in circuit current will indicate a reduction in the total value of circuit resistance. Also, a drop in circuit current indicates an increase in the total circuit resistance.

$$E = I \times R$$

$$20 = 0.002 \times 10000$$

This shows the normal values set for this circuit. When the value of one of the resistances or the ohmic value of the transistor is increased, the total circuit current will decrease, as shown in the following equation:

$$E \div R = I$$

$$20 \div 15000 = 0.0013$$

Circuit current has decreased from its original value of 0.002 A to a value of 0.0013 A, due to an increase in the ohmic value of one of the components in this series type of circuit. When one of the components' resistance decreases, the total circuit current will increase:

$$E \div R = I$$

$$20 \times 7500 = 0.0027$$

In this circuit the total resistance has decreased and the resulting current has increased to almost one and a half times its original value.

The major question to be asked is, "How does one identify these conditions when measuring circuit values?" The following examples will answer this question. The first item to discuss is the reason for any change in circuit current values. If the source voltage is maintained at a constant value, 20 V in this example, then the only way circuit current is able to change in value is if the total resistance in the circuit is changed.

Start by recalling that the emitter-to-collector connections of a transistor act as a variable resistance in any circuit. In the typical transistor circuit a small change in the voltage presented between the base and the emitter of the transistor is used to *control* the internal resistance between the emitter and the collector elements. This, in turn, controls the voltage drop between the emitter and collector elements. One method of accomplishing this is to change one of the resistances in the circuit. The emitter-to-collector elements of the transistor form such a resistance. When a voltage is applied between the base and emitter elements the internal resistance of the emitter-to-collector elements changes. An increase in the base-to-emitter voltage will create a decrease in the emitter-to-collector resistance. A decrease in the base-to-emitter voltage creates the opposite effect.

The emitter and collector elements of the transistor are connected in series, as shown in Figure 8–8. The changes in base-to-emitter voltage cause a change in the resistance between emitter and collector elements. This will be reflected in a change in

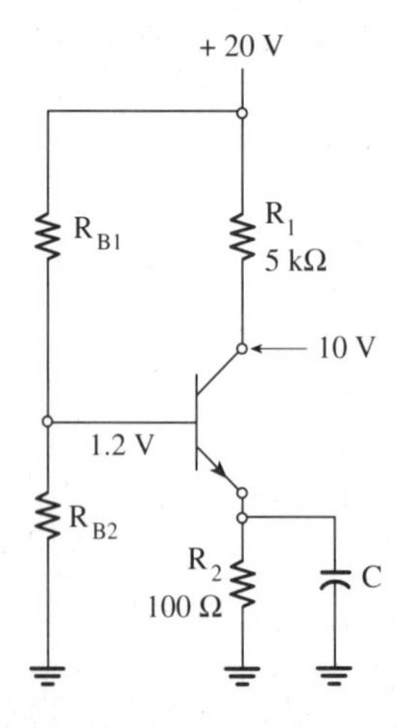

Figure 8–8 **Transistor operation depends upon the change of the base-to-emitter voltage in the circuit.**

the voltage drop that develops across these elements as current flows through the transistor. This changes the voltage drops that develop across the rest of the components in this series circuit.

The emitter and collector elements of the transistor are indicated as a variable resistance in Figure 8–9. The chart shown in this figure shows the difference in voltages that develop across each component in the circuit when the internal resistance of the transistor changes.

Let us review this chart to see what each of the lines indicates to the service technician. The upper horizontal line portrays the normal operating conditions for this circuit. This is the information shown on the schematic diagram. The service technician is attempting to validate these values with the tests. If these conditions are validated, then the circuit is operating as it was designed to and the problem must be elsewhere in the system.

Comparing the data on the middle horizontal line with the values provided in the service literature,

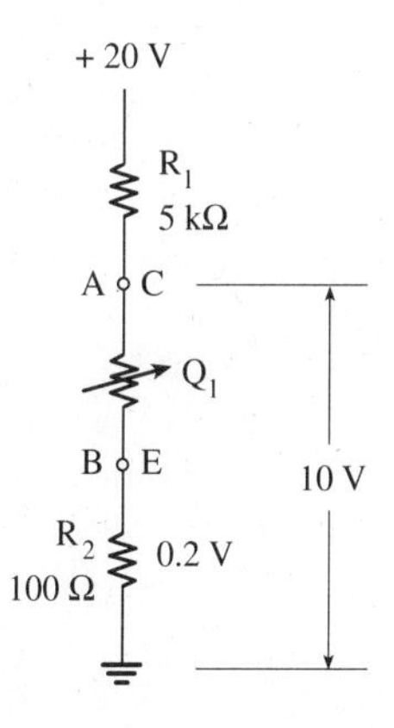

R_Q	R_T	I_T	V_{R1}	V_{Q1}	V_{R2}
4900 Ω	10 kΩ	0.0015 A	7.5 V	12.35 V	0.15 V
19.9 kΩ	25 kΩ	0.0008 A	4.0 V	15.92 V	0.08 V
1.9 KΩ	7 kΩ	0.0028 A	14.0 V	5.72 V	0.280 A

Figure 8–9 The emitter-to-collector elements of the transistor can be considered to be similar to a variable resistance for purposes of circuit analysis.

the technician will immediately discover that there is a discrepancy. This discovery will probably be obvious when voltage measurements are made and analyzed. A voltage drop developed across R_1 does not agree with that provided by the service literature. The immediate reason for this is an increase in the value of one of the resistances in the circuit. This analysis is based on Kirchhoff's rule for voltage drops in a circuit and on the rules for resistors in a series circuit. One form of this rule states that the voltage drops in a closed loop, or series circuit, will add together to equal the applied voltage. Since this circuit represents a series circuit, the laws for series circuits remind us that the only way the circuit voltage drops can change is when the resistance of one or more of the series components changes. This circuit can be described as having three resistance values in a series configuration.

Since the measured voltage drops do not agree with the values provided on the schematic diagram, the service technician must assume that one or more of these values have changed during operation of the system. At this point the service technician will stop to mentally analyze what the results of this test indicate. The schematic diagram shown in Figure 8–10 illustrates the circuit and its measured voltage drops. Returning to the rules for voltage drops in the series circuit, the greatest voltage drop will de-

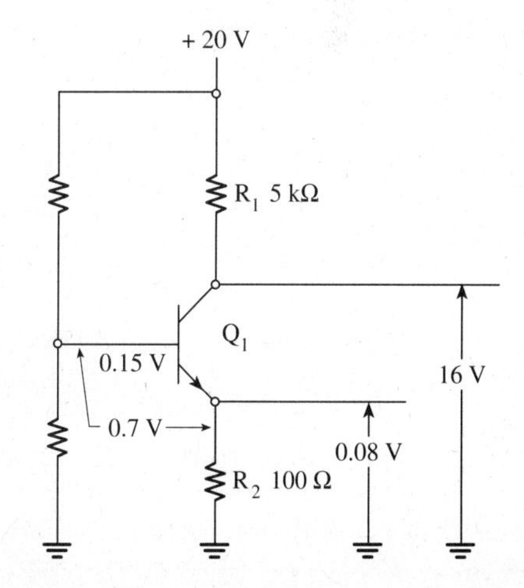

Figure 8–10 Typical emitter-to-base voltage for a silicon junction transistor is about 0.7 V for proper operation.

velop across the largest single resistance value. In this circuit the measured voltage drops are:

$$R_1 = 4.0 \text{ V}; R_{Q1} = 15.92 \text{ V}; R_2 = 0.08 \text{ V}$$

Further analysis of the measured values indicates that a much higher than normal voltage drop is occurring across the terminals of the transistor, R_{Q1}. This indicates to the technician that this value of voltage drop is caused by a very large resistance. If the resistance between the emitter and collector elements of the transistor is large enough to create this value of voltage drop, the transistor is not functioning properly. This will indicate that the area of the problem is most likely limited to the transistor.

The novice service technician would now exclaim that the problem in this circuit was located. Further, all that is required is to locate a replacement transistor and install it in the circuit. The experienced service technician would not reach this conclusion quite as rapidly. While this is *probably* the problem in this circuit, further testing is required before any components are removed and replaced. An additional test must first be made to determine if the bias voltage at the base of the transistor is correct. The schematic diagram indicates a normal value of 0.15 V at the emitter. Since this transistor is an NPN type, the voltage at the base must be between 0.6 V and 0.8 V higher than that at the emitter. This value was measured, as indicated on the schematic diagram shown in Figure 8–10, and found to be the required 0.8 V, thus falling within its correct operating values.

A visual inspection of the circuit in this example did not indicate any burned or overheated components. If you were to measure the two fixed resistances, these resistors would measure within their rated tolerances. This, of course, can only be accomplished when the circuit power is off and the resistor is physically removed from the circuit. At this point, the service technician could successfully conclude that the transistor itself was the faulty component in the circuit. Replacing it would most likely return the system to its properly operating condition. However, as a precaution, the experienced service technician would recheck the values of resistors in the circuit. This additional procedure is done to insure that the ohmic values of these resistors are correct. While this does require additional time, it might eliminate the need for a recall at a later date.

The final step in this procedure would be to turn on the system power source and measure all voltage values. Time permitting, the technician would monitor the operation of the unit over a period of several hours to be sure that other components have not failed. This would certify that the unit was operating properly.

Before moving on to another circuit and/or system, let us review what was accomplished in the previous section. First of all, the use of Ohm's law, Kirchhoff's voltage law, and the rules for series circuits were applied. The service technician first identified the circuit type as being a series circuit. Following this, the rule for voltage drops in the series circuit was followed. This rule stated that the individual voltage drops were directly proportional to the individual resistance values. Thus, the largest resistance value in the circuit would have the highest value of voltage drop.

Once this was established, the technician then considered the rules for voltage drops in a closed loop, as described by Kirchhoff. The individual voltage drops had to add to equal the voltage applied from the power source. Since these values did support Kirchhoff's law, the next step was to determine why the voltage drop developed across the emitter-to-collector elements of the transistor was incorrect for circuit operation. This referred back to the basic operation of the transistor. Once the operational values were compared to those known to be correct for typical transistors and those shown on the schematic diagram, the diagnosis of the defective component could be made. In this circuit the measurements of circuit component voltage drops and the values of the static components indicated a failure of the transistor. Its replacement would be the most likely solution.

Another rule the service technician used was the one that describes a typical transistor amplifier circuit. Values of voltage drops associated with the elements of the transistor were recalled when needed to evaluate the actual voltage values measured in the circuit.

CURRENT MEASUREMENTS

Measurement of electrical current in a circuit can be more difficult. The reason for this is that the traditional method of current measurement requires the insertion of the ammeter into the path for current flow. This procedure is illustrated in Figure 8–11. The circuit is "opened" at the junction of R_1 and R_2. A current indicating meter is inserted at the open and thus completes the path for current flow. In this circuit the current flow is equal to 0.001 A, or 1 mA. The amount of current is determined by applying the rules for series circuits and Ohm's law:

$R_1 + R_2 + R_3 = R_T$ or $10\ k\Omega + 5\ k\Omega + 20\ k\Omega = 35\ k\Omega$
(series circuit rules)

$E \div R = I$ or $35\ V \div 35\ k\Omega = 0.001\ A$
(Ohm's law)

How this amount of current is conveniently measured is one of the processes used by the successful service technician. Analysis of the circuit shows that the value of any of the resistances may be used for voltage measurement. The value of measured voltage can then be applied to Ohm's law, since the value of resistance is shown on the schematic diagram. In this example, R_1 in Figure 8–12 is used to illustrate how this works. The value of R_1 is 10 kΩ and the voltage measured across the leads of this resistor is 10 V. This value is then inserted into the Ohm's law formula to validate the value of 0.001 A. Thus, the current is measured *without* the need to open the circuit and insert the current meter.

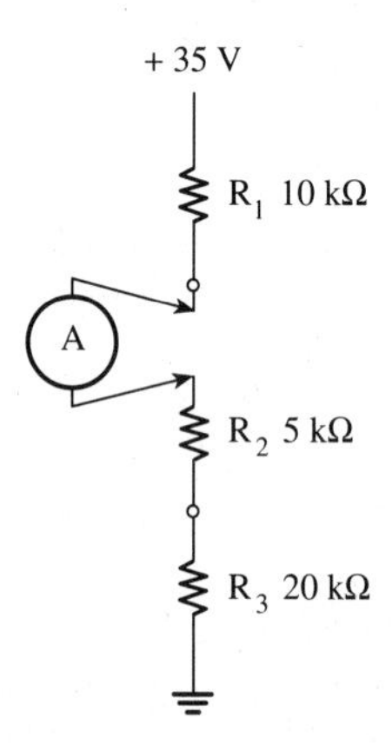

Figure 8–11 Current measurement is accomplished by inserting the current measuring meter into the current flow path.

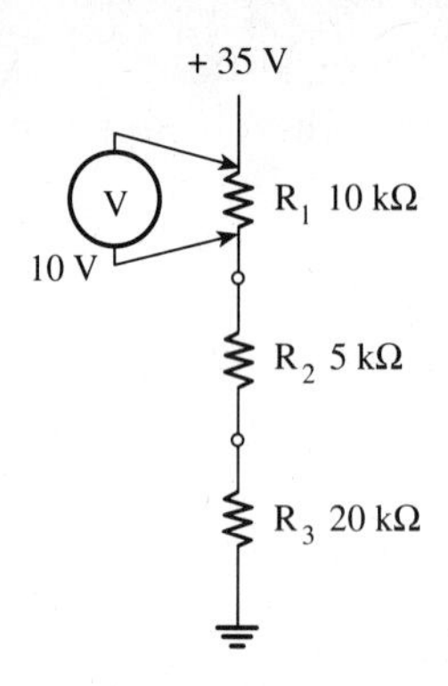

Figure 8–12 Voltage drops are measured by placing the leads of the voltmeter directly across the component, as shown here.

When the meter uses a switch to change the ranges of voltage, this system can be extended to measure almost any current value. Consider the fact that Ohm's law can be applied to any circuit when both the resistance and the voltage are known. In the previous example, the current was determined by applying these known values. Many nonautoranging meters are calibrated with ranges of 1.0 V, 10 V, 100 V, and 1000 V. A range of 0 V to 10 V is used as the reference in this example. The voltage drop being measured across the 10 kΩ in this example can be used as a current measurement. The facts known about the relationship between voltage and resistance are applied when calculating current values. Why not predetermine a value and then use it in the Ohm's law equation? For example, when 10 V is measured across a 10 kΩ resistance, the current flow is 0.001 A. The dial scale on the meter is calibrated in unit values ranging from 0 V to 10 V. Using Ohm's law, 0 V measured across a 10 kΩ resistance provides a value of 0 A. A 5 V measurement across this same 10 kΩ resistance will provide a value of 0.0005 A, or one half of the scale on the

meter's dial. This process creates a linear system of values, therefore each volt measured will be equal to 0.0001 A. The meter dial, calibrated in units ranging from 0 to 10, will then be "reading" values from 0 to 0.001 A directly from the meter dial scale. The service technician is now able to measure the values of current in the system without having to remove any component and physically insert the current meter into the circuit where it was separated.

This same procedure will work for any combination of voltages and resistance ranges. I used a very basic set of voltage and resistance to demonstrate how this will work. Many schematic diagrams will provide the voltage drop values for a specific component. One of these is illustrated in Figure 8–13. The information shown in this illustration is similar to that provided in the service literature for an industrial machine control unit. The voltage drop values shown on the diagram are those measured by the manufacturer's engineering/design staff. These values are determined to be those found under normal operation of the circuit. Consider what these will mean when voltage values differ from those provided on the service literature. As this circuit is analyzed, you must refer back to the basic procedures described earlier in this chapter:

1. What to measure.
2. What equipment to use.
3. Where to measure.
4. What to expect to find when making the measurement.
5. What the actual measurements mean.
6. The next step(s) after analysis of the measurement.

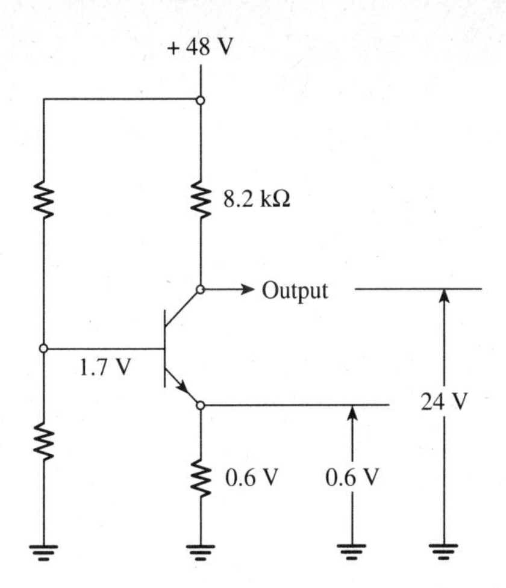

Figure 8–13 **Measurement of a voltage drop across one component can provide information about the proper operation of the circuit.**

Start by determining what to measure. Since the manufacturer provided a voltage drop as measured across a specific resistance, this may be used to determine if the circuit is operating properly. The emitter-to-common resistance value in this circuit is 200 Ω. The manufacturer's service literature indicated a voltage drop of 0.6 V for this resistance under normal operating conditions. When this circuit is determined to be the area of the problem in the unit, the service technician will be able to use this value to determine the circuit's operating condition. This completes the first step in the process.

Once the determination is made, the next step is to decide what type of test equipment should be used for this measurement. The easiest and most convenient piece of test equipment is the voltmeter. You could also use the DC voltage-measuring capability of the oscilloscope for this measurement.

Where to measure is very fundamental for this circuit. The manufacturer provided a specific place for this measurement—across the emitter resistor, or between circuit common and the emitter element of the transistor. Experience tells us that the resistor is easier to locate than the emitter element of the transistor. The voltage drop will be measured across the leads of this resistor.

The expected voltage value is provided on the schematic diagram. The measurement should indicate a voltage drop of 0.6 V when the circuit is functioning properly.

The last step in this procedure is analyzing the test results. Three possibilities can occur for this circuit. They include a measurement of the proper value as shown on the schematic. If this was the situation, the circuit would most likely be operating properly and the problem has to be elsewhere in the system. The other two situations are the voltage

measurement results that differ from those on the diagram. The measured voltage could either be higher than the recorded value or lower than this value provided on the schematic diagram.

If a value of 0.6 V is measured, there is little need to concentrate on this circuit. Return to the planning stage of your procedure and reevaluate the process. When a value *higher* than 0.6 V is measured, then there is a problem in this circuit. What does a higher than normal value indicate? Analysis of this circuit shows that the collector resistance, the emitter-to-collector leads of the transistor, and the emitter resistor form a series circuit. The laws for voltage drops in the series circuit remind us that each individual voltage drop is determined by the resistance value of the components in this circuit. Individual voltage drops are directly proportional to the specific resistance values. Two of the three components in this circuit are nonadjustable, or fixed-value, components. They are the two resistances. The third component, the transistor, is actually an adjustable, or variable, resistance as far as the circuit operation is concerned. It is an active device.

When the voltage drop measured across the 200 Ω emitter resistance is higher than expected, the most likely problem is that the ohmic value of one of the other components in the circuit has changed. The only way that the voltage drop across any component can increase is when the resistance of one of the other components has decreased. The rule for voltage drops across series resistances is that the highest value of voltage drop develops across the largest value of resistance. It is also based on the fundamental rule for voltage drops in the series circuit as developed by Kirchhoff. Using another rule, the most likely component that will fail is the transistor. The reasoning for this is simply that resistors seldom fail unless some other component has caused their failure. When servicing, the expert technician calls upon previous experience. The number of times the experience has occurred is a good indicator of which component will fail in an operational circuit. This will remind the servicer that the failure rate for transistors is much higher than that of resistors. The first component to be examined for its proper operation should be the transistor in this circuit.

If, on the other hand, the voltage value measured across the emitter resistor is *lower* than normal, the possibility of any component failing is much greater. A lower than normal voltage measurement indicates an increase in the resistance value of another component in the series circuit configuration. Either the collector resistance or the transistor could be at fault in this situation.

Probably the first consideration for identifying the specific component that failed is a visual inspection of the circuit. Often an increase in the ohmic value of a component is indicated by the image it creates. This image is the charring or burning of the component. In extreme cases the component will actually split into two parts. This occurrence is related to the work of Watt. Each electronic component has its own power rating. This rating, in units of the watt, indicates the *maximum* amount of electrical power the device is able to handle under normal operating conditions. Electrical power is determined by the formula:

$$\text{P (power)} = \text{E (volts)} \times \text{I (current)}$$

Excess power used by the component will lead to its failure. In this circuit the collector resistance has a rating at 0.125 W in addition to its 8.2 kΩ ohmic value. Normal operation of this circuit will create a wattage of just under 0.07 W. If the transistor should short circuit and fail, there would be an increase in both the voltage drop across the collector resistor and the current flow through it. A short circuit of the transistor would reduce total circuit resistance to 8.4 kΩ. With an applied voltage of 48 V, as indicated on the schematic diagram, the current will increase to just over 0.0057 A, or almost double the normal current flow value.

Consider the power used by these resistors when the elements of the transistor short circuit. The circuit now appears as shown in Figure 8–14. Total circuit resistance is now 8.4 kΩ and total current in this circuit is now almost 0.006 A. The voltage drops developed across each of the components has also increased due to the effective removal of the transistor's resistance value. Voltage drop across the collector resistor has increased to almost 48 V and the drop across the emitter resistor has increased to 1.14 V. Power consumed by the collector resistor has

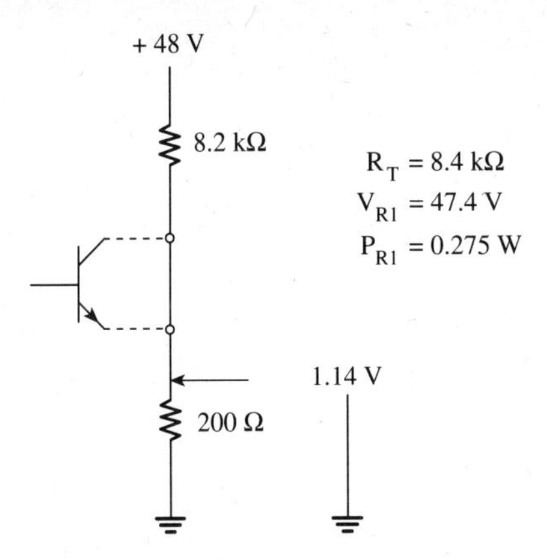

Figure 8–14 **When one component is shorted, additional current flow occurs and the remaining components dissipate more heat.**

increased from its original value of 0.07 W to a much higher value of 0.275 W. The wattage, or power, rating of the collector resistor was 0.125 W. When this greater current flow occurs, the heat generated by the current flow will be much higher than the rated power value of the resistor. This extra heat will either char the resistor or cause it to break into two pieces. A visual inspection of the circuit should show this result of excess heat and help localize the problem area.

Keep in mind that resistors can fail and create a lower than normal value of resistance. This usually happens during the failure cycle of the resistor. The end result, the one normally observed by service technicians, is when the failure has caused the resistor to overheat. When an overheated or broken resistor is located, the service technician must locate the cause of this failure. Often some other component in the circuit, such as the transistor in the example, fails and creates the visual signs.

Many of the newer types of film resistors do not have this built-in deficiency. These have been observed to be literally red hot during their operation and, when cooled off and measured, still retain their original resistance value. The specific failure, of course, depends upon the type of resistor used in the circuit.

PARALLEL CIRCUITS

Analysis of the parallel circuit uses the same rules as those for series circuits. The rules presented by Ohm, Kirchhoff, and Watt are still applicable for this analysis; the difference is in the methodology of approach and how the information is analyzed. A block diagram for a two-stage audio amplifier is shown in Figure 8–15. The signal is injected at the input of Stage One. The signal output of this stage is directly connected to the input terminal of Stage Two. The output of Stage Two is then connected to a following stage. The power source is common for both of these stages. It probably is common to all stages in the unit, as well.

This example is kept simple to present the process of analysis in a parallel path circuit. The unit could easily use more stages. The process of analysis will apply regardless of the number of stages.

In this example, assume the main fuse from the power source has failed. This indicates an overcurrent problem in the unit. Further analysis has localized the area of trouble to one of these two blocks. The next problem is how to approach and localize the area of trouble. Use of the voltmeter, as described in the section on series circuit analysis, is not appropriate since the power source is not functional. The first thing to do is study the schematic diagram to see if there is another way to approach this problem. The schematic diagram shown in Figure 8–16 shows the circuit to analyze.

This schematic shows four series circuits connected in parallel with the power source. The com-

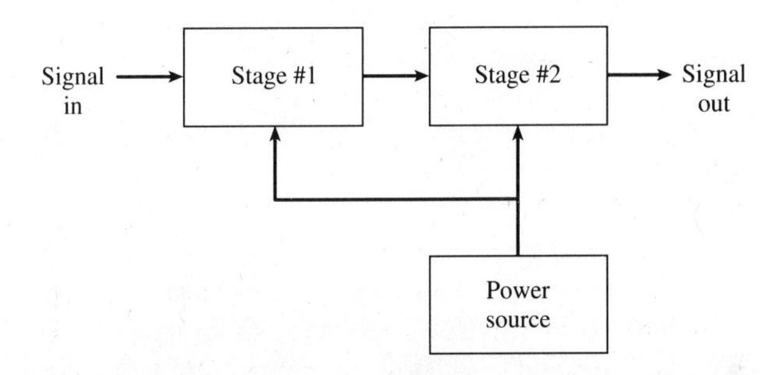

Figure 8–15 **The block diagram for a two-stage amplifier is shown here to illustrate signal flow processing.**

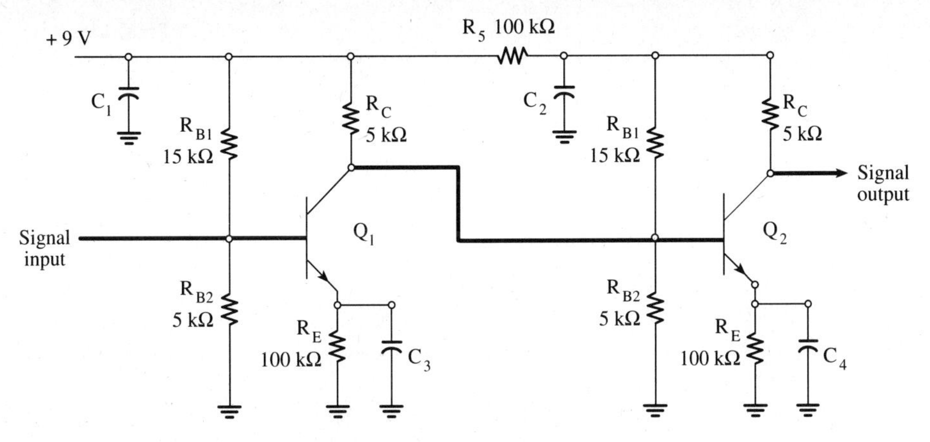

Figure 8–16 The actual circuit for the block diagram in Figure 8–15. Signal flow processing is shown by using a heavy, dark line.

ponent values for each transistor circuit are identical. A two-resistance series circuit is used to develop the operating bias of each transistor. These resistors are identified as R_{B1} and R_{B2} in both circuits. Their values are 15 kΩ and 5 kΩ, respectively. Each of the transistor circuits also has the same component values. The collector load resistors are identified as R_C and the emitter resistors as R_E for both transistor Q_1 and transistor Q_2. A series resistance of 100 Ω identified as R_5 is used to isolate the two amplifier stages. In addition, each of the transistor circuits has a decoupling capacitor connected to the positive source line. These are identified as C_1 and C_2 on the diagram. The emitters of each of the transistors also have a decoupling capacitor. The final component is the signal coupling capacitor located between the collector of Q_1 and Q_2.

The question remains: how to locate the specific component that has failed and created the overcurrent problem. The first step is to review the six basic steps identified earlier in this chapter:

1. What to measure.
2. What equipment to use.
3. Where to measure.
4. What to expect to find when making the measurement.
5. What the actual measurements mean.
6. The next step(s) after analysis of the measurement.

The lack of circuit power dictates an alternate approach to measuring values in these circuits. Since there is no voltage present, there is no current flow. The only type of measurement that can be performed without removing any circuit components is a resistance measurement.

The instrument capable of measuring resistance values is the ohmmeter. The service technician made the obvious choice by selecting this instrument.

Where to measure is the question. It would appear that a measurement at the load side of the fuse would indicate a lower than normal resistance path due to the decrease in the design values of resistance in the circuit. If this were not true, the fuse protecting the circuit would not have opened. One method, one that is not initially recommended, is to cut the wires or circuit board paths to each of the circuits while measuring the resistance at the load side of the fuse. When the resistance in this circuit increased quickly, the service technician would have located the abnormal current path. The error of doing this is that it is time-consuming and does not always provide the correct information. There is a better method described in the following paragraph.

This better method does not require destruction or modification of the circuit, only the ability to analyze the circuit. The service technician has to stop to consider what he or she knows about the circuit. As stated previously, there are four series paths all connected in parallel to the power source. An isolation resistor is found on the power source line between the two stages.

A resistance measurement at the collector of the transistor should be close to infinity. The reason for this is that the transistor is not operating and does not have any bias on it at this time. The other measurement that can be made is from the base to circuit common. The resistance should be close to the value of R_{B2} since there is no power in the circuit and the transistor is not biased at this time. These are the places to make the initial set of resistance measurements.

The expectations for resistance measurements are the same for both of the transistors. A measurement at the collector element should indicate infinity. Measurements at the bases should be close to 5 kΩ. The other place to make a measurement is at the Q_2 side of resistor R_5. A resistance measurement at this point should give a value of 20 kΩ, since this is the sum of the individual resistances of the bias network.

Finally, what do these measurements indicate? There are several possibilities for a circuit like this. Let us examine each of the suggested measurements and describe the results of the measurements.

FROM COLLECTOR TO COMMON

A resistance measurement here should indicate a value close to 20 kΩ. The transistor is not active and its internal resistance places it in cutoff. This measurement is only valid to show that the transistor is not shorted. An open transistor will not create an overcurrent condition, the cause of which we are trying to locate. Should the resistance reading be close to 100 Ω, then the transistor is shorted and the problem area is located. This is true for either of the transistors.

This test should be made early in the process of troubleshooting. The chance that the bias resistors are defective is minimal. Their current path is only 2 to 5 percent of the total current in the transistor's circuit and should not cause the overcurrent condition.

FROM LOAD SIDE OF THE FUSE TO CIRCUIT COMMON

A test at this point may provide some valuable information. For example, if the reading is zero ohms, the probability of the decoupling capacitor, C_1, being internally shorted is likely. This would be the only reason for a zero ohm reading. Shorts in either of the transistors would still provide a resistance value reading since the fixed resistors are still in the circuits. A short in C_2 would provide a reading of 100 Ω since R_5 is in the circuit between the test point and circuit common.

A resistance measurement at C_2's connection to the power source line should provide a reading of at least 20 kΩ if the circuit is valid. A short in the transistor would reduce this reading to just over 4 kΩ if the transistor were short circuited. The reasoning for this is that the two resistance paths of 20 kΩ and 5.1 kΩ are in parallel and their equivalent resistance must be less than the value of the lowest single resistance path.

Finally, where should you go when the results of the initial tests have been completed? These results must be analyzed. The best way to develop the next step is to analyze what you have learned and apply it as if you are considering each path as an independent series circuit. With the exception of the decoupling capacitors, each bias path and transistor emitter-to-collector path (with associated resistances) is an individual series circuit.

A second parallel circuit is shown in Figure 8–17. This circuit as a single stage represents the output circuits of many electronic systems, including deflection circuits for video displays and industrial equipment. This system is also used as shown as the output circuit for audio amplifiers. One of the helpful features of the portrayed circuit is that, should one stage fail, the other is still available to use as a reference to test the failed stage. In essence, the circuits consist of two complementary symmetry power transistors. One is an NPN, the other is a PNP type. The two transistors have similar characteris-

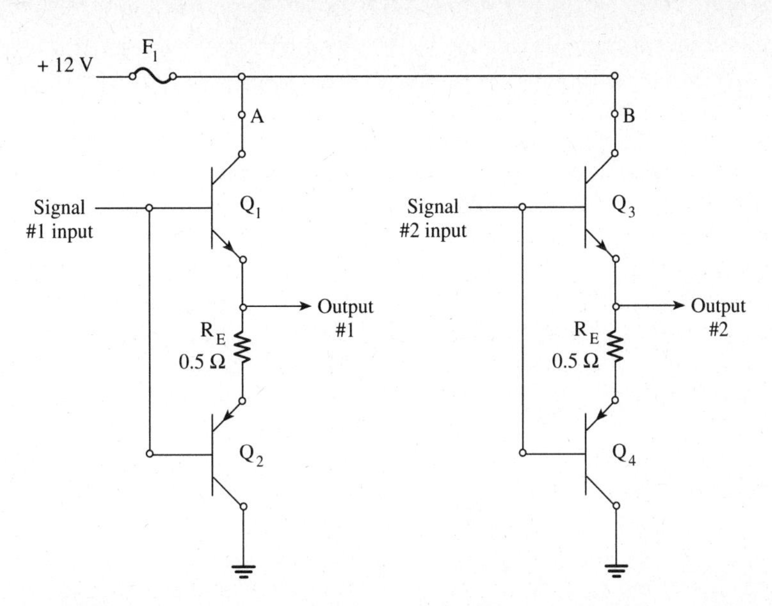

Figure 8–17 **These audio output amplifiers are connected in parallel with the power source.**

tics with the exception of their polarities. The fundamental circuit is a series circuit. Current flow starts at the common connection, flows through the collector and to the emitter of the PNP transistor Q_2, through the emitter resistor R_E, then through the emitter and the collector of the NPN transistor Q_1, and completes its journey at the positive terminal of the power source. The output connection of this circuit is at the midpoint, on either side of the emitter resistor. The output is connected to a low-impedance device, such as a deflection yoke, a speaker voice coil, or a motor.

Power output transistors are among the hardest-working components in any device. The only components that work harder are in the power source section. The output circuit is where most of the heat in the unit is created. This is also where the output components are normally mounted on heat sinks. A rule of thumb is that in any unit the hardest-working components are the ones that fail most often. It would stand to reason that, since there is more heat created in this section, it would be more apt to fail.

In this example both of the output circuits are connected to the same power source through a common fuse. When one fails, the other will also stop functioning, since the power will be disconnected from both at the same time. As an aside, it would be better to provide two fuses for this circuit, one each at point A and point B on the diagram. The question for the service technician, once one of the channels fails, is how to approach diagnosis and identification of the malfunctioning component(s).

The first question is what to measure. Since both of the stages are in parallel, a resistance measurement will provide information if neither channel has a short-circuited component. Under a "no power" condition, none of the transistors is biased in an on condition. The internal resistance of all of the transistors is very high in this condition. If one of the transistors was short circuited, then the total resistance could be very close to zero ohms. It still could be infinitely high if the measurement was taken from either point A or point B and either of the lower two transistors was open. Perhaps a resistance measurement across each of the transistors would indicate a short-circuit condition. Since the fuse had blown, this is most likely the better place to start to measure.

Possibly one of the best measurements to make at this time is the resistance measurement. This, of course, is done with an ohmmeter.

Where to measure is the next question. A resistance measurement taken between the emitter resistor and circuit common could indicate a short in one of the transistors. A similar measurement could also be taken from the emitter resistor to points A and B, respectively.

The expected measurements at each of the four components should be on the high resistance side. This would be true when the transistors have not short circuited.

A low resistance measurement for any of the four transistors would indicate the reason for the overcurrent condition and the ultimate failure of the protective fuse. The emitter resistor should be measured, since it, too, was subjected to excess current flow and its resistance value may have changed.

If the resistance measurements are close to normal for transistor emitter-to-collector readings then, the problem is elsewhere. It might be related to one of the output connection circuits. In any case further work is required.

All of the circuit analyses described in this chapter use the same format—the specific six steps identified as a part of the analysis procedure for both series and parallel type of circuits. The major activity in all examples requires the ability to plan how to approach the problem and consider what information is provided by each of the steps in the analysis process. This six-step system is applied continuously, constantly reducing the area of the problem to smaller sections of the device. Using this procedure, the ultimate result is identifying the specific component(s) that have failed. This is what the entire process related to troubleshooting is all about. The successful electronic service technician learns this process and applies it to all service-related problems. These are the steps required to become the world's best electronic technician.

REVIEW

Recognition of the basic circuits used in electronic devices is essential for the service technician. In addition, the technician must be able to remember the rules for the specific type of circuit involved. These rules are then applied during the planning process involved prior to the actual testing and analysis. The basic rules, identified by Ohm, Kirchhoff, and Watt, are applied over and over again as the area of the problem is constantly redefined and reduced in size.

You must be able to recognize series circuits and apply the basic rules about them. These rules remind the service technician that current flow in the series circuit is equal throughout the circuit, regardless of the values of the individual resistances in the circuit. Voltage drops in the series circuit are directly proportional to the individual resistance sizes, with the largest resistance value creating the largest value of voltage drop. Individual resistance values add up in the series circuit to equal the total amount of resistance. These three factors are then used in Ohm's law to determine total circuit values of voltage, current, and resistance.

These same concepts apply to parallel circuits. The service technician has to be able to recognize parallel circuits and apply them during the analysis process. Since the rules for series and parallel circuits differ, the service technician has to be able to recognize which of the two circuits is being analyzed. Then, and only then, can the proper rule or relationship be used.

In addition, the relationships established by Kirchhoff for voltage drops in the closed loop, or series circuit, also apply. These relationships are the same as those described for voltage drops in the series circuit. Kirchhoff's relationship rule for current at any junction is also necessary for the analysis process.

The final basic rule discussed in this chapter is Watt's law—current flow in the circuit creates heat. Components used in the circuit must have a sufficient power rating or they will fail. This heat must be dissipated as the circuit is operating. Excess heat will cause the component to fail. Often a heat-related failure in one component is due to the failure of a second component. Service technicians must recognize this when attempting to locate a fault.

One additional factor for the service technician to recognize is the possibility of incorrect information in the service literature. Every effort is made by the manufacturer to ensure that this does not happen. But, as we all recognize, there are times when a little misinformation does occur. The service technician must know what to expect to find in most circuits relating to typical voltage values for common circuits. When a value is provided that does not appear to be normal, further analysis is required. This will determine whether the circuit has unique values or an error was made on the diagram.

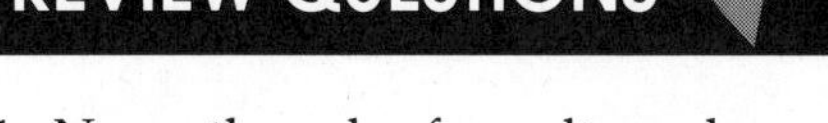

REVIEW QUESTIONS

1. Name the rules for voltage drop, current flow, and total resistance in the series circuit.
2. Name the rule for voltage drop as proposed by Kirchhoff.
3. What is the relationship of voltage drop and resistance ohmic value in the series circuit?
4. Why should you refer to the schematic and block diagram before using any test equipment for circuit analysis?
5. What is the purpose of a visual inspection of the unit during servicing?
6. Name the six basic steps used by the service technician during the planning stage.
7. What does a higher than normal voltage measurement in the series circuit indicate?
8. What does a lower than normal voltage measurement in the series circuit indicate?
9. Name the indicators of a higher than normal power condition in a component.
10. What is the typical emitter-to-base voltage of an NPN silicon transistor?
11. What does a zero emitter-to-base voltage in a transistor indicate?
12. Explain how current measurements can be made without removing any components from the circuit.
13. Why is measuring resistance in the parallel circuit difficult?
14. What is the most prevalent condition of a failed resistor?
15. What does an ohmmeter reading of zero ohms indicate?
16. What value resistance should you expect to measure across the emitter-to-collector elements of a transistor that is not connected to a power source?
17. Using the rules presented in this chapter, what voltage drop should you expect to measure across a fuse's leads when the fuse is functioning properly?
18. Explain the importance of the planning activity during the service process.
19. Explain the importance of listening to the customer's remarks as a part of the initial service process.
20. Why is it necessary to understand how the system is supposed to function as a part of the service procedure?

Chapter 8 Challenge One

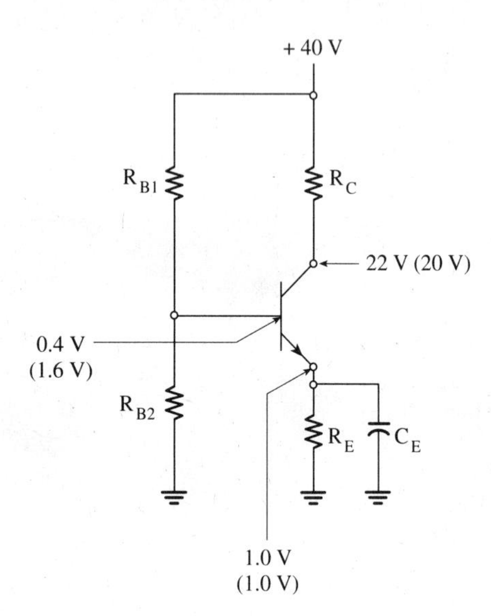

The schematic diagram provides the operating voltages for the above circuit. The actual, measured voltage values are shown in parentheses ( ). Which set of voltage values is incorrect? Explain the reason for your answer. Use electronic laws and rules to support your answer.

Chapter 8 Challenge Two

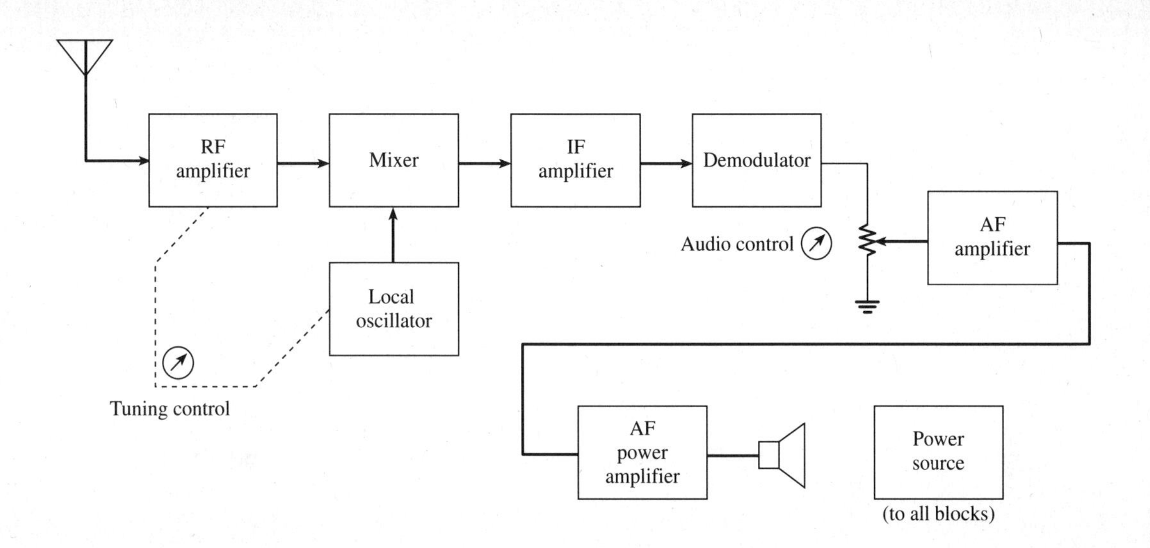

The AM radio shown above has no audio output. A schematic diagram is available. Develop a plan you can use as you start to identify problem areas in this radio. Be very specific. Use this format:

Test # ____________________

Where to test ____________________

What physical parts to use for the test point ____________________

Expected result of the test ____________________

Analysis ____________________

Chapter 8 Challenge Three

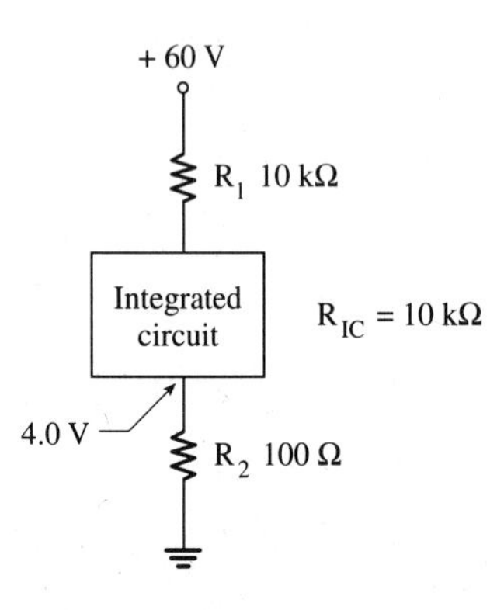

A value of 4.0 V is measured between circuit common and the upper end of resistor R_2. What does this voltage measurement indicate to you? Explain your answer in terms of the basic laws for electronics.

Chapter 8 Challenge Four

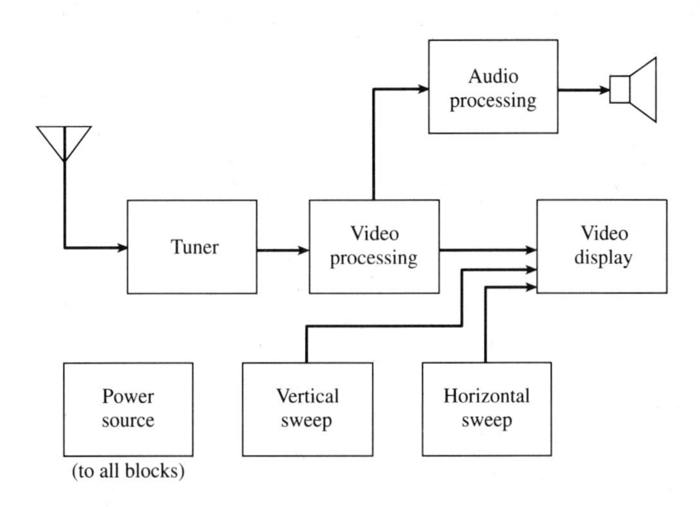

The television receiver shown above does not have a display (the screen is dark). The sound is normal and changing the channels produces different audio information. Which of the blocks would you eliminate during your initial analysis of the system? Explain why you made your decision for each of the blocks you chose to eliminate.

Chapter 8 Challenge Five

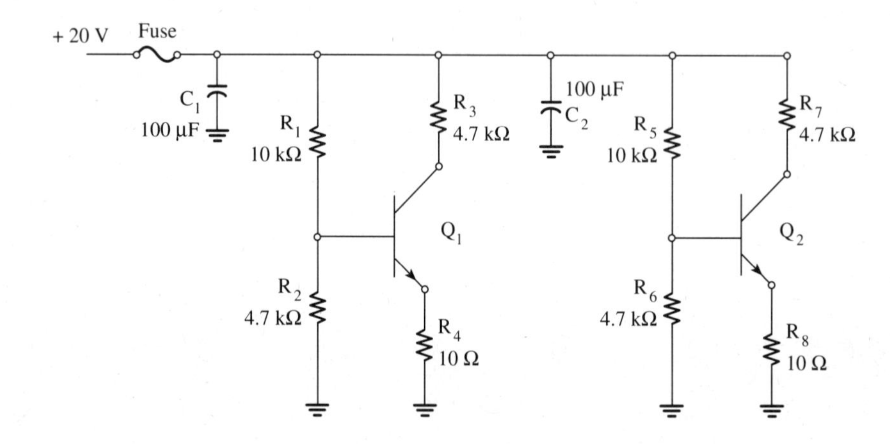

In the above circuit the fuse has opened, indicating an overcurrent condition. Develop a plan to locate the defective component(s). Use the planning sheet format. Explain why you selected each test, what you should find when making the test, and why you selected the specific piece of test equipment.

Chapter 8 Challenge Six

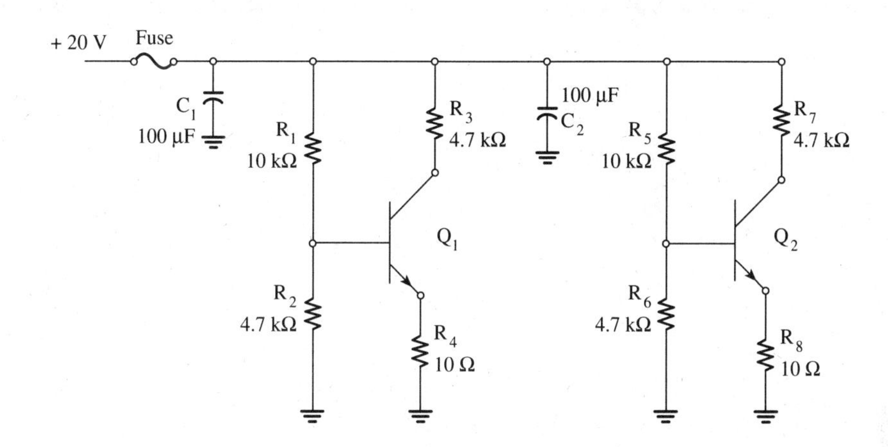

Referring to the information in Challenge Five, the circuit resistance measured at the upper end of capacitor C_1 is close to 7.4 kΩ when capacitor C_2 is removed from the circuit. What does this tell you? Explain your answer using electronic rules and laws.

CHAPTER 9

Testing Specific Components

INTRODUCTION

Once you have reduced the area of the problem to one section of the device, the next question is: "What do I do now?" The answer has to be to use some specific piece of testing equipment to localize the problem to one or more components. Selecting the proper piece of test equipment is as important as analyzing the system before making any tests. The correct test will identify a defective component. An incorrect test, or use of the wrong piece of testing equipment, can damage components. An incorrect test can also produce answers that do not lead to identification of the defective component. In all situations, the service technician must be able to select the proper piece of test equipment, know how to use it, and also know what the results of the test indicate. This chapter will assist the learner in selecting the correct test equipment and will also provide some basic information on how to use it. Specific details on use, of course, should be obtained from the manufacturer's operating manuals accompanying each test unit.

OBJECTIVES

Upon completion of this chapter, the student/reader should be able to:

1. select the proper piece of test equipment;
2. understand why and how specific test equipment is used;
3. understand the results of using specific test equipment; and
4. understand the impact of instrument input resistance ratings.

HOW TO TEST COMPONENTS

Each component may be tested in a specific manner. The efficient service technician will recognize this statement and understand its meaning. Specifically, resistors are best tested by use of the ohmmeter. An oscilloscope, for example, will seldom produce the answers necessary to determine whether a resistor has failed. Also, a voltmeter with a low input resistance value can damage circuits when used incorrectly. Knowing which piece of test equipment to select and use is the major theme of this chapter.

Electronic testing equipment can be categorized into general types and component-specific types. In addition (and not covered in this chapter) there are signal-specific test units. These are used to create signals for processing through such devices as video display terminals, shortwave receivers, and computers as well as other complex electronic equipment.

TEST EQUIPMENT CONSIDERATIONS

Input resistance is perhaps one of the most important items to discuss when considering testing equipment. Specifically, the beginning topic should be the input resistance of the tester. Almost every test instrument has an input resistance rating. This important value must be considered as one part of the selection and use process. Figure 9–1 is typical of how a test instrument is connected to evaluate the circuit. In almost every situation the instrument is connected in parallel with the circuit. One of the few exceptions to this statement is when a current measuring meter is inserted into the circuit. When measuring circuit current, the meter is placed in series with the rest of the components. In this manner all of the current flowing in the specific circuit will also flow through the meter.

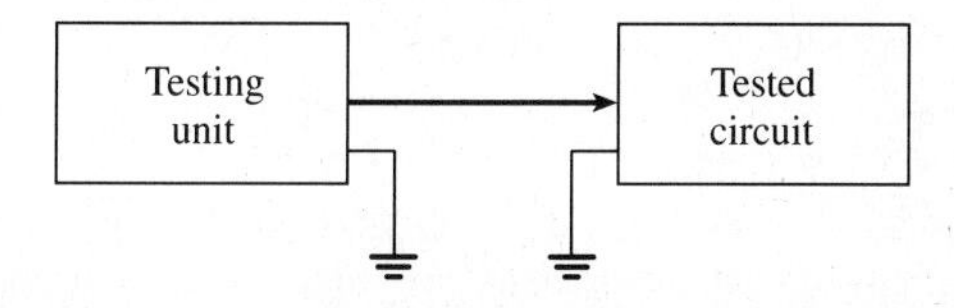

Figure 9–1 **Typical method of connecting testing equipment to the device being analyzed for repair purposes.**

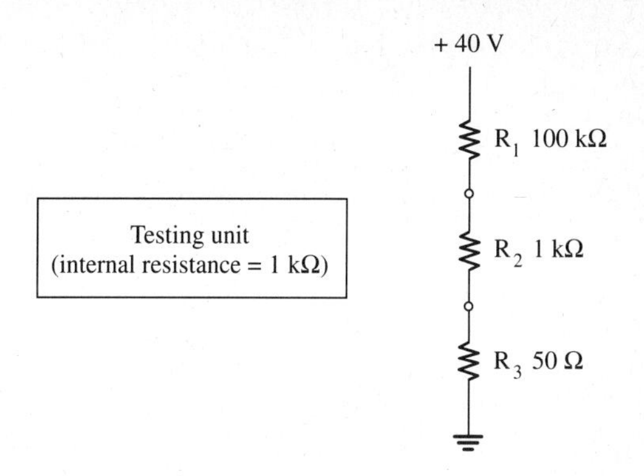

Figure 9–2 **The test unit's input resistance is 1 kΩ in this sample circuit, the same as the value of one of the resistances.**

You must consider what occurs when the test instrument is placed in parallel with an existing circuit. When the input resistance value of the test instrument is in the order of a few megohms, the effect on the test circuit is minimal. If the input resistance of the test instrument is low—less than 1 MΩ—insertion of the instrument may affect the circuit dramatically. Consider the circuit shown in Figure 9–2. This circuit consists of three resistances in a series configuration. The applied voltage is 40 V and the total resistance is 101,050 Ω. Under these conditions the current flow in the circuit is just under 0.0004 A. In this example the testing unit has an input resistance of 1000 Ω. When this value is placed in parallel with either R_1 or R_2, it has a major effect on the circuit. When a 1000 Ω resistance is parallel connected to the 100 kΩ value of R_1, the effective resistance of the meter and the fixed-value resistor is 991 Ω. This is supported by the rule stating that, when two resistances are placed in parallel, the total resistance will always be less than the lowest single value of resistance. When the value of 991 Ω is added to the other values of 1000 Ω and 50 Ω, the total resistance of the circuit will be reduced to slightly over 2 kΩ. Total circuit current will increase to almost 0.02 A, or 50 times the original value. Excess current flow, caused by the insertion of the test meter, could damage some of the original components in this circuit.

Excess current flow is one of the problems that can be created by incorrect selection of testing equipment. Another related problem is the change in the level of voltage applied to the individual components when a tester with the wrong input resistance is used. Using the same example, if the value of R_1 was reduced to 991 Ω in the process of measuring its voltage drop, then the normal voltage drop of close to 38 V would change to about 20 V. The voltage drop normally developed across R_2 would increase from its normal value of less than 4 V to close to 20 V. The possibility of damage to the device represented by R_2 is very high when this occurs.

The same type of conditions could easily occur when the test instrument is placed in parallel with R_2. Two equal value resistances in parallel create an effective value of one half of the value of one of the resistances. This effect would not be as dramatic as the one previously described, but the possibility of component damage still exists. The other major factor is that the insertion of this value of resistance and the ultimate change in circuit conditions will provide a false reading of circuit values. The technician would think that something was wrong with the circuit, when, in fact, the only thing wrong was insertion of the low-resistance test instrument.

Now consider the effect on the circuit when an instrument having an internal resistance of 5 MΩ is placed in parallel with any one of the three resistances. The effective value on R_1 would be minimal, since the total resistance of the two values in parallel is just over 98 kΩ. None of the values of voltage and current would change enough to affect circuit operation or component values. The message in this example is important to the service technician—you should locate the area of the problem, not create it. Using an improper piece of test equipment can cause damage and provide false information. Neither of these is productive; both may create additional work for the service technician by providing incorrect information and possibly damaging otherwise good components.

Another technician-created problem is illustrated in Figure 9–3. Many pieces of test equipment obtain

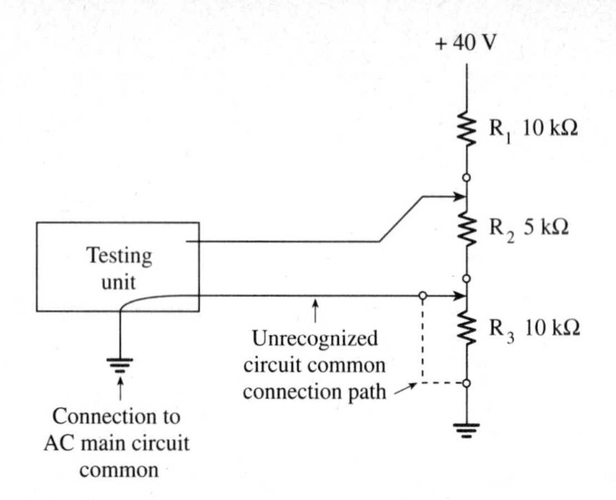

Figure 9–3 A common internal connection for the negative leads and the power source can provide false readings for the unsuspecting technician.

their operating power from the AC power lines. One of the wires on the power plug, as well as the "ground" lead on the plug, is connected to the "ground" terminal of the building's wiring. Often the negative lead of the test instrument is also connected to this common grounding. In effect, the negative lead of the tester is then always at common, ground potential, or zero volt potential.

When the tester's negative lead is connected to the common point of the circuit being tested, the effect on the measurement is valid. However, should this negative lead be connected to some other point in the circuit, the readings obtained will be in error. In this figure, the negative lead is connected to the junction of R_2 and R_3. This, in effect, places a short circuit across R_3, removing it from the circuit. The result will provide incorrect voltage readings and will also increase circuit current. If the service technician had used a meter with an isolated or battery power source, this problem would not have happened. The only method of measuring the voltage drop across R_2 when the negative lead of the instrument is connected to circuit common is to make two measurements, one at each end of the resistor, and then subtract the lower value reading from the higher one.

A final thought on this subject is that, under some conditions, this unseen circuit common, or "ground loop," could damage either the test instrument or the unit being tested. I have learned from experience that a ground loop such as this has lifted the printed circuit foil in the meter from its base material and actually melted it. This does delay further service work and is costly, as the instrument or the damaged parts, or both, must be replaced before diagnosis of the nonoperating system can be completed. In addition to being costly, the time involved to identify the components that have been damaged by the improper use of test equipment cannot be charged to the customer. All in all, the best approach is to (a) know your test equipment and (b) know how to use it properly.

METER OPERATION

One of the topics taught early in the field of electronics education is the operation of the basic meter. As you progress through the learning cycle, this information is often forgotten or overlooked. The knowledge of how the basic meter movement functions has become less important as the electronic service industry has moved away from use of the analog meter movement to the digital meter movement. Even though this is more "state of the art," the concept of how the meter measures voltage, current, and resistance is still fundamental for the successful service technician. A review of this process is very important if you are to be successful in the electronic service industry.

The basic meter movement is shown in Figure 9–4. This meter consists of a coil of wire, or armature, suspended in a permanent magnet's field. A magnetic field develops when an electrical current flows through the armature. This field reacts with the fixed magnetic field created by the permanent magnet. The interaction of the two magnetic fields creates a rotary motion and the armature rotates. The strength of the magnetic field developed in the armature determines how much armature rotation, or deflection, occurs. An indicating needle is attached to the armature to show how much deflection is occurring. The amount of current flow through the armature is normally very low, on the order of 1 milliampere, or less. This is the maximum current flow that will create a full-scale deflection on the meter's dial scale. The movement of the armature is linear as current flows through it. One half of the total allowable current will create a deflection of one half of the dial scale, etc.

The example meter shown in Figure 9–4 has an internal resistance of 100 Ω. It is designed to develop a full-scale deflection when 0.001 A of current flows through its armature. Using Ohm's law, the voltage required to create a 0.001 A current flow through 100 Ω is 0.1 V. The concept of these values is critical if you are to use the meter without damaging it.

Figure 9–5 illustrates how the meter is inserted into a circuit to measure current. The total resistance of this circuit is 15.1 kΩ. The addition of the meter's

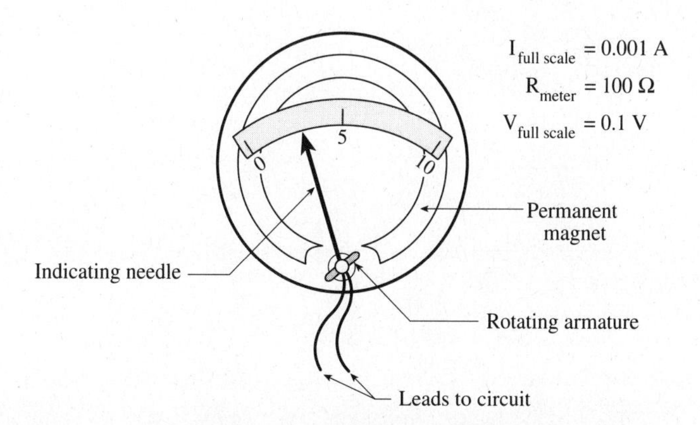

Figure 9–4 The basic electrical meter consists of a permanent magnet and a moving armature with a needle, or indicator, attached to the armature.

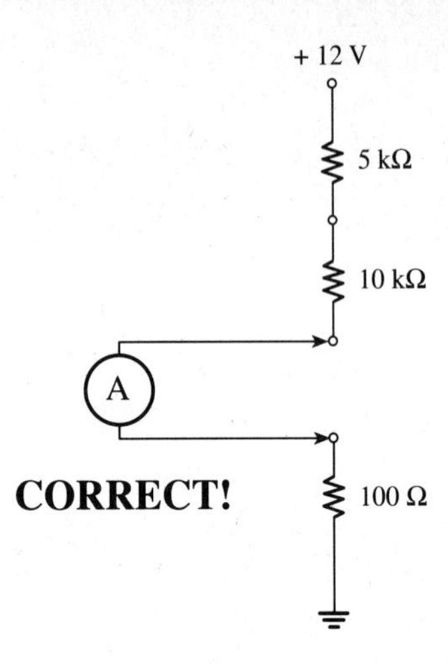

Figure 9–5 **The correct placement of an ammeter when evaluating circuit current.**

100 Ω resistance will have little effect on the total current flow in the circuit. This, of course, is one of the correct methods of measuring current.

Now, look at the circuit shown in Figure 9–6. In this circuit the meter is incorrectly applied to a circuit. Current cannot be measured with the meter adjusted to indicate a range of amperes of current when the meter's leads are applied in parallel with a component. The voltage drop that develops across the 10 kΩ resistance is about 8 V. When this amount of voltage is applied directly to the meter's armature, two things happen. One of these is that the total resistance in the parallel combination of 10 kΩ and 100 Ω is effectively just over 99 Ω. This will reduce the total resistance in the circuit and increase the total current flow to a level that will probably damage the components in the circuit. At the same time, this 8 V applied directly to the meter's armature will damage the armature. Current amounting to 80 mA will try to flow through the meter instead of its design value of 1 mA. Obviously, the meter will be damaged when this is attempted.

This same concept is used for the digital meter operation. The block diagram for a digital multimeter with autoranging and autopolarity is shown in Figure 9–7. The block marked "input signal conditioning" is used to adjust the value of the input signal to one that the meter is able to evaluate. This signal is then applied to an A/D (analog-to-digital) converter. The output of the A/D converter is applied to a microprocessor. At the same time, another output from the A/D converter is sent to a counter. Its output is also applied to the microprocessor. The microprocessor is designed to supply two output signals; one of these is used to develop a digital display, the other provides information for an analog bar graph. A small voltage difference is presented to an integrated circuit.

Measurement of current flow in the circuit is ac-

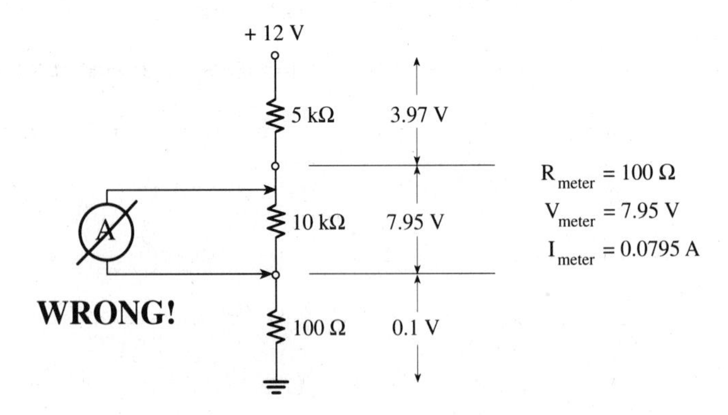

Figure 9–6 **Placing the low-resistance ammeter in parallel with a voltage drop source will damage the meter. Current flow is 0.0795 A while the current for full-scale deflection is 0.001 A.**

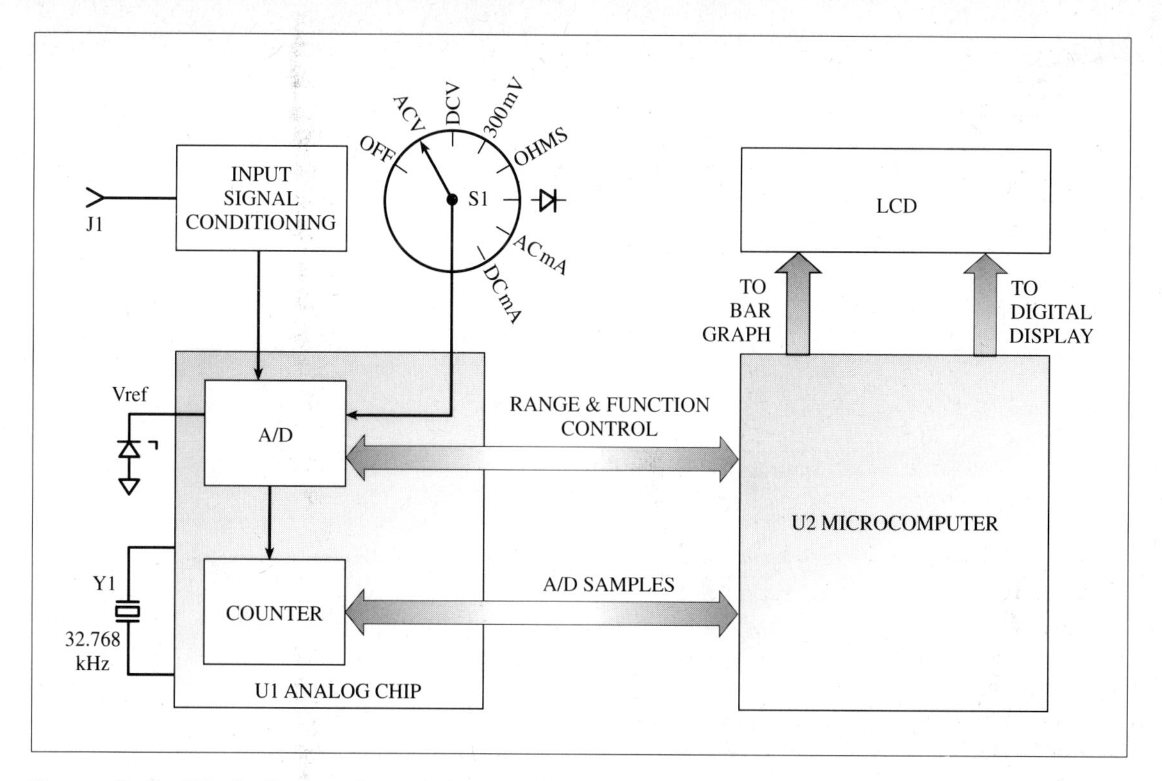

Figure 9–7 **Block diagram for a digital multimeter with autoranging and autopolarity capabilities.**

complished by evaluating the quantity of voltage drop across a resistance. This is the same process as the one described in Chapter 8, which demonstrated how a voltage drop measured across a fixed resistance can be converted into a current reading when you use Ohm's law. The designers of the digital meter have done all of the mathematics and programmed them into the microprocessor IC. The end result is a digital display of the current flow through the circuit at the point of measurement.

UNIVERSAL TEST EQUIPMENT

The term "universal" may be misleading. It is used here to represent a general group of test equipment including analog and digital multimeters, signal generators, and oscilloscopes. This chapter deals with component testing; almost all of the material presented here will relate to component testing. In this section we will concentrate on the use of the analog and digital multimeter for component testing. Two of the more commonly found multimeters are shown in Figure 9–8. These instruments were originally called V-O-Ms, or volt-ohm-milliammeters. Today, the term multimeter is used since it better describes the function of the instrument.

The specific layout of the front panel of these instruments will vary, depending upon the manufacturer. Some of them are more sophisticated than other models produced by the same manufacturer. In a way, they are similar to "lines" of items produced by companies that produce automobiles, home appliances, and power tools. Many instrument manufacturers have a basic model. In addition, upgraded models of the same units are also available. These upgraded units have additional operating characteristics making them more adaptable to the service industry.

One such instrument is shown in Figure 9–9. This device is able to measure AC and DC voltages, cur-

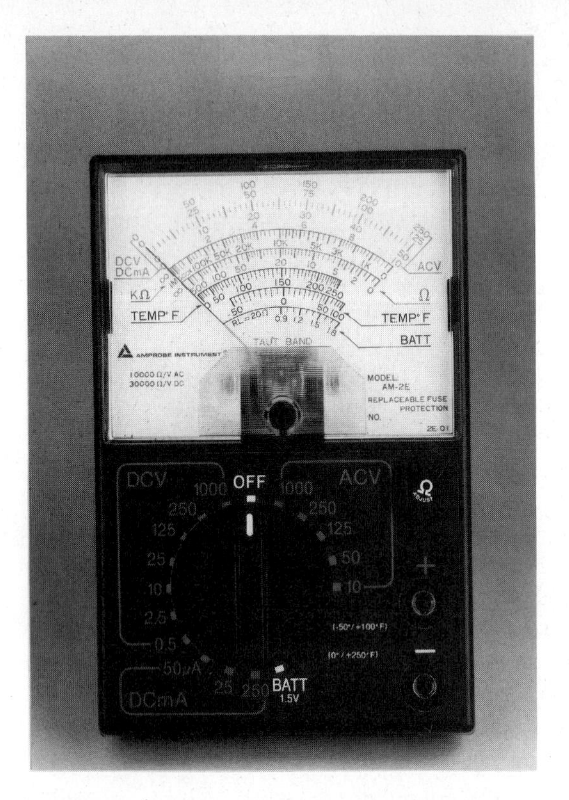

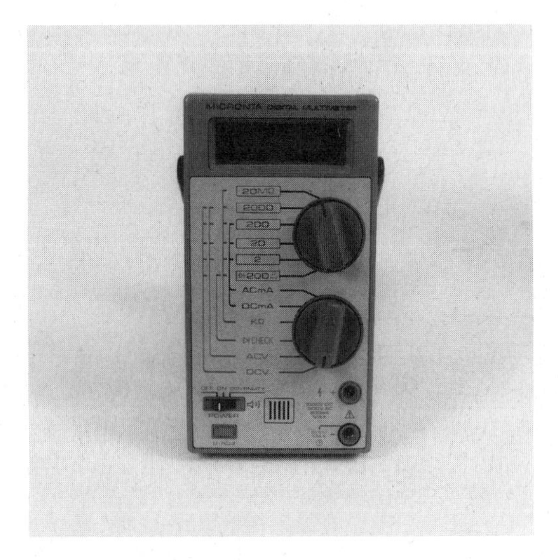

Figure 9–8 **Both the analog and the digital multimeter are shown in these photographs. (*Analog meter photo courtesy Amprobe Instrument; digital photo J. Goldberg*)**

rent, and resistance. In addition, the manufacturer has added some "nice to have" features. These include a logging scale with a zero center marking on the dial, a mirror backing on the dial, and a scale for reading AC peak-to-peak values. The logging scale is designed to monitor changes in values ranging either higher or lower than the center range setting on the meter. The mirror backing is used to minimize parallax conditions and thus provide a more accurate measurement. The AC peak-to-peak ranges are mathematically calculated to convert the AC rms values into peak-to-peak readings.

The best test instrument for evaluating peak-to-peak voltages is still the oscilloscope. While the voltmeter will provide a peak-to-peak reading, it cannot indicate whether this reading is a valid waveform. Waveform distortion can best be observed with an oscilloscope.

The digital multimeter shown in Figure 9–10 has some other interesting features. The manufacturer of this instrument makes a basic model as well as some units with additional features. One of these is shown in this figure. One additional scale is included in addition to the basic digital readout common to these meters. This scale is shown as a horizontal line at the bottom of the digital readout, the purpose of which is to provide a sort of analog readout for the digital instrument. The horizontal bar graph at the bottom of the display is used for this purpose. When attempting to adjust some circuits for maximum voltage or current or minimum values, the digital readout is often too slow. It tends to lag behind any changes in circuit values. The analog readout has a faster reaction time and is better able to display the changes that occur during the adjustment operation. The bar graph display corrects this lack of instant feedback for adjustment purposes.

Some digital instruments have one or two addi-

Figure 9–9 **The analog multimeter.** (*Courtesy B&K Precision Instruments Division, Maxtec Corporation*)

tional functions that make them easier to use than the basic models of the same device. One of these is known as "auto polarity." This term indicates that the meter will display the polarity of the measurement, regardless of the placement of the test leads into the circuit. Typically the instrument's negative test lead is connected to circuit common or to the most negative point of the circuit being evaluated. If, for some reason, the test leads are incorrectly placed, either the instrument will be damaged or it will not provide a proper reading. The autopolarity function eliminates the need to be concerned about the polarity of the test leads. When the polarity is positive, the readout indicates a numeric value. If a negative value is encountered, a minus sign is also displayed on the readout.

The second additional function is called "autoranging." Again, the basic meter uses a mechanical switching circuit to change the range of the value being measured. If the value exceeds the range setting, the instrument could be damaged. The autoranging function changes the range setting automatically and acts as a range selection switch in this manner. Both the autopolarity and autoranging

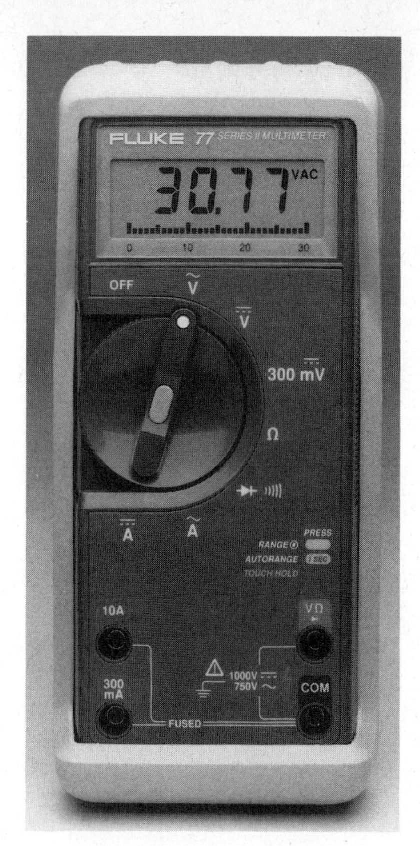

Figure 9–10 **One of the more commonly used digital multimeters.** (*Copyright 1991 John Fluke Mfg. Co., Inc. Reproduced with permission from the John Fluke Mfg. Co., Inc.*)

functions are convenient additions to any instrument, and they do not make the basic meter much more expensive. They are recommended additions.

COMPONENT-SPECIFIC EQUIPMENT

The second set of test equipment is categorized as component specific. Each of the following test instruments is designed to evaluate a specific electronic component. Some of them are essential for the normal types of electronic service requirements of today. Others could be classified as being nice to have, but not essential. I leave the decision as to which of these are essential up to the individual. The decision, of course, will depend upon several factors, including need, price, availability, and the requirements of the manufacturers of equipment being serviced under factory warranty authorizations. Some of the more common component-specific types of testing equipment are described in this section and illustrated in the following photographs. Please recognize that the units I have included are only samples of the excellent equipment available from several different test equipment manufacturers. Individuals wishing to purchase such equipment should examine as many types as possible before making any final decision as to which one to purchase.

Bridge Testers One of the older types of test equipment designed for evaluation of resistors, capacitors, and inductors is the Wheatstone Bridge shown in Figure 9–11. This piece of testing equipment requires the removal of at least one end of the component to be tested. The leads of the component are connected to the terminals of the tester. The large dial is rotated until the indicating device shows that the system is in balance. At this point the value of the specific component is "read" from the face of the large dial.

This process is slow and cumbersome. The need to remove one or both of the leads of the component is not always an efficient method of testing. However, the Wheatstone Bridge provides one of the more accurate methods of evaluating these components. This tester also has the ability to check for capacitor dielectric voltage breakdown and leakage.

Advanced Capacitor Testers One of the more advanced types of testers for capacitors and inductors is shown in Figure 9–12. This tester can be set to automatically evaluate capacitors or inductors. The tests for capacitors include measurement of the component values, leakage, and the ability to respond to the specific evaluation needs of a large variety of different types of capacitors. Inductor testing includes the ability to identify shorted inductor winding leads and the quality of the inductor.

A second type of capacitor tester is shown in Figure 9–13. These are digital types of instruments. Both perform capacitor tests in the same manner; the basic difference between the two is that one requires selecting the range of capacitance, while the other contains more sophisticated circuitry enabling it to automatically select the proper range for the reading.

Testing Capacitors First of all, consider what normally goes wrong with the capacitor. It will develop one of the following conditions: open, shorted, or

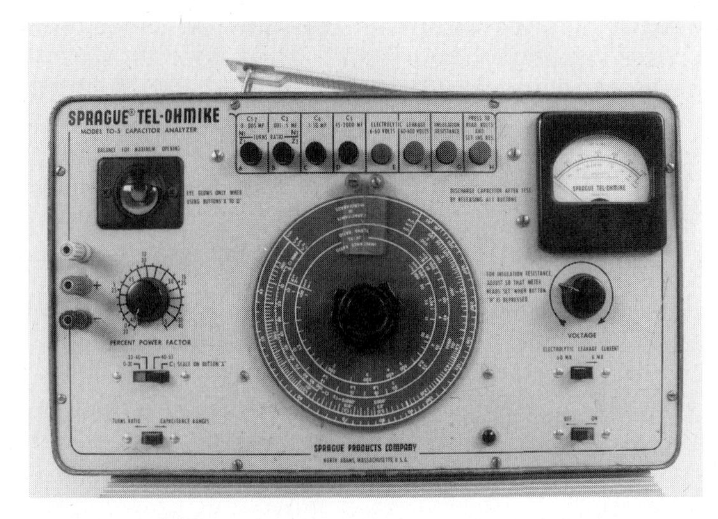

Figure 9–11 **This older piece of test equipment is still able to evaluate resistance, capacitance, and inductance values. (*Photo by J. Goldberg*)**

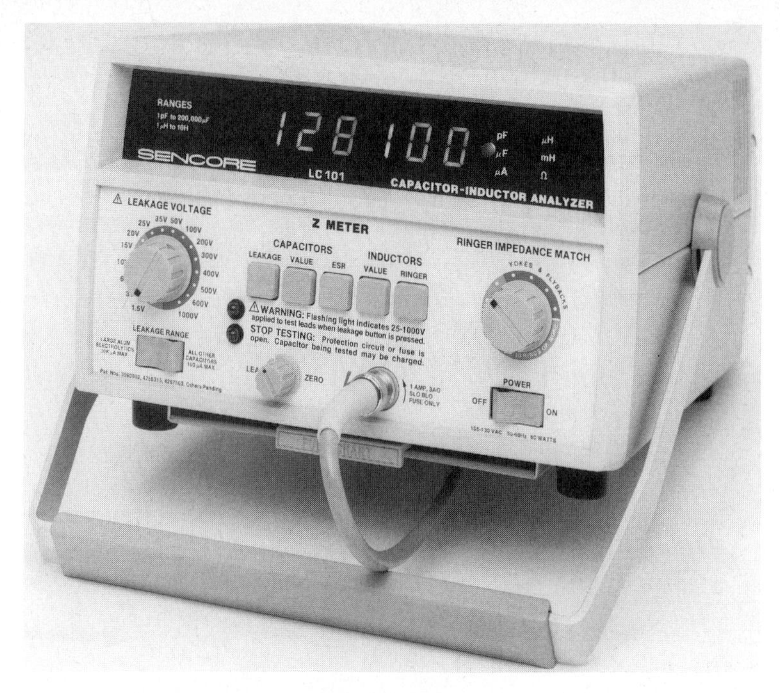

Figure 9–12 **A modern version of the RCL meter. This unit has additional capabilities, including evaluation of coils and transformer windings. (*Courtesy Sencore Inc.*)**

dielectric absorption. Some semiconductors have a high level of dielectric leakage. Any one of these conditions will degrade the operation of the equipment in which it is installed. Due to the effects of

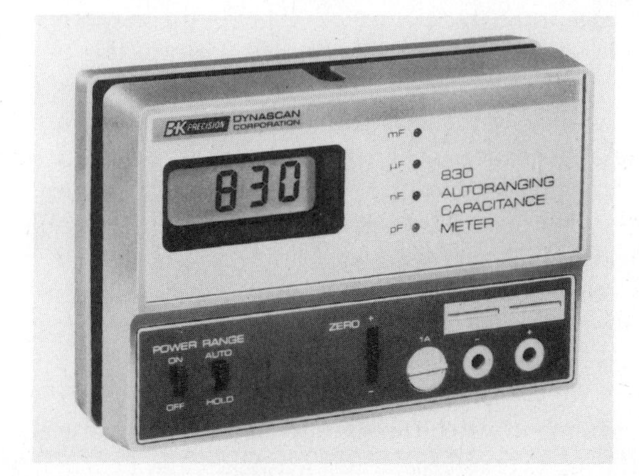

Figure 9–13 **Portable inductance-capacitance meter for field use. (*Courtesy B&K Precision Instruments Division, Maxtec Corporation*)**

other components in the circuit, the really best method of testing any capacitor is to remove it from the circuit. Once removed, the capacitor can be tested with an ohmmeter or with a capacitor tester.

Ohmmeter testing of the capacitor will place a low voltage charge on its plates, which will charge the plates of the capacitor. The ohmmeter readout will show a steady change in the value of resistance being measured in the process as the capacitor charges. In truth, the value of resistance is not being measured. The charge being placed on the plates of the capacitor will change the quantity of current flow in the instrument. This will be displayed as a change in resistance. If this measurement indicates an increase in the resistance value, the probability of the capacitor being "good" is high.

When a resistance reading of close to zero ohms is measured, the capacitor is shorted. Its internal dielectric has failed to the point where it no longer acts as an insulator. The capacitor must be replaced when this measurement is observed.

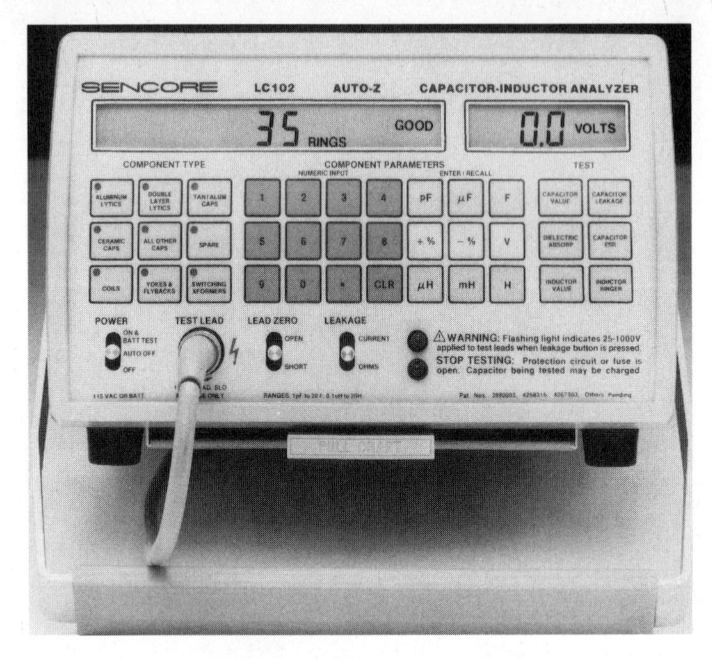

Figure 9–14 **Another version of the portable inductance-capacitance-resistance meter. (*Courtesy Sencore Inc.*)**

The more difficult measurement to make is one that indicates internal leakage. When a capacitor's dielectric starts to fail, it will no longer accept and hold a voltage charge. The ability to charge to its fully rated value is reduced. The technician will observe a reduction of filtering action or a loss in the ability to pass signals between stages.

Measuring Inductors Inductors will fail and become totally open, totally shorted, or partially shorted. The last condition occurs when one or two turns of wire are short circuited. Fully open inductors can be tested for continuity with an ohmmeter, but the ohmmeter will fail to provide adequate information about shorted turns in an inductor. Since many inductors are one component of a resonant circuit, a shorted turn or two will change the point of resonance. Unfortunately for the service technician, the resistance of many inductors is often close to zero ohms. If it is not zero ohms, then the actual value of DC resistance is often very low and less than two or three ohms. Remember, current flow opposition in a resonant circuit, or an AC circuit, is based on reactance and not only on the value of DC resistance. The values of reactance are also provided in units of the ohm. This may be confusing unless you recognize the difference.

One of the better methods of testing inductors uses a bridge type of tester. Many of the newer testers using integrated circuits will perform this type of test. The tester shown in Figure 9–14 is one of several currently on the market. This instrument is capable of measuring resistance and capacitance in addition to inductance values. The ability to measure all of these components makes it a very versatile instrument.

An evaluation of the various types of test instruments presently available for the testing of capacitors and inductors is difficult to perform. The service technician has to consider what is to be tested and how it is to be accomplished. Almost all capacitors and a great many inductors must be tested out of the circuit. It is possible to test capacitors by temporarily placing another one in parallel with the unit suspected to be open or leaking. However, if the unit is shorted, the only method of testing is to remove it from the circuit.

Many inductors can be tested while in the circuit. One of the better methods of doing this is using the bridge type of tester. Today, the bridge tester is found in the portable and battery-operated units illustrated in this section. These are not as large and cumbersome as earlier models; many are small and can be held in one hand. They are operated from small replaceable battery power sources. This miniaturization of testing equipment offers the service technician the ability to include one in the portable testing equipment normally carried on service calls. Their relatively low cost permits their inclusion on the test benches of service shops.

SEMICONDUCTOR TESTING

Evaluation of the large variety of semiconductor devices can sometimes be very confusing. It is almost impossible to test the integrated circuit devices using the ohmmeter. On the other hand, basic devices, such as diodes and transistors, can be tested with an ohmmeter. When you attempt to evaluate some of the integrated circuits, you will need to use the oscilloscope or an IC device-specific tester. The oscilloscope can be used to evaluate signals at the input of the IC, and then these signals can be compared with those at its output. A dual-trace oscilloscope is best for this because it can display both the input and output voltage waveforms at the same time and show any phase relationships.

Reference to a block diagram is almost a requirement for servicing these devices. You need to use the description of the block to determine whether it is functioning as it should in the circuit. The large variety of integrated circuit types and functions requires the use of manufacturers' service literature or other reference sources for proper analysis.

It often is convenient to use a logic probe when you have the need to evaluate the output of a digital integrated circuit. This probe, shown in Figure 9–15, is designed to indicate whether the voltage at a specific connection is "logic high" or "logic low." These values are often between 3.5 and 5.0 V for the "high" and 1.5 to 0 V for the "low." The logic probe has two or more light-emitting diodes in its circuit. One of these will be red and one will be green. Sometimes, a multicolored light-emitting diode is used instead of individual ones. When the "logic high" condition is measured, one of the lights will

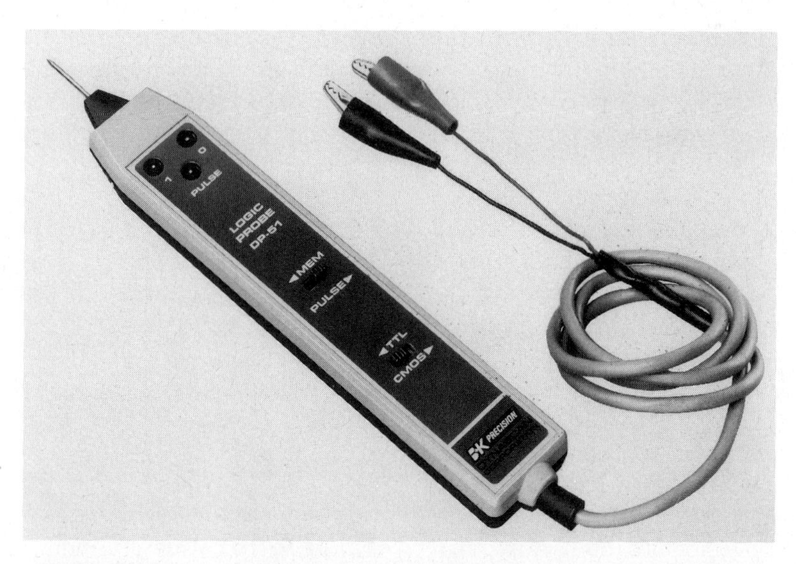

Figure 9–15 **A logic probe is used to "read" the output states of digital circuit components. (*Courtesy B&K Precision Instruments Division, Maxtec Corporation*)**

glow. This is usually the red one. The green light-emitting diode will glow when a "logic low" condition is encountered. This small piece of test equipment is convenient to use and easy to transport.

Testing of semiconductors can be accomplished by removing them from the circuit. I do not recommend this procedure unless there is no other way to obtain this information. There are many excellent out-of-circuit semiconductor testers on the market today. Experience has shown that no matter how careful you are, it is possible to damage the semiconductor while attempting to remove it for testing. Component leads may break off or the circuit board foil can be damaged due to heat lift from the base material or breakage during the removal attempt. It is possible (and does happen to the best technician) to reinstall a working semiconductor incorrectly. When this happens, the circuit still does not work properly and the technician has to retest and re-evaluate the system to locate this new problem. An alternate method of testing transistors in circuit is presented later in this chapter.

Some semiconductors are very sensitive to static electrical charges. Many people fail to realize the potential voltage charge at their fingertips. The static electrical charge created by the human body can be very high. It often is created when you wear clothing made with polyester fibers. This charge may be very high, often into kilo- or megavolt ranges.

Picking up one of these semiconductors without taking the necessary antistatic procedures will damage a new unit. It is impossible to observe whether the new unit has failed due to handling. If a semiconductor must be removed, the manufacturer's procedure for safe handling must be followed.

DIODES AND OTHER TWO-ELEMENT SEMICONDUCTORS

Diodes and other two-element semiconductors can often be evaluated with an ohmmeter. A diode is one of the easiest devices to test in this manner. The ohmmeter's internal power source is used to bias the diode. You must know the polarity of the ohmmeter's test leads when this test is being made. Almost all of the current group of ohmmeters have their red or (+) lead connected to the positive terminal of the internal power source. You should check this for the instrument being used to make this measurement to be sure.

The drawing in Figure 9–16 shows how this is done. The upper meter on this drawing has its positive lead connected to the anode of the diode. This will provide a forward bias on the diode, and its resistance measurement should be between 500 and 1000 Ω. When the diode is reversed biased by the ohmmeter, as seen in the lower half of the drawing, its resistance should measure close to infinity.

A second method of testing the diode is shown in Figure 9–17. When a silicon diode is forward biased, as shown in this drawing, the voltage drop that develops across its terminals can be evaluated. If the diode is functioning properly, this value of voltage drop should be around 0.7 V. If the diode's base-to-emitter junction is leaking, the voltage drop will be greater. When it is open, the voltage drop across these terminals will be much higher. When the diode is open, the voltage drop across its terminals will approach the value of the applied voltage.

In the circuit being discussed, if the diode is open, the voltage reading will be an AC voltage. The reason for this is that the only method of converting the AC input voltage to a DC voltage is when the diode is functioning correctly. Using an oscilloscope and measuring from circuit common to the anode of the diode will show a value equal to the applied

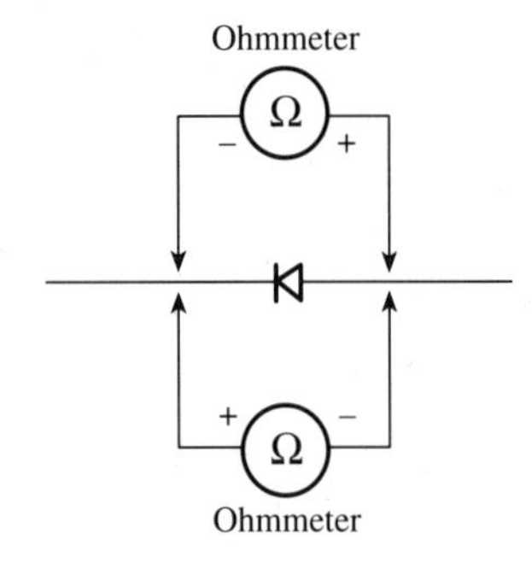

Figure 9–16 Using an ohmmeter to evaluate the condition of a semiconductor diode. The power source in the ohmmeter will bias the diode into either its forward or reverse condition.

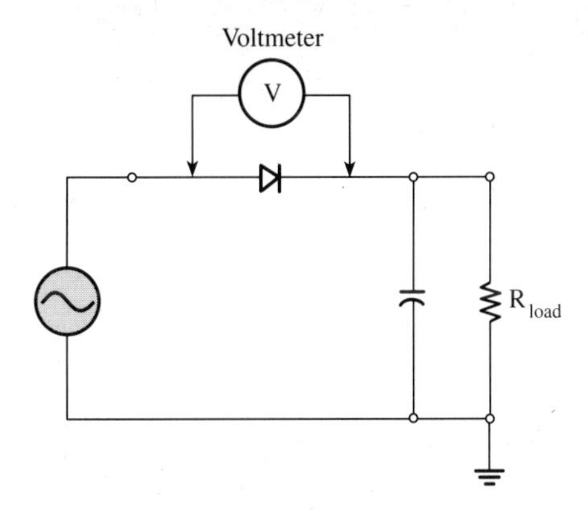

Figure 9–17 Semiconductor diodes can also be tested under operational conditions. The voltage drop that develops across its terminals will indicate its operational state.

AC voltage. When the diode is shorted, this AC voltage will develop across the load. The oscilloscope reading will indicate this condition. A reading with a DC voltmeter measured across the load will show a value of zero volts when the diode is open.

TRANSISTORS

Transistors may be tested with a voltmeter while they are in the circuit. Figure 9–18 shows a typical transistor amplifier circuit. One method of evaluating the transistor is to measure the operating voltage at its elements while it is in the circuit and under power. Perhaps the prevalent use of the transistor is in its application as an amplifier. With the exception of output stages, most amplifiers operate as Class A devices. The Class A amplifier faithfully reproduces the input signal. It is biased at the midpoint of its operational curve.

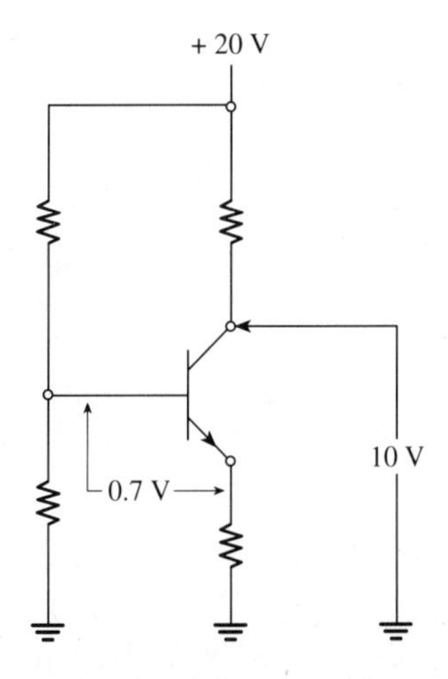

Figure 9–18 The emitter-to-base voltage reading on a transistor will indicate its operational condition.

The information about the Class A amplifier shows that often the collector voltage is close to half of the applied voltage from the power source. This is illustrated in this circuit. The source voltage is +20 V, and the collector voltage in this circuit is +10 V. In practical circuits the value of collector voltage will range between 40 percent and 60 percent of the source voltage. In the circuit shown here, this range will be between 8 V and 12 V. A measurement falling between these two extremes will be an initial indication of acceptable operation.

The other major voltage drop develops between the emitter element and the base element of the transistor. If the transistor is a silicon device, this voltage should be approximately 0.7 V. This junction in the transistor is actually a diode junction. Since this is true, the voltage measurement for a diode will apply to voltages measured between these terminals.

Another method of testing the transistor is with the use of an oscilloscope. The circuit to be tested must have its operating power connected. This type of test requires an input signal for evaluation. The operational configuration of the amplifier must be known before this test is performed. The three waveforms present in each of the three basic amplifier configurations is required knowledge for this test. For example, the common emitter type of amplifier will exhibit these qualities when evaluating the signal at its input and comparing that signal to one observed at the output: voltage gain and phase inversion of the signal between the input and the output of the circuit. A comparison of these two elements with the oscilloscope will verify whether the amplifier is operating properly.

The three configurations for amplifiers are shown in Figure 9–19. This figure includes comparison of input and output signal phases as well as the quantity of relative voltage gain. This type of information

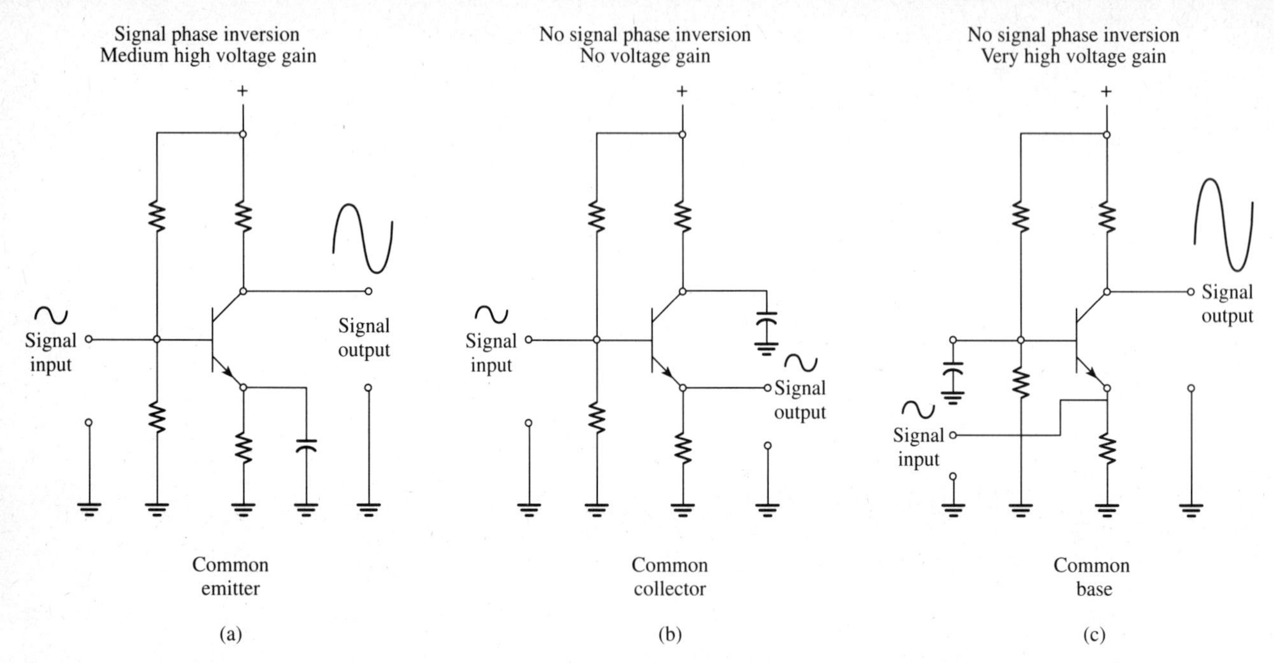

Figure 9–19 The three basic circuit configurations for amplifier are (a) common emitter, (b) common collector, and (c) common base.

is an essential component when planning how to analyze the circuit. Each of these three circuit configurations will be explained using the basic rules for circuits established by Ohm and Kirchhoff.

THE COMMON EMITTER

This circuit is used more often than any other type of amplifier circuit. Amplifier configurations are identified by the name of the one element that is used for both the input signal and the output signal. In this circuit it is the emitter element. A schematic diagram for the common emitter amplifier configuration is shown in Figure 9–20. The signal input for this amplifier is between the base element and the emitter element. This is true since signal common is the same as circuit common in this circuit and the use of the base-to-common and the emitter-to-common resistances are used for current flow, but not for signal control. The output is connected between the collector element and the emitter element. The emitter element is common to both the input and the output in the circuit.

Let us examine the operating conditions as a positive going signal is applied between the base and the emitter of the transistor. This will raise the voltage drop between the actual base and emitter elements and will result in an increase in current flow across the emitter-to-collector path in the transistor. The circuit consisting of R_{LOAD}, the collector-to-emitter elements, and R_E is effectively a series circuit. In any series circuit, current flow can increase only if the total value of resistance is reduced. The effect of the increase in input signal voltage and its increase in the base-to-emitter voltage will reduce the internal resistance of the transistor.

When the internal resistance of the transistor is reduced, the voltage measured from the collector to circuit common will be reduced. In other words, an increase in the level of voltage at the base of the transistor will reduce the voltage at the collector of the transistor. The waveforms in Figure 9–20 show this in graphic form. The input signal voltage of the common emitter transistor amplifier produces a waveform with a reverse polarity from that of the input signal. In addition, the swing in operating voltage between the collector and circuit common is

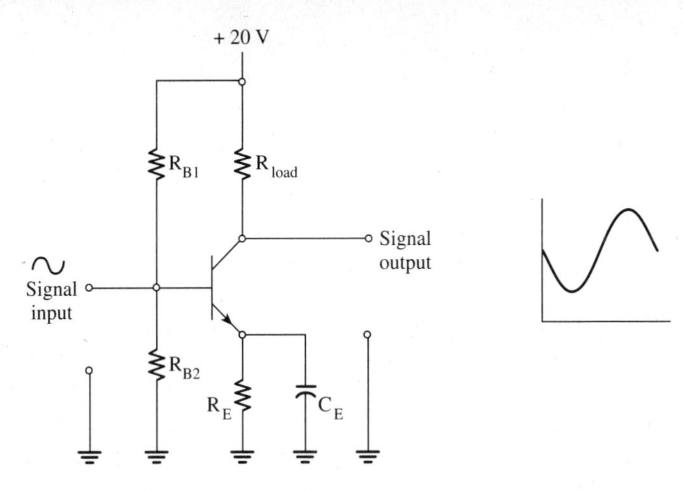

Figure 9–20 Circuit configuration on left and waveforms for the common emitter amplifier.

much greater than that of the input signal voltage. This indicates a voltage gain for the circuit.

A second factor requiring recognition is shown in Figure 9–21. In this circuit the emitter bypass capacitor, C_E, is removed. The purpose of this capacitor, or any capacitor, is to attempt to remove or reduce any changes in the level of voltage at that specific point in the circuit. In series circuit a reduction in the value of resistance for any one component will reduce the voltage drop across that same component. This reduction in voltage is added to the voltage drops that originally developed across the other components in the circuit.

The use of a sine wave signal voltage does exactly the same thing to the circuit when the emitter bypass capacitor is removed. The waveform shown at the emitter of the transistor in this figure has a polarity that is opposite to that of the output signal. These two opposite polarity voltages add, and the result is a reduction in the total output signal voltage level. Emitter bypass capacitors are used to reduce this effect of signal degeneration.

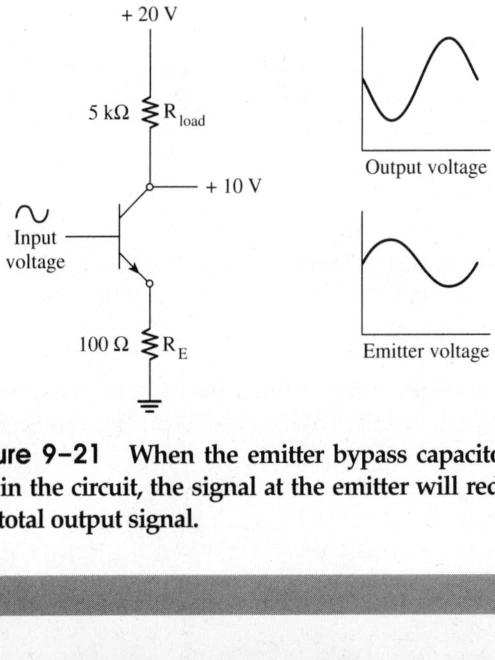

Figure 9–21 When the emitter bypass capacitor is not in the circuit, the signal at the emitter will reduce the total output signal.

Troubleshooting Troubleshooting the common emitter amplifier configuration is best done using a dual-trace oscilloscope and a constant signal source from a function generator. The signal at the input of the circuit should be compared with the signal at its output. One should expect to find a voltage gain at the output as well as phase inversion of the signal.

COMMON BASE AMPLIFIER

The common base amplifier circuit is used for a very high level of signal gain. The circuit for this configuration is shown in Figure 9–22. It can be recognized using either one of two factors—the bypass capacitor at the base element or when the signal in-

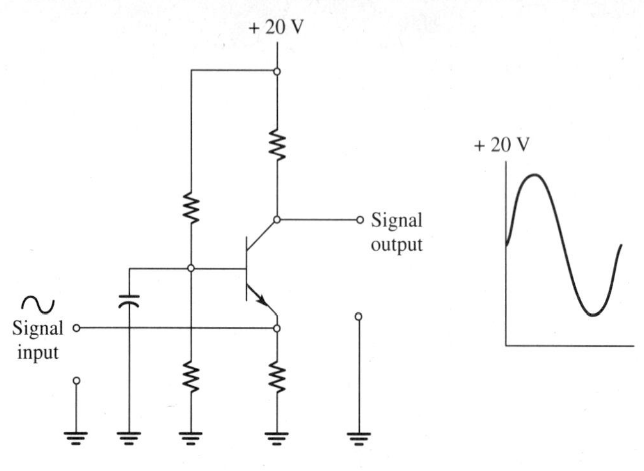

Figure 9–22 **Signals at the input and output of the common base amplifier are in phase and provide the highest voltage gain.**

put is at the emitter and the signal output is at the collector.

The concept of the internal resistance between emitter and collector elements also applies to this circuit. The difference between the common base amplifier and the common emitter amplifier is based on which of the input elements is held at a constant value and which has a varying voltage applied to its terminals.

The common base amplifier configuration will "lock" the base element with a fixed value of voltage. When the input signal voltage is applied to the emitter, the voltage drop between emitter and base changes. An increase in the base-to-emitter voltage will reduce the voltage drop between these same elements. This, in turn, will increase the internal resistance between the emitter and the collector of the transistor. Thus, an increase in this voltage value will create an increase in the voltage between the collector and circuit common. Both voltages are in phase and the voltage gain is very high for this circuit. This concept of the increase in base-to-emitter voltage creating an increase in the emitter-to-collector voltage is the basis for the high level of voltage gain. Both the input voltage and the output voltage are in phase and, as a result, the total output voltage is the sum of these two values.

Troubleshooting Troubleshooting the common base amplifier configuration is best done using a dual-trace oscilloscope and a constant signal source from a function generator. The signal at the input of the circuit should be compared with the signal at its output. You should expect to find a very high level of voltage gain at the output as well as a lack of any phase inversion of the signal.

COMMON COLLECTOR

The common collector amplifier is the final configuration in this series. This circuit is shown in Figure 9–23. In this circuit the collector element is bypassed. There should be no signal voltage present at this element, even though the operating voltage is present. The input signal voltage is presented between the base and the emitter of the transistor. The output signal is taken from the voltage that develops across the emitter resistor. Amplification occurs in this system by an increase in the base-to-circuit common voltage. This increases the voltage drop between base and emitter and reduces the internal resistance of the emitter-to-collector terminals of the transistor.

When the emitter-to-collector voltage is reduced, the voltage drop that develops across the emitter

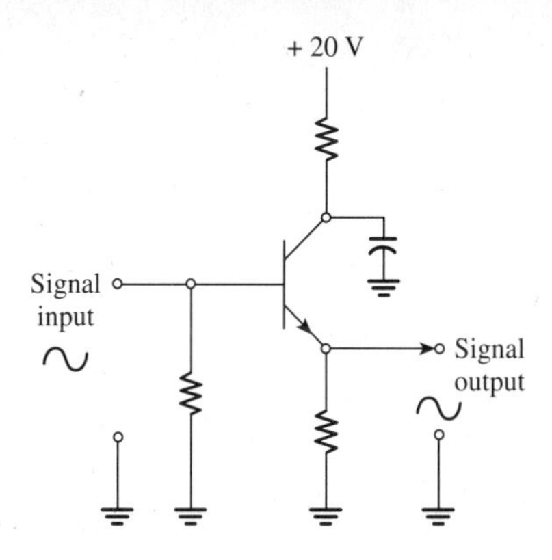

Figure 9–23 **The input and output signals for the common collector amplifier are in phase and their voltages are almost equal.**

resistor will increase. Therefore, both signals increase at the same time and are considered to be in phase. Since the voltage measured at the emitter of an NPN transistor is always less than that at its base, the output signal for this configuration will also always be less than the output signal at the base element.

The characteristics of this common collector amplifier configuration are a slight voltage loss and a high current gain. Often this circuit is said to have a voltage gain factor of unity, or one. It is used as an impedance matching circuit or as a power output stage used for operating low-voltage and high-current-source devices such as loudspeakers, deflection yokes, or motors performing some function on an industrial machine.

Troubleshooting Troubleshooting the common collector amplifier configuration also is best done using a dual-trace oscilloscope and a constant signal source from a function generator. The signal at the input of the circuit should be compared with the signal at its output. You should not expect to find a voltage gain at the output, and there will not be any phase inversion of the signal.

Testing the circuit consists of making the base-to-emitter voltage test. When this voltage is close to 0.7 V, the input circuit of the transistor can be considered to be functional. A voltage test of the emitter-to-collector circuit should show values of less than that of the applied source voltage. The specific amount of DC voltage measured between these two elements will depend on the values of any associated voltage-dropping resistors in the circuit.

CONNECTORS AND CABLES

These devices are subjected to vibration, motion, pulling, and just normal wear and tear during their operating life. How you determine that a problem is occurring in a cable or a connector often is very simple. The process of twisting, pulling, or distorting the cable will usually provide information about its ability to perform. When intermittent conditions occur it is wise to test the interconnecting cables. Specifically, those cables that are external to the equipment should be tested first. It is also possible for one of the internal cable connections to work itself loose during the normal operation of the system. The process of visual inspection and checking the integrity of the cable connector is important.

There are times when the terminals of the circuit board on which the connector is attached will oxidize. These can be cleaned with the proper chemicals. The process of removing and replacing the connector is often sufficient to solve the problem. You should always remember to keep your hands away from the actual connectors on circuit boards. Skin emits an oil, which can contaminate the surface of the board. In some extreme cases the oil will etch the surface of the connector on the board. This results in development of a higher than normal resistance between connector and board. This additional resistance produces a voltage drop when current flows, as does any resistance in a series circuit. The result is a loss of voltage across this added resistance. The loss may be in operating voltage or it may be in the amplitude of the signal being processed at that point.

Another problem that occurs with connectors and cables is the actual chafing or rubbing of the wires against the frame or case of the unit. This can be found during a visual inspection of the system and its connecting devices. An emergency repair may be made on the existing cable if possible. The best way to repair this type of problem is by replacing the faulty cable with a new one. It may be necessary to reroute the cable run to minimize future chafing and destruction.

Individual cable connections can be checked for continuity with an ohmmeter or other resistance-measuring device. This is relatively simple when there are just a few individual wires in the cable. However, if the cable contains many pairs or individual wires, you may need to use a commercial cable tester. These devices test cables for shorts, opens, and partial shorts. Whether you should own a tester of this type depends upon how often you need it or if using it will save you hours of troubleshooting and expedite the repair.

CIRCUIT BOARD TROUBLESHOOTING

The method of diagnosis for circuit boards often requires some physical activity. The first step is to locate the suspected board. Often removing and reinstalling it will clear up any problem because the problem may have been due to a buildup of oxidation on the connector for the board. If the board is permanently mounted in the device, you must use a different set of diagnostic techniques.

The first diagnostic step for permanently mounted boards is visual inspection. Look for components that have been subjected to excessive heat. These will have scorched surfaces, broken and scorched parts, or the board surrounding the component may be scorched. A keen sense of smell will often help you to locate these damaged components. When you do find them, use traditional testing equipment to help pinpoint the location of the specific component or components that have failed. Do not forget that, in many situations, the scorched component is the result of another component in the same circuit having failed. Replacing the damaged component may not always return the system to normal operating conditions. Always make additional tests once the damaged component has been identified.

Another method of locating a defective board is the pressure test. Often using the eraser end of a pencil and a gentle push on the surface of the board will modify the existing condition. The system will either return to its normal operating condition or stop functioning. This test is used to locate components whose leads may have become unsoldered during the normal operation of the system.

Most circuit boards are produced by machines. The individual components are also mounted by machine insertion. A final step in the process is known as "wave soldering." The board is held just above and parallel to the surface of a tank of molten solder. A wave is created in the solder tank, which is used to solder all of the connections on the circuit board and the components. At times, and particularly after much use, it is possible for one or more of these solder connections to fatigue. Often you cannot see this failure with the naked eye. The process of gently probing the surface of the board often will help in locating one area of the board where this failure has occurred. An inspection of the suspected area with a magnifying glass should reveal the connection that has failed. Once identified, the connection is resoldered and the board should be able to operate properly.

REVIEW

The experienced service technician knows which of the many types of testing equipment should be used for the most efficient test and analysis. This decision is one of the many to be made during the planning period prior to making the actual measurements. In addition, this technician also has some idea of what to expect to find when the actual measurement is made. The service technician must choose between general types of testers, such as the volt-ohm-milliammeter and the oscilloscope, and the more component-specific types, such as capacitor and inductor testers.

Knowing how to connect the testing equipment is also critical. With the exception of a current-measuring meter, almost all measurements are made by placing the test leads in parallel with the circuit or component to be tested. The technician also must be aware of what effect, if any, the addition of the tester will have on the circuit or component. The input resistance of the tester is a critical factor. If it is low—on the order of 100 Ω to 1000 Ω, then it will change circuit currents and voltage drops. This could damage some of the components that were operating correctly prior to the test. The experienced technician will use testing equipment that has an input resistance of at least 1 MΩ or higher.

Another item to be aware of when connecting testers into a circuit is whether the circuit common of the tester and circuit common for the device are the same. If they are different, a false reading could be obtained. You also must be aware of the "ground loop" condition that happens when the circuit common, or grounds, are not at the same electrical potential.

The basic types of volt-ohm-milliammeters are available with either an analog or digital display. Some of the more sophisticated meters include autopolarity and autoranging circuits. One thing to keep in mind when selecting any multipurpose meter is its ability to provide an analog readout. Many of the digital meters currently available also include a horizontal bar graph display. This will help in adjusting controls or in tuning circuits when either a maximum or a minimum circuit value is required.

There are many different component-specific testers on the market. The more commonly used ones measure capacitors and inductors. These also are available in either autoranging or switch-selectable ranges.

Capacitors are tested for leakage value, shorts, opens, or dielectric absorption. An ohmmeter can be used to test for the capacitor's ability to accept a voltage charge. This test will indicate whether the capacitor is not open or shorted. A capacitor tester is required for value measurements and other tests.

Inductors are tested for opens or shorts. The ability of an ohmmeter to indicate a partial short is minimal; this type of test must be done with more specialized equipment.

Semiconductors can be tested with some very exotic equipment or they may be tested with the ohmmeter and oscilloscope. A semiconductor diode's resistance can be measured. If the diode is forward biased, its resistance is on the order of 500 Ω to 1000 Ω. When the diode is reverse biased, its resistance is very high—often too high to measure with standard testing equipment. An oscilloscope is used to evaluate a powered circuit for diode action.

Transistors may also be tested with an ohmmeter. A better method is using a voltmeter to measure the base-to-emitter voltage. If the transistor is a silicon type, this voltage should be around 0.7 V. The oscilloscope is also used to measure signal processing in transistor circuits. The best type of oscilloscope is a dual-trace type, which enables the technician to observe both the input and the output signal and its phase relationship. This relationship provides operational information once the technician recognizes the type of circuit being used as an amplifier.

Testing integrated circuits is more complicated. You can use a logic probe, which will indicate whether the signal at a specific terminal is logic high or logic low. Service information is required for testing analog types of integrated circuits. Waveform analysis with the oscilloscope is often required for this evaluation.

Connectors and circuit boards also can create problems. These are evaluated first with a visual inspection. Next, you should gently wiggle, twist, and prod to attempt to locate a specific problem area. Often the soldered connections become loose and require an application of fresh solder. This activity will often solve an intermittent operating problem.

REVIEW QUESTIONS

1. What is the first step the service technician must take in servicing a product?
2. What is meant by the term "component-specific equipment"?
3. Why is the input resistance of a test instrument an important factor in the selection and use of the tester?
4. How are voltmeters connected to the circuit in order to measure a voltage value?
5. What type of fundamental electrical process is used with the analog meter?
6. Explain the term autoranging and its use.
7. Explain the term autopolarity and its use.
8. How is a capacitor tested with the ohmmeter?
9. What does an ohmmeter test of a capacitor indicate?
10. What is the limitation of an ohmmeter test of an inductor?
11. Explain how a semiconductor diode is tested with the ohmmeter.
12. How can an oscilloscope be used to test a semiconductor diode in a power supply system?
13. What does a voltage drop of 0.7 V DC measured across a diode's terminals indicate?
14. What is the typical base-to-emitter voltage drop for a silicon transistor?
15. Describe the input and output waveform relationships for a common emitter amplifier.
16. Describe the input and output waveform relationships for a common base amplifier.
17. Describe the input and output waveform relationships for a common collector amplifier.
18. What is the purpose of a logic probe and where is it used for testing purposes?
19. Why should you keep your hands off the contacts on a printed circuit board?
20. Explain the reason for twisting, pushing, and prodding circuit boards.

Chapter 9 Challenge One

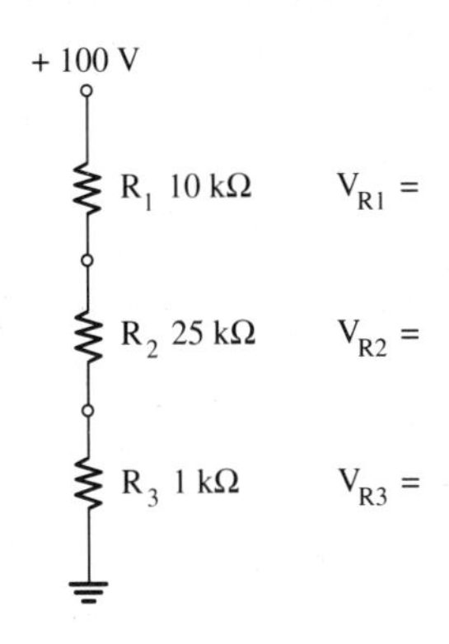

What approximate voltage values should be found in the above circuit *before* any voltage measurements are made? Explain how you made this determination using electronic terms and laws.

Chapter 9 Challenge Two

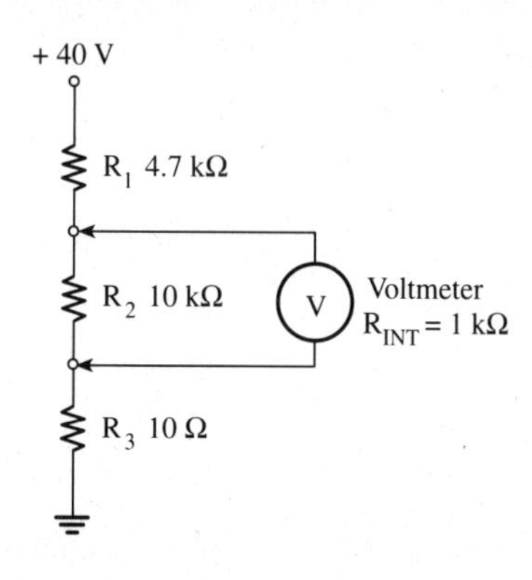

What effect on the voltages and circuits will occur when the voltmeter (with its values shown) is used to measure the voltage drop across resistor R_2? Explain your answer in terms of electronic rules and laws.

Chapter 9 Challenge Three

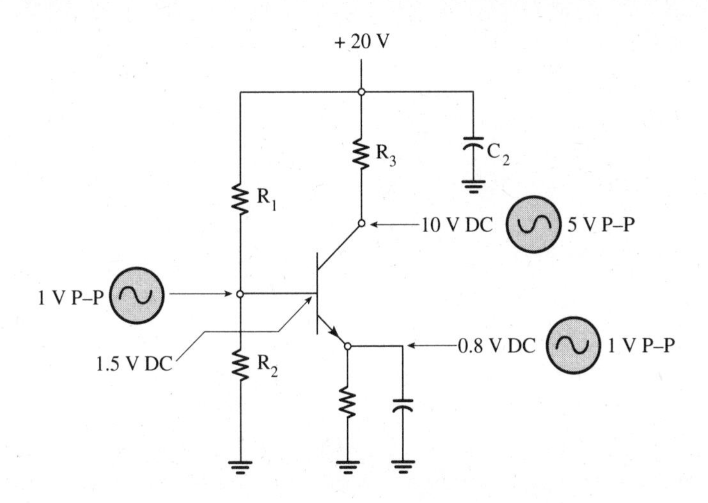

The output voltage for the above circuit is (1) out of phase and (2) slightly larger than the input signal. Develop a plan to test this circuit to identify a defective component. Use this format:

1. Type of amplifier circuit configuration
2. Expected waveform values comparing input to output
3. Abnormal indicators on the diagram (from the measurements)
4. Possible failures
5. Method of testing

Chapter 9 Challenge Four

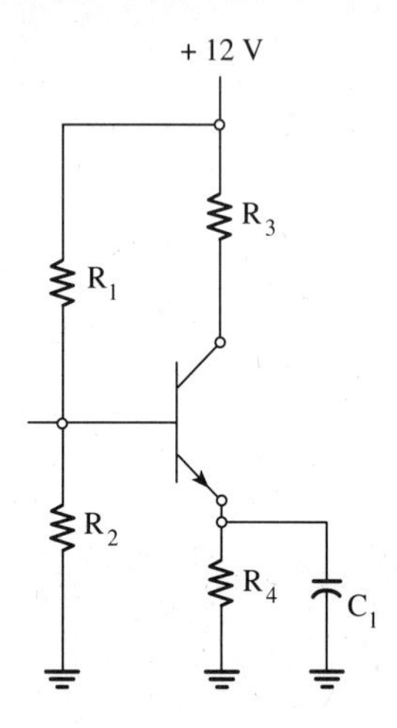

A temporary short circuit is placed between emitter and base in the above circuit. The voltage drop measured at the collector of Q_1 rises from its initial value of 5 V to a value of 12 V. Explain, using electronic terms, why this occurs and what it indicates to you.

Chapter 9 Challenge Five

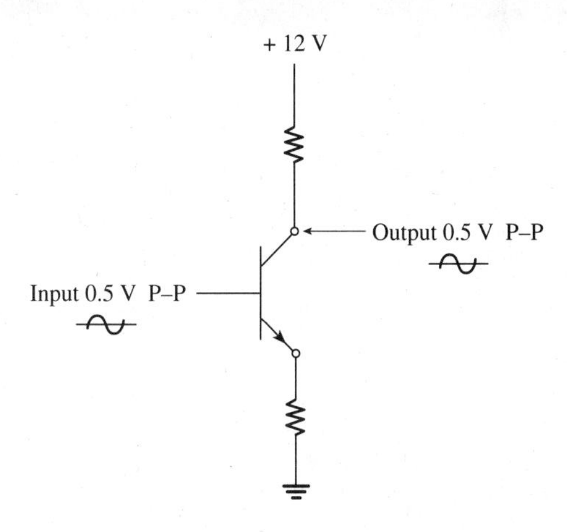

Does this circuit have the correct value of input and output signals? Describe:

1. Type of amplifier configuration
2. Expected gain from input to output for this configuration
3. Expected phase relationship between input and output signals
4. A systematic plan to locate the specific problem in this circuit.

CHAPTER 10

Testing the Power Supply I: Traditional Power Supplies

INTRODUCTION

The statement, "The hardest-working components are those that fail first," is true for electronic components and systems as well as for mechanical devices. Almost all electronic devices convert alternating current obtained from the power utility companies into direct current to operate their transistors, integrated circuits, and vacuum tubes. (Do not overlook vacuum tubes, for they are still used as cathode ray tubes in the majority of television receivers and in video display terminals of computers.) The basic rule of power transfer applies to electronic devices. The power consumed by the unit as it operates must be supplied from the power source. This rule cannot be violated, which contributes to the feeling that the power source has to be the hardest-working section of block in the system.

The design of the power supply has changed during the past years. Efforts to reduce cost, size, and weight, as well as to increase efficiency, have created supplies that meet these qualifications. As you would expect, the results of these efforts have produced some new and exciting types of circuits—known as scan-derived and switching. The sophisticated types of electronic equipment also require some rather close control over the levels of voltage and current supplied to the load from the power supply. These topics and those of the more traditional power supply types require two chapters in this book. This chapter describes the more traditional power supply types, including half-wave, full-wave, and full-wave bridge circuits. Chapter 11 presents materials on the less traditional, but more current, scan-derived and switching supplies and material about regulation circuits.

OBJECTIVES

Upon completion of this chapter and the following companion chapter, the student/reader should be able to:

1. understand how a variety of electronic power supplies function;
2. recognize typical power supply problems;
3. understand how to test the power supply;
4. recognize which piece of test equipment to use to test the power supply; and

5. understand what the specific tests indicate as they relate to service and repair.

WHY THE POWER SUPPLY?

Perhaps the best answer to this question is another question: "Why not?" The introduction to this chapter indicated that the power supply, or power source, is the hardest-working section of any electronic device. The rule for determining total power in any circuit, whether series or parallel circuit, states that the total power is the sum of each load's power. When you consider that each stage or section of any electronic system acts as a load on the power source, this rule has even greater meaning. Figure 10–1 illustrates this point. Each block of this typical AM radio uses a small amount of power. Typically, the output stage requires the greatest amount of power. Total power used by this AM radio is equal to the sum of the power used by each of its stages. The total of the power used by each stage in this radio is equal to 3.4 watts. The power source block has to be able to provide this quantity of power at both the proper voltage and the proper current levels for the radio to operate correctly.

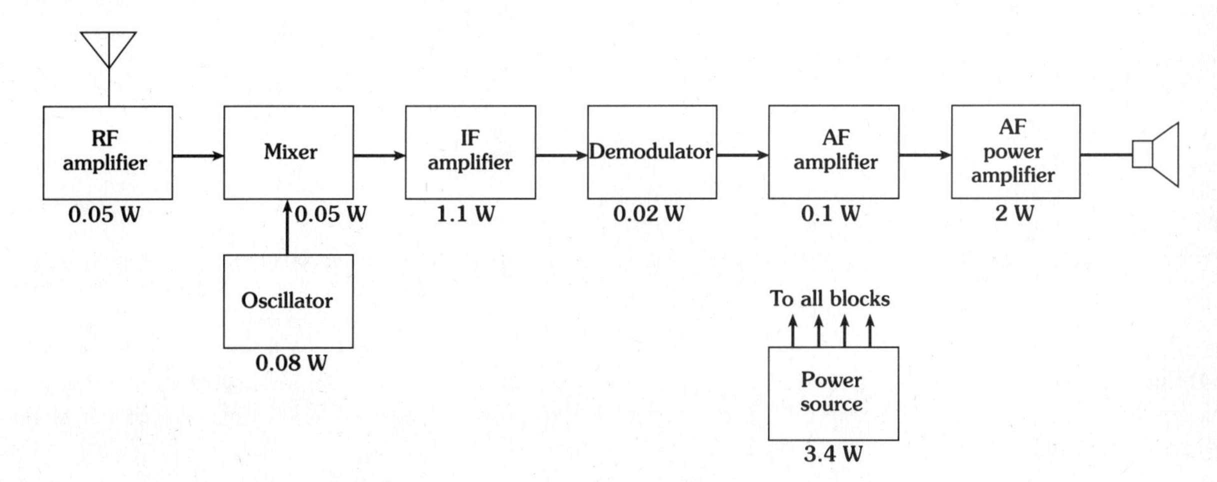

Figure 10–1 Individual block power requirements for a typical transistorized AM radio.

If, for any reason, the power source is unable to produce the full 3.4 watts of power, then the radio will not operate properly. Any one of several things might occur: there could be an output level less than that of the original radio design; unwanted noise, or a low-frequency hum, could be heard on the information being processed by the radio; or the components in the power supply section could fail as they attempted to produce the required levels of electrical power. Any of these could result in a lack of output from the device.

POWER SUPPLY TYPES

There are several types of power supplies available at this time. These include chemical, photovoltaic, and electronic. The chemical world has provided us with batteries. Electronic manufacturers have produced a variety of types of electronic power supplies, several of which are described in this chapter. In addition, the electronic industry now offers photovoltaic cells, a spin-off from the space industry. The emphasis in this chapter and the next is on the electronic power supply because that is what powers most electronic equipment.

CHEMICAL CELLS

Batteries consist of one or more individual chemical cells. These individual cells can be wired into series, parallel, or series-parallel combinations. Perhaps the most commonly used battery is one made up of six individual 1.5 V cells wired in a series connection. This battery provides the 9.0 V used by many electronic devices.

Actually, the battery's cell combination can create any value of voltage and current required to operate an electrical or electronic device. The fundamental rules for cell combination are the same as those used to analyze series and parallel circuits. In the series circuit the current is equal everywhere in the circuit and the voltage drops across each load add to equal the applied voltage. In a series-wired battery the voltages developed by the individual cells add to equal the terminal voltage of the battery. Current flow in this series voltage-aiding arrangement is limited to the amount of current that the lowest-rated cell is able to process. This means that, should one cell be limited to 0.5 A of current, while all of the other cells are capable of producing 1.0 A, the available level of current is limited to the smallest single value of 0.5 A.

When cells are wired as a parallel combination, the voltages remain equal. This is the same as the rule used for voltage in the parallel circuit. Load currents in the parallel circuit add, as do the current capacities for parallel connected cells.

ELECTRONIC POWER SUPPLIES

The greatest number of power supplies in use in business and industry are those that obtain their operating power from commercial AC power company wiring. International standards for AC power sources are either 120 V or 240 V sources. Their operating frequency is either 50 or 60 Hertz. These sources provide alternating current power to the user. Almost all electronic components require direct current. Therefore, a system has to be used to convert the alternating current source into a direct current source. These conversion devices are called power sources, or power supplies.

A block diagram for a typical traditional power supply system is shown in Figure 10–2. This system consists of several blocks. Each one has a specific function in the system. As a first step in the chapter, be certain that you understand each of the blocks and their specific functions.

AC source: This block is the input to the power supply system. It is connected to the AC power company wiring. Typical components in this block include the power cable, fuses, and power control (on-off) switches.

transformer: This block is used to change the values of voltage and current. It also provides isolation between the AC power line and the system. The transformer's input is either 120 V or 240 V from the power company wiring. The output voltage and current from the transformer is dependent upon the requirements of the rest of the system. Almost all electronic devices containing solid-state components operate at voltage levels

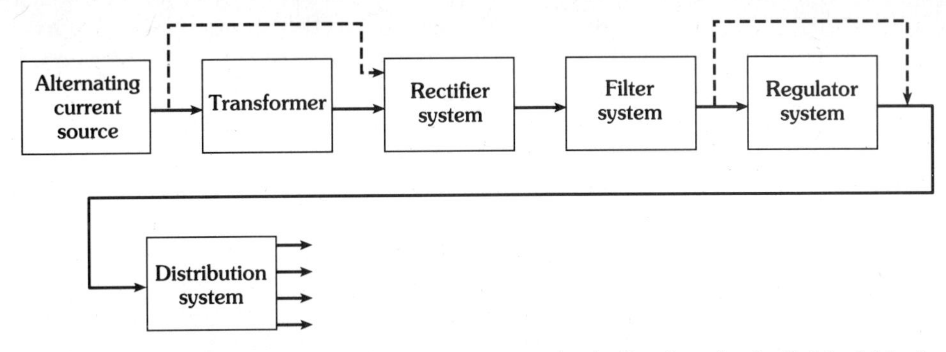

Figure 10–2 Block diagram for a power supply, indicating the individual blocks found in these units.

much lower than the 120 or 240 V AC source values on the transformer primary winding. The transformers used for these devices are called voltage step-down transformers. The output voltage from these transformers is often in the range of 12 or 24 V at the transformer's secondary, or output, winding.

There are systems in use that require voltages much higher than the 120 or 240 V input values from the power company. These, too, can be supplied from the output of the transformer. When the secondary winding voltages are raised above the level of the input voltage value, the transformer is called a voltage step-up transformer.

Some electronic systems use both the voltage step-up and voltage step-down capabilities of the transformer. Typical of these systems are the older television receivers and many communications transmitter/receiver units. The transformer can be manufactured to supply more than one voltage from its output. The specific amount of voltage and current available from each winding depends upon the needs of the system. The power rating of the transformer is equal to the sum of each of the secondary winding power requirements.

rectification system: The third block in this system is identified as the rectification system. This block changes the alternating current voltage into direct current voltage. Electronic devices, called rectifiers, permit current flow in only one direction. When properly installed, the rectifier changes the AC with its constantly changing polarity into pulsations of single current flow direct current.

filter system: The function of this block is to smooth out the pulses of direct current and provide a constant level of direct current. The ultimate would be to provide the same consistent level of direct current as you would obtain from a chemical source.

regulator system: This block's purpose is to keep the amplitude of voltage at a constant level. The output voltage of many power supplies will change as the demands of the load change. The regulator is designed to maintain a constant value of output voltage under varying load conditions.

distribution system: Once the direct current is produced at the proper values and filtered to be equal to chemical cell sources, it must be distributed to the various sections of the device. This must be accomplished with a minimum of power losses.

Note the dashed lines going around the transformer block and the regulator block in Figure 10–2. Some pieces of electronic equipment are designed to operate directly from the AC power source without the need for a transformer. Often one of the two wires from the power source is connected directly to the chassis, or frame, of the system. Particular care must be taken when working on a system of this type, since there is a potential for electric shock if the wiring from the system and the wiring from the test equipment do not have the same common connection. Other equipment does not have voltage regulation requirements. When either or both of these conditions exist, the respective two blocks are eliminated from the system.

SAFETY

One hundred twenty volt alternating current can be dangerous! The service technician cannot immediately observe this voltage when making test equipment connections to the system. Of course, the current can be checked with a voltmeter or other electrical testing device. In practice, very few, if any, electronic technicians make this type of test when starting to make circuit analysis measurements.

It is possible to connect the power cord plugs of two electrical devices into the same AC power line and have their individual chassis, or common, connections connected to two different wires in the AC power line. This is illustrated in Figure 10–3. The load is connected to the 120 V AC power source and has its internal common wire connected to the chassis or frame of the system. This is the upper part of the illustration. The lower part represents the testing equipment. It, too, is connected to the 120 V AC power source. However, unknown to the service technician, the common wire from the 120 V AC source that connects to the tester's chassis and frame is connected to the *other* side of the 120 V AC power source. The end result is a difference of 120 V when measured from the common on the system to the common on the test equipment. If the technician places one hand on the system and the other on the test equipment, a current would flow between these two units. The only current path available would be through the technician's body. Needless to say, this experience could be a killer! If the experience is not fatal, it certainly could cause severe electrical burns on the body.

There is a method available to minimize this type of dangerous condition; it requires learning to use one additional piece of test equipment. This piece is called the isolation transformer. A schematic diagram for an isolation transformer is shown in Figure 10–4, along with a photograph of commercially available units. The isolation transformer does exactly as its name implies. Its purpose is to isolate the unit being tested from the 120 V AC power source. The basic circuit is shown in this figure. Some of these transformers offer additional assistance to the technician. This is shown as a switching arrangement on the transformer's top. The switch is

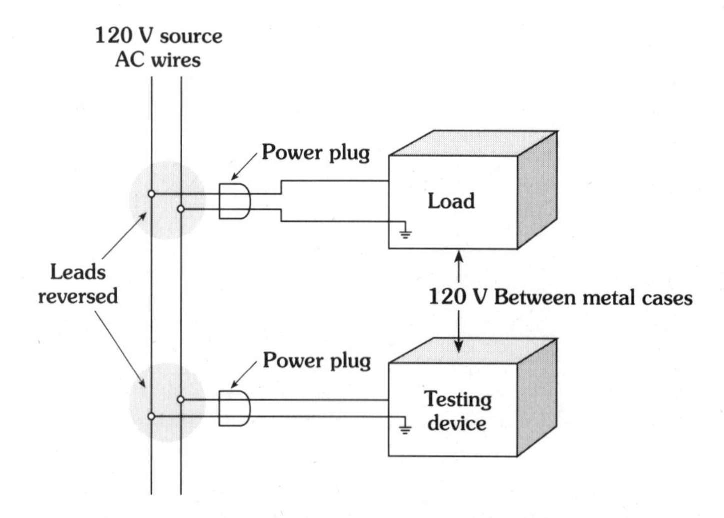

Figure 10–3 The possibility of an electrical shock exists when the AC power cords for two units are not connected to the same AC common wiring.

Figure 10–4 **Isolation transformers, as shown here, will minimize electrical shock and equipment damage. (*Photos courtesy The Vector Group, Instrument Division*)**

connected to some taps on the transformer's windings, permitting the selection of a variety of output voltages for testing purposes.

TESTING POWER SUPPLIES

Once the system's problem area has been localized to the power supply section, the first step is to locate and read the service literature for the device. This service literature should provide the output voltage(s) of the power supply, the circuit diagram for the supply, a parts list, and, with a little luck, its physical layout.

The next step is to develop a plan of approach for diagnosing the power supply. At this point the service technician needs to be able to recognize the type of power supply and how the power supply functions. There are a few typical types of electronic power supply systems presently in use, including the half-wave, full-wave, full-wave bridge, switching, and scan-derived systems. The first three of these are described in this chapter; the others are described in Chapter 11.

POWER SUPPLY PROBLEMS

The specific type of problem associated with the unit has to be recognized once the type of power supply has been identified. Typical types of problems for most power supplies include:

1. *No output*: The unit is totally "dead." None of the systems has any indication of the presence of operating power.
2. *Limited output*: The unit shows some indication of a willingness to operate, but not all sections are

functioning properly. Possible conditions include low audio output, poor RF signal processing, and, if it is a transmitting unit, low signal output.

3. *Poor filtering*: All sections seem to have operating power, but there are indications of erratic operation. Audio devices will produce a low-frequency humming noise at a rate of either 60 Hz or 120 Hz due to poor filtering. Video devices will display wavy lines that appear to have a sine wave quality, either horizontally or vertically, or both, due to poor filtering. At times these sine wave-like lines may move, or float, across the video information being displayed. At other times they may appear at the edges of the display.
4. *Lack of regulation* (if used): Output conditions will change as the load changes. The output should be constant with a consistent level of input signal. A lack of voltage or current regulation may create a condition where an increase in the audio level controls will not provide an increase in the level of the output signal.

It is possible to have similar problems in other sections or blocks of the device. The problems described here are typical of those created in the power supply section.

POWER SUPPLY COMPONENTS

The traditional power supply consists of a minimum of three components—these are the transformer, the rectifier, and the filter capacitor. Power supplies may use more than one of each of these items. The fundamental rules for servicing are the same, regardless of how many of the components are used.

RECTIFIERS

Rectifiers act as current-controlling devices in this circuit. This statement is based on the rule that the total resistance in any circuit is one of the factors that determine the quantity of current flow. The rectifier acts as a resistance in the power supply circuit. Rectifiers are, in a sense, devices capable of exhibiting two levels of internal resistance. One level is

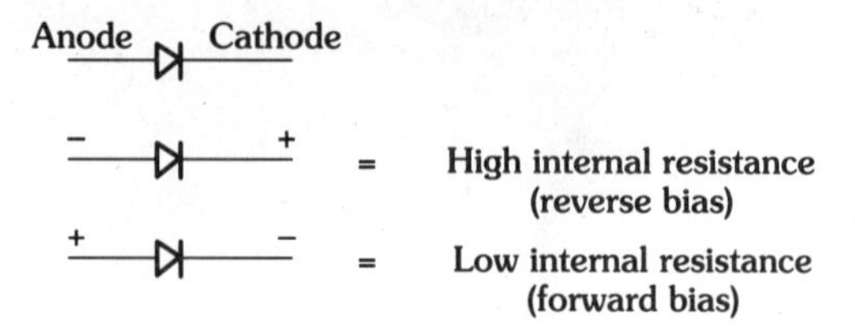

Figure 10–5 **A diode in a power supply can be tested with an ohmmeter. The diode will have a low forward bias resistance value and a high reverse bias resistance value.**

an extremely high value and the other level is an extremely low value of resistance. Rectifier diode action is illustrated in Figure 10–5. The graphic symbol for a solid-state diode is shown in the upper part of this illustration. The cathode of the diode is indicated by the vertical line, and the anode is shown on the left, at the wider part of the arrow-like figure.

The diode operates under one of two conditions: they are either forward biased or reverse biased. Both of these conditions are shown in the remaining parts of this figure. When the diode is forward biased by the voltage applied from the circuit it is in, its internal resistance is very low. It acts as a very low value of resistance and permits current flow through it. The actual measured value of resistance will vary, depending upon the type of meter used to measure it. A digital meter will often read between 500 Ω and 600 Ω when measuring the forward bias resistance of a silicon diode. Forward bias occurs when the diode's anode element is positive with respect to its cathode element. This polarity is shown in the lower part of the figure. Some digital meters are able to indicate the forward bias voltage of this junction.

The diode can also be reverse biased. This condition occurs when the voltage at its anode is more negative than that at its cathode. Under this condition the diode acts as a very high value of resistance. A limited amount, if any, of current will flow through its terminals under a reverse bias condition. In a sense you can say that the diode acts as a voltage-sensitive switch in an electrical circuit. It is either "on" or "off" as it processes current flow in the

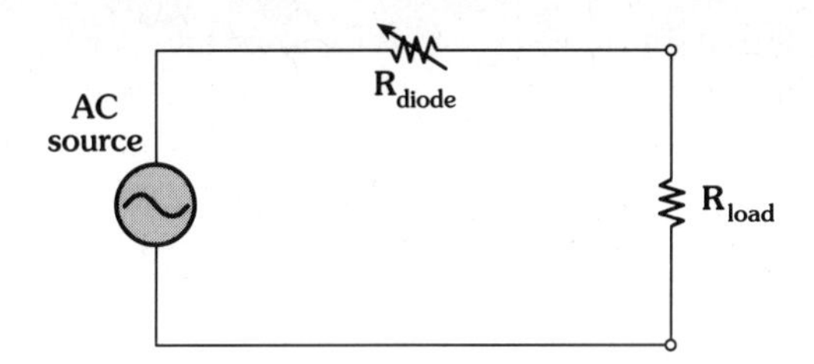

Figure 10–6 The diode can be considered as a variable resistance in a circuit. Its internal resistance is determined by the polarity of the bias voltage on its elements.

circuit. These conditions are based solely on the polarity of the applied voltage to its terminals.

Now, let us look at an application of the diode. Figure 10–6 shows an AC voltage source, a variable resistance marked "diode," and a load. The load and the diode are represented as resistances in this circuit. The AC source provides a constantly reversing polarity of voltage to this series circuit. The diode acts as a dual-resistance device in the circuit. A voltage is applied to the circuit from the AC source. In addition, the voltage has a constantly reversing polarity.

The components identified as R_{DIODE} and R_{LOAD} are connected in series with the AC power source. The rules presented by Kirchhoff for voltage drops in the closed loop, or series circuit, apply here. The rules for basic series circuits also apply. Using Kirchhoff, the voltage drops that develop across the two resistances will *always* equal applied source voltage. The rule for voltage drops in the series circuit as they relate to the ratio of the individual resistance values also apply to this circuit. When forward bias is applied to R_{DIODE}, its internal resistance becomes very low. The voltage drop that develops across it is also very low, since these two factors are directly related to each other. Under these conditions the majority of the applied voltage is developed across the resistance of the load.

The bias is reversed each half-cycle of the applied AC voltage. When this occurs, the internal resistance of the diode becomes very high. The individual voltage drops that develop in this series circuit are still dependent upon the size of the individual resistances. Since the internal resistance of R_{DIODE} has become very high, the voltage drop across the diode becomes close to the value of the applied voltage and the voltage drop across the load is minimal.

The application of the one half of the AC voltage in a circuit containing a diode and a load is shown in Figure 10–7. This figure shows the AC voltage waveform being applied to the diode and load resistor. The output voltage waveform is also shown. The two waveforms are similar in shape and should also be similar in amplitude. A positive-value voltage of the AC waveform is applied to this circuit during the waveform's first half-cycle. The diode is forward biased and current will flow through the diode and through the load. This illustrates the effect of diode action in a rectifier circuit when the diode is forward biased by the applied AC source voltage.

The polarity markings on Figure 10–7, which are shown in parentheses ( ), indicate the polarity of the applied AC source voltage during the second half

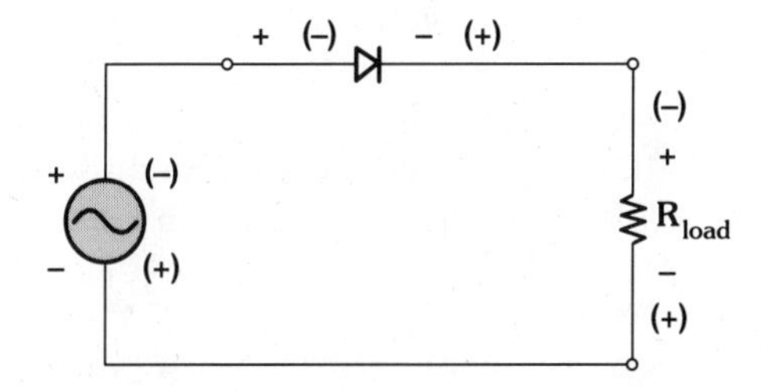

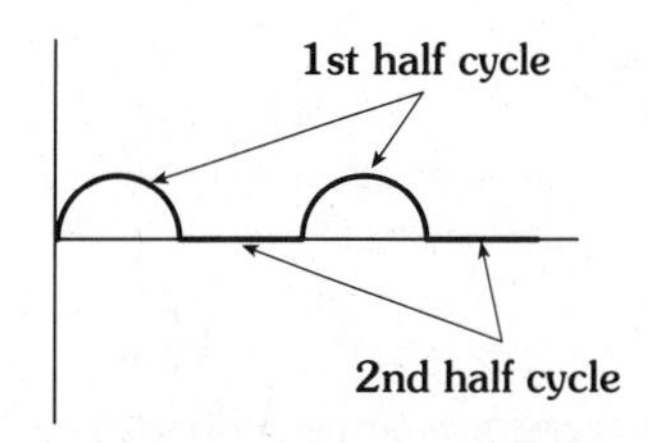

Figure 10–7 The half-wave rectifier system produces an output that is half of the input AC voltage waveform.

of its cycle. Now the diode is reverse biased. Its internal resistance is much higher than that of the load resistance. Almost all of the applied AC source voltage now develops as a voltage drop across the diode. Little, if any, can develop across the load. The reason for this is apparent when you consider the rules for the values of voltage drops in a series circuit as they apply to the individual values of resistances in the same circuit. The output voltage during this second half-cycle, as shown on the same figure, indicates a zero volt output value.

CAPACITORS

Capacitors act in a similar manner to that of the sponge. Their action is limited to changes in the amplitude of the voltage in the circuit. If, for example, the voltage rises in the circuit, the capacitor will attempt to absorb the increase. This increase charges the plates of the capacitor. A time lag in the change of voltage level occurs during the period when the plates are charging. When the voltage in the circuit decreases, the charged capacitor plates will give up their voltaic charge in an effort to maintain the previous level of voltage.

This action is very beneficial to the operation of the electronic power supply. Diode action produces half of the applied AC waveform across the load. This fluctuation is not desirable for the operation of electronic components, since they function better on direct current. Capacitor action will "fill in" the voltage during the time when the rectified AC voltage is either rising or falling. This reduces the changes in the amplitude of the rectified AC voltage. The result is a value that closely represents direct current voltage. There are some variations in the level of this voltage, called "ripple." A quality power supply will have less than a 1 or 2 percent ripple factor.

Capacitor failures do occur. They include a short circuit, an open circuit, and a partial failure known as "leakage." The open capacitor can be diagnosed by using the oscilloscope. The waveform is observed by measuring the DC voltage level at the load or across the capacitor's terminals. If the observed waveform appears as one half of the sine wave, and it rises from zero volts to the maximum amplitude, then the capacitor is open and not functioning.

A shorted capacitor will open the circuit protective devices in the power supply section. The shorted capacitor places an extremely low resistance value in parallel with the load's resistance. When two resistances are placed in parallel, the total resistance is always less than the value of the lower value resistor. When a resistance of less than one ohm (short circuit) is placed across the load, the total resistance presented to the power source is also less than one ohm. Shorted capacitors are found by using the ohmmeter.

The third type of failure—partial failure of the capacitor—is harder to locate. When a capacitor develops the condition known as leakage, its ability to react to changes in the amplitude of the voltage is weakened. This condition is identified by using the oscilloscope when an increase in the value of the ripple voltage is observed.

HALF-WAVE POWER SOURCES

A circuit similar to those illustrated for diode action is shown in Figure 10–8. There are three differences between this illustration and the one previously described. The first is the identification of the load resistance as an output resistance; the second is the addition of a capacitor placed in parallel with the output resistance; and the third is the identification of three points in the circuit labeled (a), (b), and (c). These points are shown on the schematic diagram as well as with corresponding waveforms below the circuit.

An examination of the operation of the circuit offers some interesting information. The input voltage for this circuit has an alternating current value of 120 V. This value is measured in its most common mode as an rms value. The rms value is the same one normally measured with the AC voltmeter. However, when this same 120 V rms value is measured with an oscilloscope, a different voltage value is observed. The oscilloscope's measurement will provide a reading of just under 340 V! The reason for this is clear. The oscilloscope readings are not using the same system of measurement as does

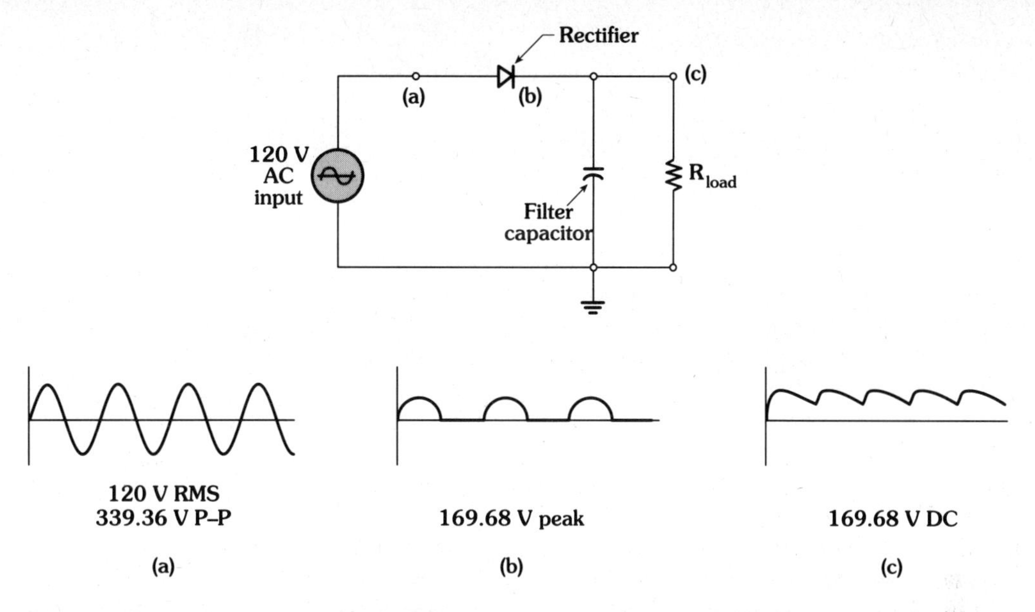

Figure 10–8 Use of a parallel-connected filter capacitor will "fill in" the mission voltage values in the half-wave rectifier system.

the voltmeter. The oscilloscope measurements are provided as peak-to-peak voltage values. These, by definition, are 2.828 times higher in value than the same reading as an rms voltage. Remember, we are describing the same value of voltage under two different systems of measurement. You might consider this as similar to describing the measurement of a liquid in gallons and liters. The same quantity could be measured but the difference in number values would depend upon the system describing the measurement.

These values are measured at point (a) on the circuit. Both the rms and the peak-to-peak values are provided. The reason for giving both sets of voltage values will be apparent when you consider what occurs during the diode's rectification process. This process literally cuts off one half of the applied peak-to-peak voltage. The result is shown in part (b) of Figure 10–8. This voltage is also obtained by using the oscilloscope at point (b) of the circuit. The voltage output from the diode is one half of the peak-to-peak value of 339.36 V, or 169.68 V. This value, described as a peak reading, is the amount of voltage that is presented to the output of the circuit when the diode is forward biased.

The voltage at this point in the circuit is still not usable by the load. It requires further conditioning before it can be applied to operate the load. The additional conditioning requirement is to filter, or eliminate the variations in it. The variations in the operating voltage at this point are known as the ripple voltage. This voltage often appears in sawtooth form on the DC voltage created by the power supply and filter system. The amount of ripple voltage that is acceptable is normally about 1 to 2 percent of the total DC voltage.

Reducing the amount of ripple voltage is accomplished by using a capacitor. The capacitor is connected in parallel with the output resistance. (In some circuits the output resistance is nonexistent; it is replaced by the actual load.)

In this circuit the voltage that develops across the output resistance during the time when the diode is forward biased also charges the capacitor to this same value of voltage. When the diode is reverse biased during the second half of the input AC voltage cycle, the capacitor will give up its voltage charge into the load. The reason for this is that the diode circuit appears to be an open circuit when it is reverse biased. The only source of voltage for the load at this time is the voltage charge developed in the capacitor. The result is shown in Figure 10–8 (c). This illustrates the result of the diode action and the capacitor's action in the circuit.

Another way of looking at the capacitor's action is shown in Figure 10–9. This is the same circuit, except that the diode is replaced by the symbol for a switch. When acting as a rectifier, the diode is essentially a switch as far as current flow in the circuit is concerned. When the diode is forward biased, it has a low internal resistance; it acts like a switch that is closed and current flows through the completed circuit. This will provide an operational voltage for the load and also charge the capacitor at the same time. Since the capacitor is connected in parallel with the load, its voltage charge will be the same voltage level.

When the diode is reverse biased, its internal resistance is extremely high and the switch representing the diode appears to be open. There is no current flow from the AC source through the diode and to the load. The capacitor cannot charge; it can, however, act as a source of voltage since it is still connected in parallel with the load. This is exactly what it does as it discharges its voltage across the load.

SERVICING THE HALF-WAVE POWER SUPPLY

This circuit may be considered to be a closed loop, or series circuit. This being the case, the rule for linear, or in-line, types of circuits can be used. This is the half-split rule. The procedure for testing this circuit is:

1. Determine if the output voltage is proper. The type of equipment to use for this measurement can be either the multimeter or an oscilloscope. The oscilloscope will provide more information because it will tell you whether the capacitor is functional in the circuit. One of four conditions should exist. You should have (a) a no-voltage condition; (b) a low-voltage condition; (c) a high level of ripple, or (d) the system will be working properly.
2. Determine if the input is proper using the oscilloscope or the voltmeter. If there is no input voltage, then the output voltage will not be present. You should not assume that, just because there is no output, the input system is operating correctly.
3. If the input is good and there is little or no output, use the half-split rule. Locate a point at or near the center of the circuit for the next test. In the circuit shown in Figure 10–10, this point is identified as point A. Use the oscilloscope for this

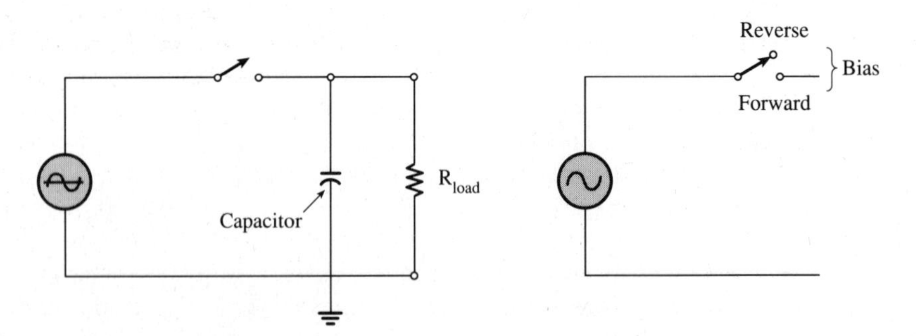

Figure 10–9 This circuit illustrates the action of the capacitor during each half of the input AC voltage cycle. When the diode is conducting, the capacitor is charged. When the diode is not conducting, the capacitor gives up its charge into the load.

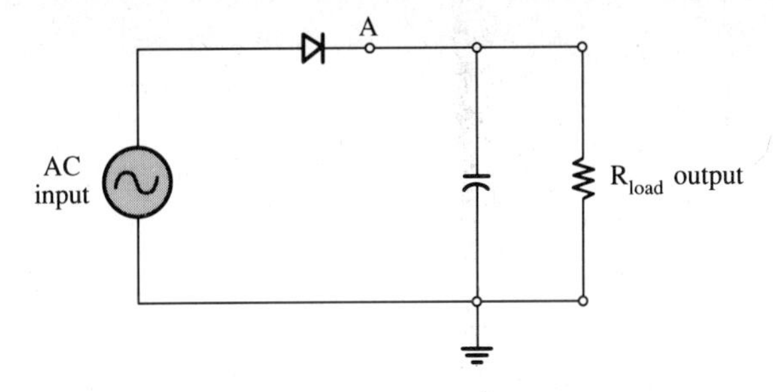

Figure 10–10 **The midpoint of this circuit is identified as point A on the schematic diagram. This is the starting point for diagnosis.**

measurement because the effect of the capacitor is included in the operation of the system. You cannot isolate its action when it is operating properly.

4. If the diode is working properly, then the voltage and waveform at this point should agree with the manufacturer's specifications in the service manual. This being true, the left-hand bracket placed at the start of the diagnosis procedure is now moved to point A. This tells us that all is working properly between point A and the input to the system.
5. The best test to make at point A is to measure the DC output voltage with an oscilloscope. If the amplitude of the ripple voltage present at this point is above that provided by the manufacturer, then the problem is most likely limited to the capacitor. It may be tested using one of the methods and pieces of test equipment described in Chapter 9. Replacing it should solve the problem.
6. If the test at point A does not provide the proper voltage amplitudes and waveshapes, then the problem is between point A and the input to the system. The right-hand bracket is then moved to this point. There are only two components to be tested. One of these, the input voltage source, was previously tested and found to be correct. This leaves the diode as the only remaining component in this part of the circuit. Testing the diode is accomplished by using the ohmmeter or a diode tester. The diode's resistance should be relatively low when the meter forward biases it. The resistance should be infinitely high when the meter reverse biases it. Test results indicating low resistance in both directions or high resistance in both directions tell the technician that the diode is defective and must be replaced.

THE FULL-WAVE POWER SUPPLY

The half-wave power supply operates at a 50 percent efficiency rate. This supply relies on the action of the capacitor to "fill in" for the other half of the wave. This, of course, does not make the most efficient type of power supply. There are two or more circuits that use both halves of the AC wave cycle effectively. One of these requires the use of a transformer having a connection to the center of its secondary winding. This system is known as the full-wave center-tapped transformer system. The other power source is called the full-wave bridge system. It does not use the center-tapped winding transformer. Each of these systems is in use in present-day electronic equipment.

THE FULL-WAVE CENTER-TAPPED TRANSFORMER SYSTEM

A diagram of a full-wave center-tapped winding transformer is shown in Figure 10–11. The secondary winding of this type of transformer has a connection at the midpoint of its winding. If, for example, there were 500 turns of wire on the secondary winding, then the center tap connection would be made at the 250th turn. In this example two resistances are connected in series across the full secondary winding and their midpoint is connected to the center tap of the transformer. The resistances are used to show a completed electrical circuit.

The output voltage from the center-tapped transformer can be measured as the total voltage measures either from point A to point C, or it can be measured from the common point in the center, point B, to each of the other connections. In this example you could say that the output voltage was 24

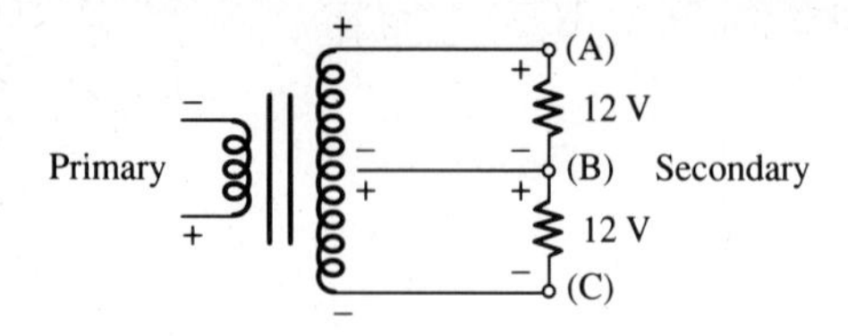

Figure 10–11 **The full-wave center-tapped transformer develops polarities at its output connections during its operational cycle.**

v.c.t. (center-tapped) or that the output voltage was 12-0-12 volts. This second method indicates the values of voltage measured from the center tap connection to each end of the windings. Both methods provide the same information.

Also note the polarity markings in this figure. Transformer action creates a 180-degree phase shift between input and output windings of a transformer. The polarity markings for the secondary winding are shown. Since the center tap will be established as the common point, the polarity of the output voltages will differ for each of the windings. One will be positive, the other negative. This, of course, reverses during each cycle of the input AC voltage.

Figure 10–12 shows how this center-tapped transformer is used in the full-wave rectification system. The center tap connection of the transformer is connected to circuit common. The anodes of two diodes are connected, as shown, one to each end of the transformer's secondary winding. The cathodes of these two diodes are connected to each other and to the filter capacitor and the load resistance.

During the first half-cycle of the AC voltage at the input to the transformer, the polarity shown on this drawing develops. Transformer action will provide a reversal of the polarity of the voltages between primary and secondary windings. Electron current flow is from circuit common at the center tap connection, through the load, through diode D_1, and then to the (+) terminal of the transformer. The polarity of the applied voltage forward biases this diode. This action creates a half-wave rectification system. Its waveform, without capacitor action, is shown to the right of the circuit as the D_1 waveform. Diode D_2 is reverse biased and does not conduct. Therefore, it has a zero output voltage during this half-cycle of the input voltage.

During the second half-cycle of the AC voltage at the input to the transformer, the polarity shown on the drawing is reversed. The voltage polarity presented between diode D_1 and the center tap connec-

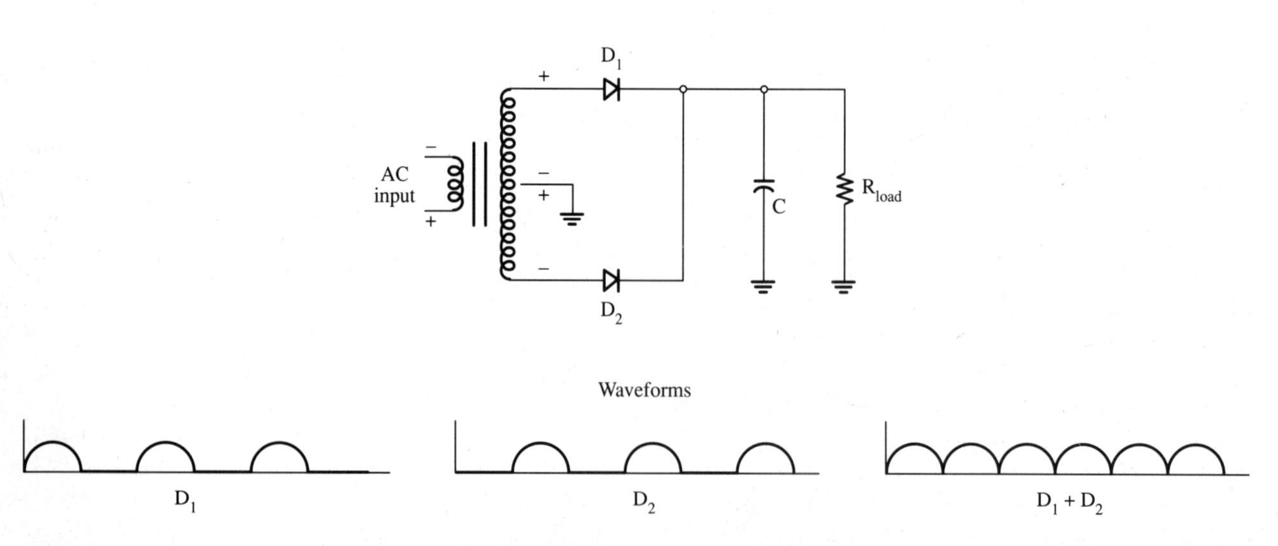

Figure 10–12 **Current flow action creates the voltage waveforms in the full-wave center-tapped rectifier system illustrated here.**

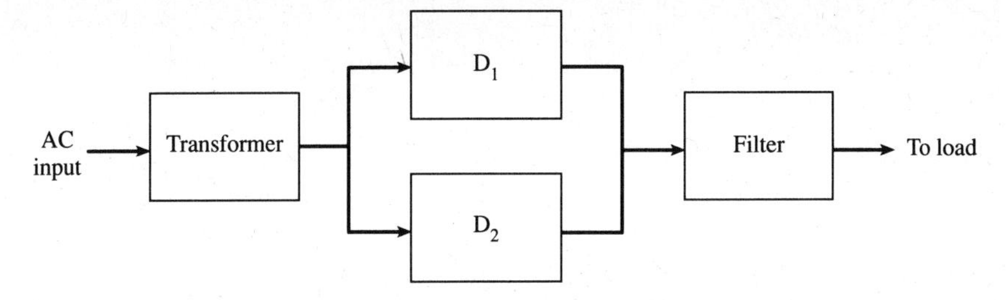

Figure 10–13 **A partial block diagram for the full-wave center-tapped rectifier system.**

tion to circuit common reverse biases the diode. The voltage presented between diode D_2 and the center tap connection of the transformer forward biases this diode and current flow occurs through this diode and the load. The path for current flow during this second half-cycle is from circuit common, through the load resistor, through diode D_2, and then to the other terminal on the transformer. (At this time it is indicated by a (−) sign, but when the polarity of the input voltage is reversed, the polarities of all of the terminals on the transformer also are reversed.) This action reverse biases diode D_1 and it has no current flow.

The polarity reversal of the AC voltage source continuously switches the diodes' forward and reverse bias conditions. They are out of phase with each other and function at different times. There is only one load resistor and it is connected to circuit common and the meeting points of the anodes of the two diodes; therefore, all of the current flow in this circuit is through the load. Current flow is through the load in the same direction no matter which of the two diodes is conducting. This staggered action develops the combined waveform shown as $D_1 + D_2$ in Figure 10–12. Capacitor action fills in for the low spots between waveforms.

Servicing the Full-wave Center-tapped Power Supply This servicing is accomplished in the following manner (the block diagram for this system is shown in Figure 10–13). It is essentially a linear system with one exception. The exception is the action described for the two diodes. They do act in a manner that is similar to the one described for the meeting, or joining, block diagram. The difference is that this meeting occurs at different times. Since the characteristics of this system are mostly linear, it will be treated as such in the initial stages of analysis.

1. Determine whether the output voltage is proper. The type of equipment to use for this measurement can be either the multimeter or oscilloscope. The oscilloscope will provide more information because it will inform you whether the capacitor is functional in the circuit. One of three conditions should exist: You should have a no-voltage condition, a low-voltage condition, or a high level of ripple.
2. Determine whether the input is proper using the oscilloscope or the voltmeter. If there is no input voltage, then the output voltage will not be present. You should not assume that, just because there is no output, the input system is operating correctly.
3. An analysis of the circuit indicates that its midpoint occurs at the junction of the anodes of the two diodes. This is an excellent place to make the next test. It also should be fairly easy to locate this point on the circuit. This test is best made with the oscilloscope. You should expect to find a filtered DC voltage with a slight amount of ripple voltage present.
4. If the amplitude of the waveform is acceptable, but the ripple voltage is excessive, then the left-hand bracket is moved to the midpoint of the circuit. The only component left in the circuit is the capacitor. Its partial or full failure would create

the pulsating DC voltage with a high level of ripple voltage.

FULL-WAVE BRIDGE CIRCUITS

These circuits use four diodes and do not require the use of a center-tapped transformer. One of these circuits is illustrated by the circuit shown in Figure 10–14. The diodes work in pairs in this circuit. Both D_1 and D_2 conduct during the first half-cycle of the AC input wave. Diodes D_3 and D_4 conduct during the second half-cycle of the wave. The electron flow path during the first half-cycle starts at the negative terminal of the transformer's secondary winding. Current flow is through diode D_1, to the load and through it, then through diode D_2 and to the other end of the winding on the transformer. This action creates the same half-wave waveform as shown for the full-wave center-tapped system.

When the polarity of the applied AC source voltage is reversed, the other two diodes conduct. Current flow during this half-cycle is from the upper winding connection of the transformer, through diode D_3, through the load, then through diode D_4 and to the other end of the transformer's secondary winding. The output waveform from this half-cycle is also similar to that of the full-wave center-tapped system. It is out of phase with the other half-cycle waveform.

Since all current flow through the load is in the same direction and has the same polarity, the output voltage waveform is also the same as observed in the full-wave center-tapped system.

Servicing the Full-wave Bridge System The block diagram for this system is very similar to that of the full-wave bridge system; therefore, the service approach will be the same. The oscilloscope will display the waveform of the output voltage. If this waveform has a high level of ripple voltage, then the problem is most likely the filter network and its capacitor. If the waveform has peaks with two different amplitudes, then one set of the diodes is probably defective.

The waveforms obtained from a system with a defective diode set are shown in Figure 10–15. The lack of amplitude for one of the two peaks is typical of that observed for a failure of this type. Part (a) of this figure illustrates the waveforms when both diode sets are working properly and the capacitor circuit is not connected. When one pair of diodes is malfunctioning, the waveforms will appear as shown in part (b) of the figure.

Consider this type of circuit to be similar to two series loops. One of the two loops is shown in Figure 10–16. It consists of the secondary of the transformer, diodes D_1 and D_2, the filter capacitor, and the load (represented by the resistor symbol). During the half-cycle of the AC input when the polarity of secondary winding of the transformer is as shown, both of the diodes are forward biased. Each diode will exhibit a voltage drop of about 0.7 V,

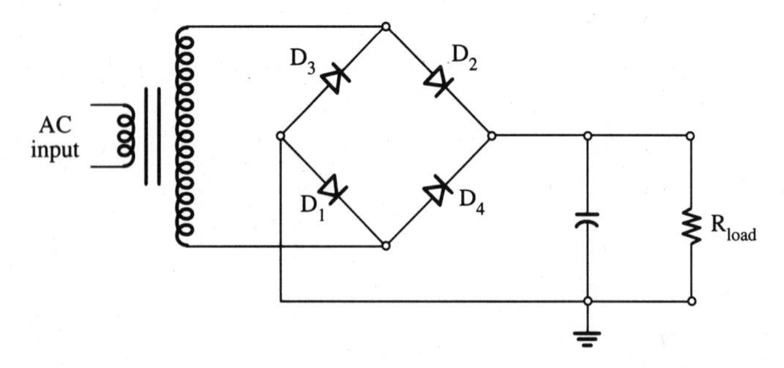

Figure 10–14 A full-wave bridge rectifier system uses four diodes and does not require the use of the center-tapped transformer.

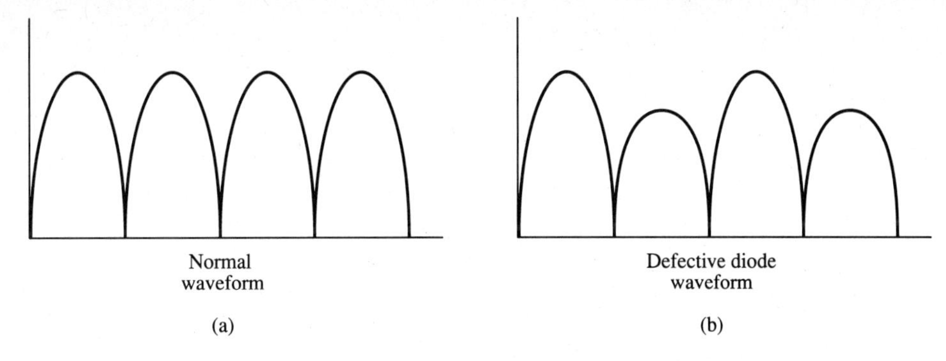

Figure 10–15 **Output waveform for a properly functioning full-wave bridge rectifier shown in (a) and (b) when one diode is defective, but has not totally failed.**

for a total of 1.4 V. The major amount of voltage drop will then develop across the load. This will also charge the capacitor to the value of the applied voltage.

If either or both of the diodes opened, the result could be an open circuit or a very high value of resistance. This condition would be observed as a high voltage drop across the diode instead of across the load resistance. In either case current flow would be minimal, if at all, and the load would not function.

One other type of condition could also exist—development of a short circuit in one or more of the diodes. When this condition exists, current flow is no longer directed to one of the two pairs of diodes. Current flow when one diode is shorted occurs in both directions at the same time. Look at Figure 10–17. Let us add some resistance values and determine

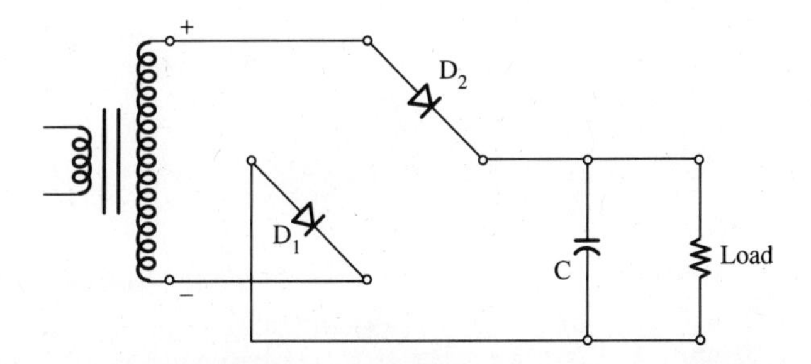

Figure 10–16 **One half of the operational components of the full-wave bridge rectifier. These two diodes work as a pair to provide half of the output waveform.**

what the total circuit current would be when the system functioned properly. The 12 V rms value of the transformer secondary winding will create close to 18 V DC at the load. When 18 V DC is placed across the 5 kΩ load, 0.0036 A of current will flow through the load.

Consider the current paths if all four diodes were replaced by 100 Ω resistances. The result of this condition is illustrated by the diagram shown in Figure 10–18. The circuit effectively consists of two branches of 100 Ω resistances connected in series across the transformer secondary winding. The effective total resistance is 100 Ω for the entire circuit.

Actually, the total effective resistance for this combination would be slightly less when you include the value of the load resistance. This is based on analysis of the circuit. Two 100 Ω resistances connected in series are equal to 200 Ω. Two additional 100 Ω resistors also connected in series equal a second value of 200 Ω. The effective resistance for these two 200 Ω resistances is 100 Ω. When this combination is placed in parallel with the 5 kΩ load, the total effective resistance is 98 Ω. Current flow would now attempt to reach a value of almost 0.184 A. Under most circumstances, this 67 percent increase in current flow would cause the circuit protective fuse to open. An open fuse is typical for shorted diode conditions in the circuit.

Possibly the best way to test for short-circuited diodes is to use an ohmmeter. The internal power source in the ohmmeter will place a bias voltage across the terminals of the diode. If the diode is

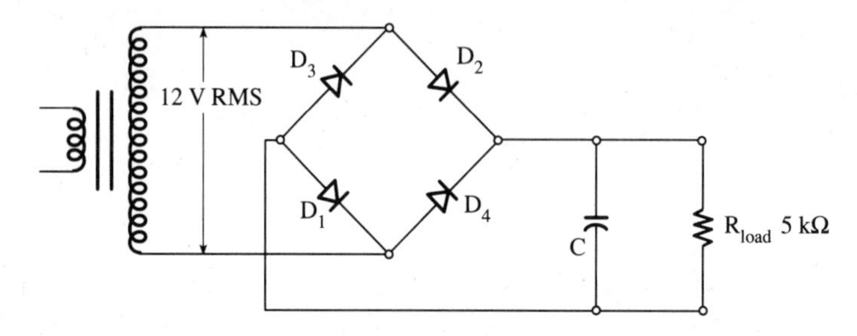

Figure 10–17 **Full-wave bridge rectifier system. Each diode has a resistance of around 100 Ω when forward biased.**

okay, then its resistance will be relatively low when it is forward biased. The silicon diodes used in power supplies often have an ohmic value of around 500 Ω when forward biased. Their reverse biased ohmic value is often too high to measure with an instrument of this type.

TESTING POWER SUPPLIES, CHEMICAL CELLS, AND BATTERIES

The phrase, "to load or not to load," should be considered when evaluating any power source. The conditions created by the load will directly affect the action of the power source. An evaluation of the unloaded power supply will often provide false information about its condition. Load conditions will change the level of capacitor action because of factors known as RC time constants. They directly relate to the time required to charge and discharge the capacitor. This, then, affects the ability of the capacitor to fill in the voltage during those periods when the diodes are not at the peak of the AC input voltage level.

Load conditions are described in this section. While the description is directed primarily to the chemical cell, the same type of conditions apply to electronic power supplies as well. An unloaded power supply often will provide a higher than normal output voltage. Evaluation of this value may indicate that the power supply section is functioning normally. Ripple voltage is minimal with a light load. The normal loaded conditions often show higher levels of ripple voltage when capacitors are starting to fail.

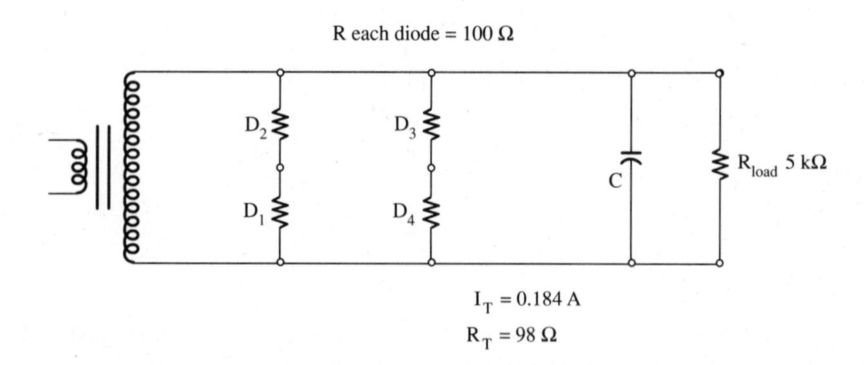

Figure 10–18 **Equivalent circuit using resistances to show diode values.**

CHEMICAL CELLS AND BATTERIES

Chemical cells and batteries should be tested under load conditions. The reason for this is very simple and reverts to the basic rule for voltage drop in the series circuit. This rule states that a voltage drop will develop across a component in the series circuit when a current flows through that component. The internal construction of the chemical cell is illustrated in Figure 10–19, which shows the basic symbol for the cell and a resistor symbol. The resistor symbol indicates the internal resistance of the cell. All power sources have a value of internal resistance. When new, this internal resistance has a very low ohmic value. As the cell ages and provides its power to the load, this value of internal resistance will increase. The increase in this internal resistance value will also increase the voltage drop that develops across it when the cell is connected to a load.

Figure 10–20 illustrates how the internal resistance of a 9 V battery affects its output voltage. One way of showing this is to use this formula for determining the load voltage:

$$V_{LOAD} = V_{CELL} - V_{INT}$$

Using this formula, let us examine the various amounts of load voltage for different internal resistance values. These are shown in the following chart. This circuit uses a 9 V battery as its power source and a load value of 500 Ω.

This chart shows how the internal resistance of either an individual cell, or an entire battery made up of individual cells, is an important factor in determining the output voltage for the load. The one ohm value is very similar to a "no load" condition. This is the value of voltage you would measure when the battery is not connected to its load. Once it is connected to the load the terminal voltage of the battery will drop due to its internal resistance.

You must remember that all batteries and cells are tested under load conditions. Commercially available testing units place an internal load resistance across the test leads for more accurate measurement of battery conditions.

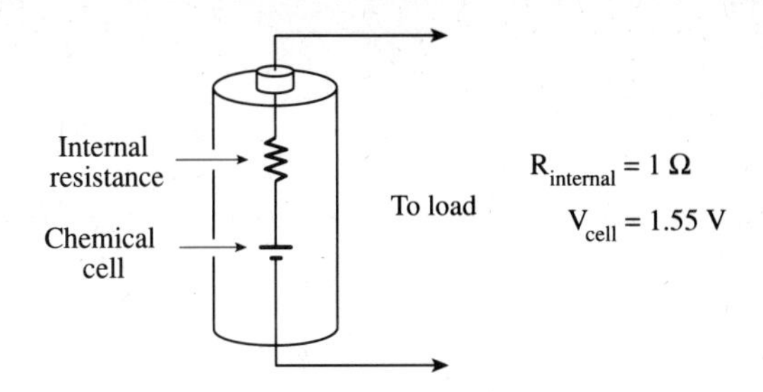

Figure 10–19 A chemical cell has an internal resistance value. This resistance is a part of the circuit and will develop a voltage drop when current flows in the circuit.

POWER SUPPLY TESTING

Power supply testing is best accomplished with the oscilloscope. The reason for this is the need to evaluate the amplitude, frequency, and shape of the waveform produced by the supply. The service technician should be able to recognize the difference between the half-wave supply waveform and the full-wave supply waveform. It will not be possible to identify whether the supply is the center-tapped type or the bridge type when observing the output waveform. What should be evaluated are the amplitude and shape of the output voltage.

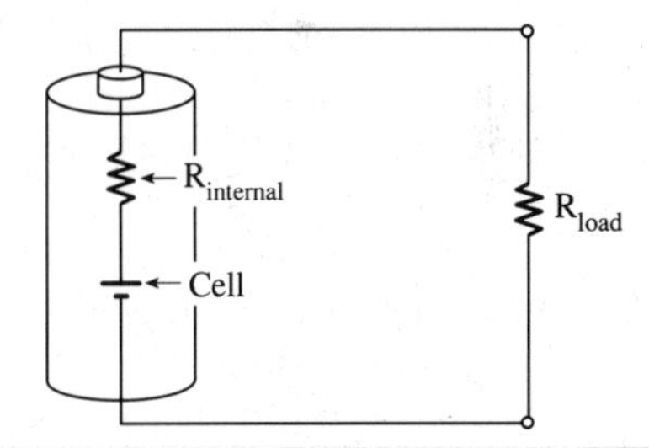

R_{INT}	R_{TOTAL}	V_{INT}	V_{LOAD}
R_{INT} 1 Ω	501 Ω	0.018 V	8.98 V
R_{INT} 100 Ω	600 Ω	1.5 V	7.5 V
R_{INT} 500 Ω	1 kΩ	4.5 V	4.5 V
R_{INT} 5 kΩ	5.5 kΩ	8.2 V	0.82 V

Figure 10–20 This circuit and the accompanying chart illustrate the effect of internal resistance on the voltage available to operate the load.

Ripple voltage should be within the acceptable levels of 1 to 2 percent of the total voltage. The frequency of the ripple waveform will show whether the output is half wave or full wave. This, in turn, will provide information about the ability of all of the diodes to perform as expected in the circuit.

REVIEW

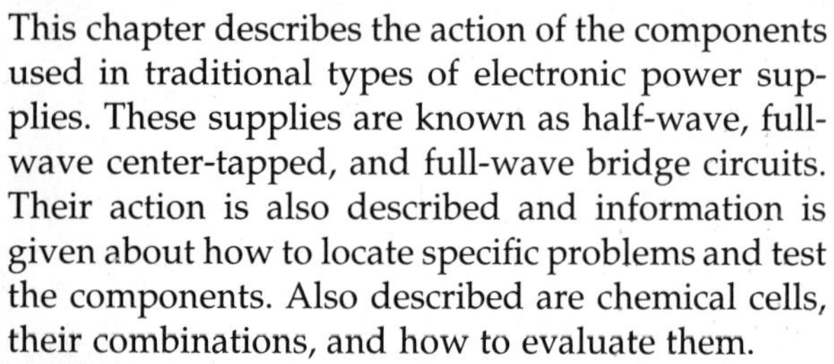

This chapter describes the action of the components used in traditional types of electronic power supplies. These supplies are known as half-wave, full-wave center-tapped, and full-wave bridge circuits. Their action is also described and information is given about how to locate specific problems and test the components. Also described are chemical cells, their combinations, and how to evaluate them.

The reason for suspecting the power supply in the system is that it is the hardest-working section. This would indicate that the section that works the hardest is often the one to fail. In addition to working the hardest, this is the one area in most electronic devices in which the greatest amount of heat is created. This, too, leads to component failure. The power supply should never be overlooked during the diagnostic process.

Chemical cells are often combined into series, parallel, or combination series-parallel batteries. The rules for series circuits: voltage drops add and current flow is limited to a single value; this applies to combinations of chemical cells. The rules for parallel circuits: voltage is constant and current flow through each component adds to equal total current flow; this also applies to those parallel cell combinations. When cells are combined as both series and parallel elements, then the individual rules apply to each section of the combination.

Electronic power supplies are used in most electronic equipment. The components used in these supplies include a transformer, rectifier diodes, and capacitors. The transformer is used to provide the desired levels of voltage and current to the rectifying system. The rectifier diodes permit current flow in only one direction. When they are placed in series with the load and the transformer, the result is a pulsating direct current. The capacitor is used to smooth out the pulsations of the direct current created by diode action.

A regulator, if used, will monitor the levels of voltage and current. The function of the regulator system is to attempt to maintain the design level of these values over a range of load conditions. The distribution system is the method of providing the proper levels of voltage and current to each of the operational sections of the electronic device.

Safety is of prime importance to anyone working around or with electrical energy. Electronic technicians must be aware of the potential danger of electricity. This is particularly true since electricity cannot be seen. Safe working habits must be formed and observed. One safety device that should be used whenever performing tests is the isolation transformer.

Power supply problems can range from a zero output condition to a lower than normal output. Often symptoms such as an audio hum, wavy lines in the video display, or low output from the unit indicate a power supply problem.

The rectifier is a diode in the circuit. This device is polarity sensitive. When forward biased, the rectifier diode has a very low internal resistance. If the diode is connected in series with a load, then the greatest amount of voltage will develop as a voltage, drop across the load. When the diode is reverse biased, it exhibits a very high level of internal resistance. When it is connected in series with a load, the largest voltage drop will develop across the terminals of the diode. There will be little or no voltage drop across the load at this time. The rules identified by Ohm and Kirchhoff apply to diode rectifier circuits.

Capacitors act as sponges in an electronic circuit. In the power supply the capacitor is connected in parallel with the load. When the circuit voltage rises, the capacitor attempts to absorb this rise. The capacitor plates accept the voltage charge. When the circuit voltage drops, the capacitor gives up this charge and attempts to maintain the circuit voltage at its previous level.

The half-wave power supply processes half of the AC wave. The basic circuit consists of an AC voltage source, one diode, one capacitor, and the load. Capacitor action is used to fill in the missing half-wave.

The full-wave center-tapped power supply processes both halves of the input AC wave. This is accomplished by using two diodes. In the two-diode system, transformer secondary winding with a center-tap connection is required. One of the diodes conducts during each half of the input AC waveform. The input AC waveform reverses its polarity continuously. The output of this circuit is connected to one load. All current through the load is in the same direction. Diode conduction creates a pulsating direct current. Capacitor action is used to fill in for the changes in the pulsating DC voltage.

The full-wave bridge power supply does not need the center-tapped transformer secondary. It does, however, require four diodes for it to function. In this system two diodes conduct during each half of the AC input waveform. The output of this system is connected to one load. All current flow through the load is in the same direction. A capacitor is used to fill in for the variations in output voltage levels.

Servicing any power supply is based on the basic rules for troubleshooting a series circuit. The best instrument to use for service is the oscilloscope. It will provide more information with one test than any other test instrument. The displayed waveform will indicate the amplitude, frequency, and shape of the voltages in the power supply. All of this can be done when the output of the supply is tested with the oscilloscope.

The basic power supply circuit may be considered to be linear in form. The rules for series circuits apply to almost all power supplies. The exception to this is the joining, or meeting, of the two diodes in full-wave systems. While these are technically not in a linear form, they may be considered to be because their actions occur at different times.

REVIEW QUESTIONS

1. Why is the power supply section identified as the hardest-working section of any electronic device?
2. Which piece of electronic testing equipment is recommended for testing the power supply section?
3. What information does the test equipment identified in question 2 provide to the servicer?
4. When chemical cells are connected in series, what rules apply to their output voltage and current?
5. When chemical cells are connected in parallel, what rules apply to their output voltage and current?
6. Describe the purpose of the transformer in the electronic power supply.
7. Are transformers tested under load or without a load connected to their secondary windings? Explain your answer.
8. Describe the method of testing a diode.
9. What does the regulator section of the power supply do?
10. What is the function of the isolation transformer?
11. What does a sine wave form along the edge of a video display indicate?
12. What is the condition of the diode when a voltage drop of 48 V is measured across its terminals?
13. Describe the symptoms observed when a capacitor is leaky.
14. What indications are observed on the oscilloscope when one diode in a full-wave center-tapped system is open?
15. What condition exists when one or more diodes in the full-wave bridge system are shorted?
16. Why is it necessary to test the output of the power supply under load conditions?
17. What type of signal processing system is used to analyze the half-wave rectifier system?
18. Describe ripple voltage and its cause.
19. What is meant by the term "internal resistance"?
20. What effect does the value of internal resistance of any power source have on its output?

Chapter 10 Challenge One

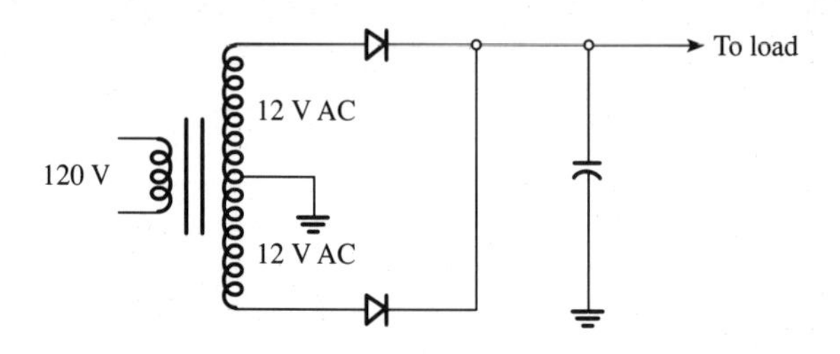

The DC output voltage for this rectifier system should be close to 18 V; however, it measures slightly more than 9 V. Describe a method or plan to locate the problem component.

Chapter 10 Challenge Two

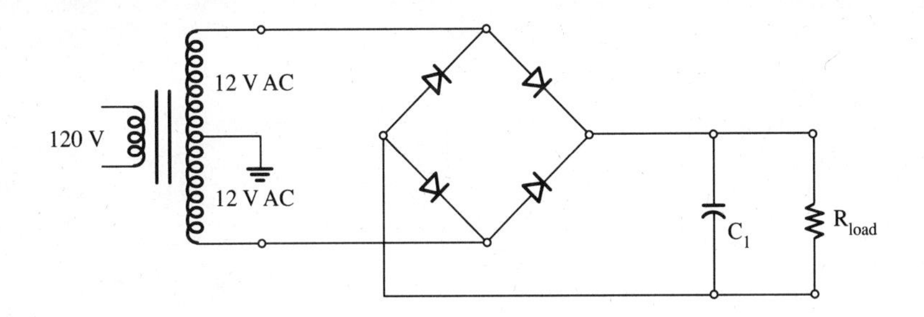

The power supply shown above has zero volt output to the load. Describe a method or plan to locate the specific component that created this fault:

Open transformer primary winding

Chapter 10 Challenge Three

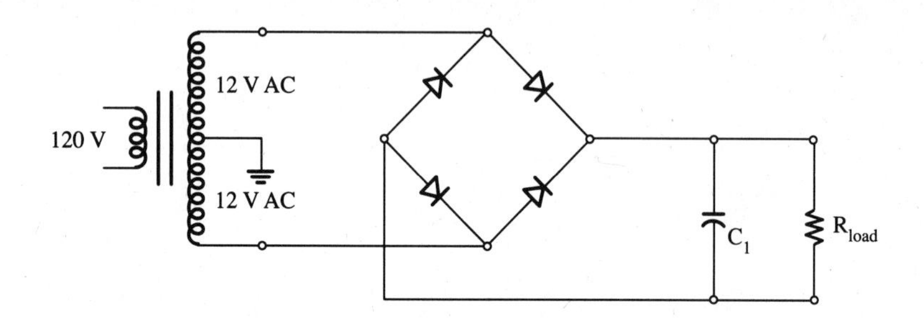

The power supply shown above has a very high level of ripple voltage at its output to the load. Describe a method or plan to locate the specific component that created this fault:

Open filter capacitor C_1

Chapter 10 Challenge Four

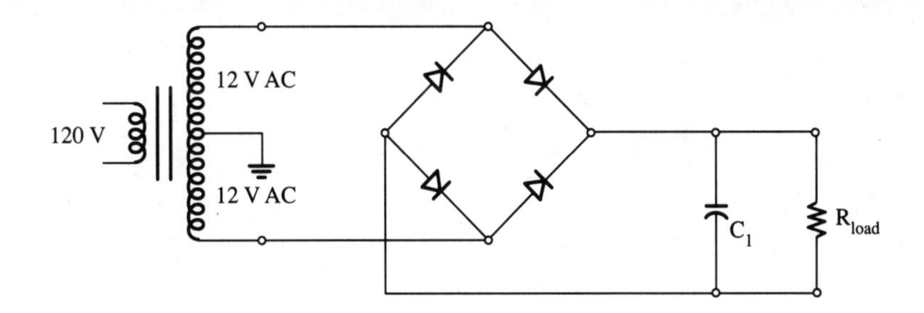

The power supply shown above has a lower than normal voltage output to the load. There is little or no ripple observed with an oscilloscope. Describe a method or plan to locate the specific component that created this fault:

Open one half of the transformer secondary winding

CHAPTER 11

Testing the Power Supply II: Nontraditional Power Supplies

INTRODUCTION

The introduction of electronics and radio communications for aircraft presented some problems for the designers of this equipment. One major concern was the need to reduce the size and weight of the radio equipment. This was necessary so that the designers of the airplanes could increase their ability to carry heavier payloads. In those days, well before the knowledge of solid-state components and their ability to operate from a low-voltage direct current power source, each radio and electronic device required its own power supply. These power supplies were designed with the knowledge of that day and used the 60 Hz power frequency. Little was known about the use of high frequencies for transformer action. Almost all of the power supplies of those early days used a low-voltage DC power source and either a motor-generator to convert the DC into higher voltage values or a device known as a vibrator, or chopper, to break up the DC and enable it to work with a transformer.

The concept of electromagnetic induction is still used in almost all power supplies. This concept is based on the theory that a moving magnetic field will induce, or create, a voltage on a wire in its field. The opposite is also true. When a wire or a conductor moves through a magnetic field, a voltage is induced onto that wire. The frequency of this interaction is one of the factors that determine the value of the created voltage. Another factor is the number of turns of wire on a particular armature winding that interact with the magnetic field.

The early days of electrical generation often used a power line frequency of 25 Hz. This very low frequency required the use of some very large power transformers. Both cost and size demands ultimately required that these devices be reduced. The present standard for power generation on the American continent is 60 Hz.

Designers of aircraft electronic equipment found that they could increase the frequency of the electromagnetic field movement and thus reduce the size, weight, and even the cost of the power transformer components. A frequency of 400 Hz was used to meet these design requirements.

The development of the television receiver created the need for a high voltage power source. The requirements ran from a low value of around 10,000 V to more than 25,000 V. Power transformers and their associated components required for these voltages are large and heavy when 60 Hz power is used. The designers of this equipment found that they could use a frequency of just under 16,000 Hz for the high voltage power supply components. The industry initially standardized on a frequency of 15,750 Hz. This was used for horizontal scanning in addition to the high voltage power supply circuits. Later on, this frequency was adjusted to its present value of 15,734 Hz to better accommodate color television information processing.

Initial use of the high frequency power supply was limited to the development of the required high voltages for cathode-ray tube operation. Later on, the designers found they could also use this frequency and its transformers to create the necessary low voltages for solid-state operation. This is the present method of producing the operating voltages for almost all television receivers, video display terminals, and computer monitors. The process has not changed, but the name now used to describe these circuits is "scan-derived power supply," since the frequencies are related to the horizontal scanning frequency of these devices.

Designers of power supplies have continued to advance their circuits. One of the outcomes of this activity was development of power supplies using even higher operating frequencies. These frequencies operate in the range of 20 kHz to 100 kHz. Power supplies using these high frequencies are called high-frequency switching supplies. Their main advantages are their small size and low weight. Many solid-state devices require a low voltage and high current power source. Often these are in the area of 5.0 V and up to 60 or more amperes. The typical supply requirements to provide this amount of power operating at a frequency of 60 Hz would weigh well over 25 pounds. A switching supply capable of producing this power weighs less than five pounds. The third section of this chapter describes switching supplies and how to troubleshoot them.

The second section of this chapter discusses the concept of voltage regulation. Many of the solid-state electronic devices presently being produced and marketed require a constant value of voltage or current from the power source. This need is addressed with the use of voltage or current regulation circuits. These, too, need to be diagnosed and serviced. Thus, the inclusion of this topic in this chapter.

OBJECTIVES

Upon completion of this chapter, the reader/student should:

1. understand how scan-derived power sources function;
2. understand how switching power sources function;

3. understand how power supply regulators function;
4. be able to apply the basic electronic rules for troubleshooting these power supplies and regulator systems; and
5. be able to recognize typical troubles in nontraditional power supplies.

SCAN-DERIVED POWER SOURCES

High-frequency power supplies are not new to the field of electronics; they have been used for many years in a variety of equipment. These power sources are based on applications of the rules for electromagnetic induction. The quantity of voltage induced on a wire or wires in a magnetic field will increase as the speed of the movement of either the field or the wires is increased. While this may sound rather difficult to understand, the statement is made much simpler if you realize that there is an important relationship between any magnetic field and any wires moving through it. This relationship is used for all high-frequency power supplies. Several statements may be made related to this concept:

1. The quantity of voltage induced on a wire increases as the speed of interaction increases.
2. Speed of motion is directly related to the operating frequency of the system.
3. When a specific voltage is required at a frequency of 60 Hz, fewer turns of wire are required to create this voltage at a higher frequency.
4. The physical size and weight of associated components in the power supply will decrease as the frequency of the generated voltage rises.

A block diagram for a typical scan-derived power source used in either a television receiver or a computer monitor is shown in Figure 11–1. The input to the block diagram is the traditional 120 V, 60 Hz AC system. Once the voltage is applied to the unit, it is processed by one of the more traditional power supply types described in Chapter 10. One of the blocks in these devices is a high-frequency oscillator. The operating frequency of this block is often 15,734 Hz, the same frequency as that used for the horizontal scanning circuit. (This is how the name "scan-derived power source" was established.) The output of this oscillator block feeds into a high-frequency power supply block. This block is often called the "flyback" circuit, since the process of creating the high voltage occurs during the period of time when the cathode-ray tube's beam rapidly returns, or "flies back," from the right side of the tube to the left side.

Included in the high-frequency power source, or flyback system, is a multiwinding transformer. This transformer's windings are designed to create the desired values of voltage and current. Among these are the high voltage required for cathode-ray tube operation and the operating voltages for other sections of the device.

The operating voltages required to operate a television receiver or computer monitor range from values of 5.0 V up to more than 500 V. These are shown as individual blocks on this diagram. The number of individual voltage blocks will depend upon the design characteristics of the unit. The output of

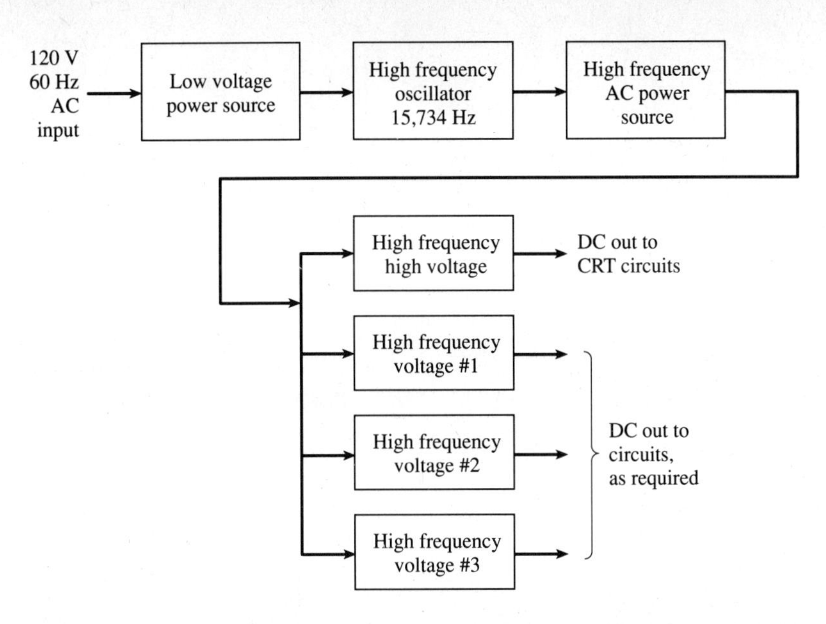

Figure 11–1 Block diagram for a power supply using the horizontal high frequency available from a television receiver.

these blocks is then used as a source for the operating power of the solid-state components in the system.

A partial schematic diagram for a scan-derived system used in a video display terminal is shown in Figure 11–2. The schematic shows its input to be the 15,734 Hz signal from the horizontal oscillator circuit. These waveforms are shown in the form of pulses having an amplitude of about 150 V. This is also shown on the schematic diagram.

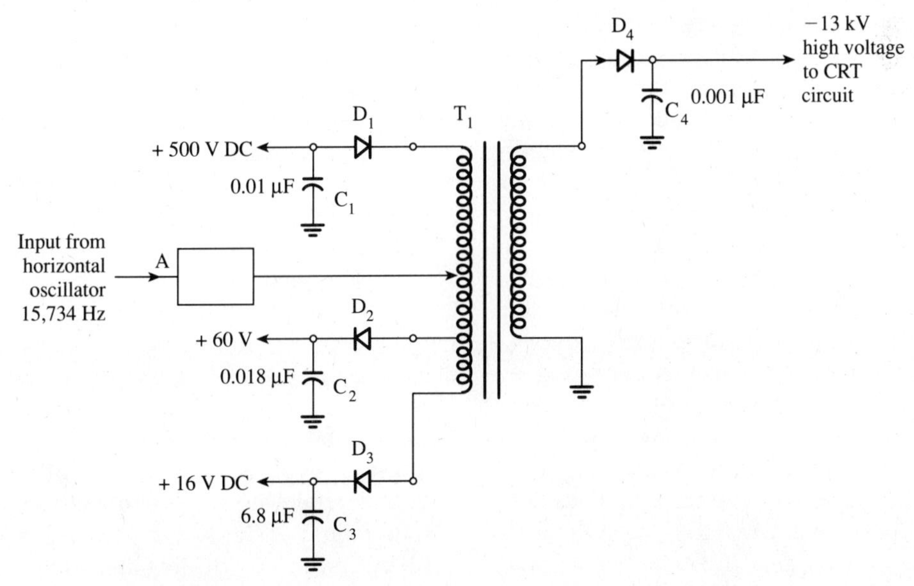

Figure 11–2 Schematic diagram for the scanning circuit-derived power supply.

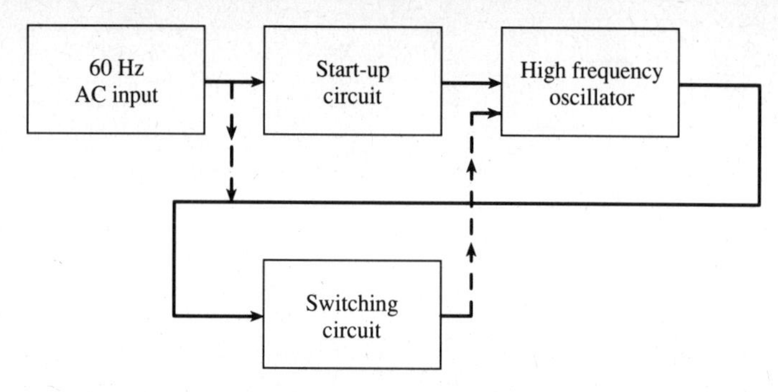

Figure 11–3 **This block diagram illustrates the concept of a start-up system for the scanning circuit power supply.**

There are four diodes acting as rectifiers in this system. Each of the diodes is used for a discrete secondary power source. The four sources are:

1. +16 volts, using D_3 and C_3;
2. +60 volts, using D_2 and C_2;
3. +500 volts, using D_1 and C_1; and
4. −13 kV, using D_4 and C_4.

Each of these sources is actually an individual half-wave rectifier circuit. The diode acts as the rectifier and the capacitor acts as the filter for these circuits. Look at the values of the capacitors used for filtering. Three of the four capacitors have values of less than 1.0 μF. These values are very much lower than the capacitance values used in 60 Hz power supplies. In addition, they are much smaller. The reason for this ability to use smaller values of capacitance is due to the shorter RC time constant necessary for operation at the higher frequency.

Figure 11–3 illustrates a block diagram for the type of circuit configuration that is used once the system is up and running. This circuit uses a start-up circuit to obtain the initial power required to operate the system. Once the system is operating, an electronic switching circuit turns off the start-up system and provides the required levels of power to the balance of the circuits in the system.

The concept of using the higher-frequency scan circuit-derived power supply makes a lot of sense. First of all, the components involved are smaller, less costly, and weigh a lot less than those used for the traditional 60 Hz supplies. Second, all of the individual circuit power is derived from this one source. It places a much smaller power demand on the input circuit and any other power source.

TROUBLESHOOTING THE SCAN-DERIVED POWER SOURCE

The rules established earlier for analysis and troubleshooting also apply to these circuits. The block diagram shown in Figure 11–4 illustrates the type of distribution system used for this power source. It is similar to that described in an earlier chapter as a separating, or splitting, type of system. Because of this similarity, the same rules apply to this system as those that would apply if the system were used for signal processing. The output of the horizontal oscillator section is a signal voltage operating at a frequency of 15,734 Hz.

This signal voltage is injected into the primary of a high-frequency power transformer. There are several output circuits connected to the secondary of this transformer. These are shown on the block diagram along with their respective output voltage levels. Each of the output blocks provides operating power to one or more sections of the monitor.

One of two major failures can occur with a system of this type. These are either a partial failure or a total failure of the system. In the partial failure one or more blocks will be unable to provide the required levels of voltage and current. The first step

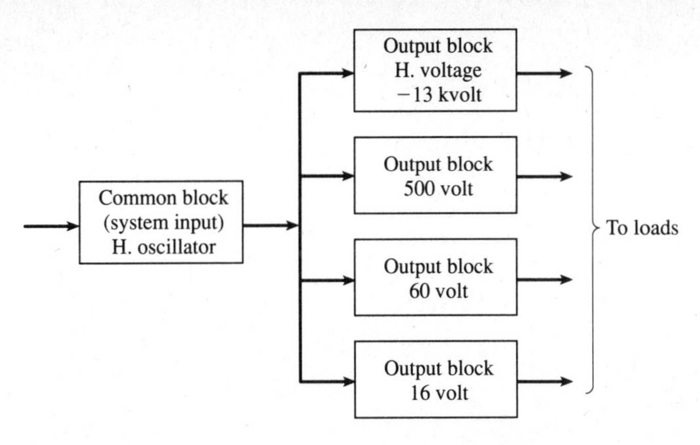

Figure 11–4 **The power distribution system used with the scan-derived power supply has a single source and four output sections.**

is to determine the type of failure. Perhaps the easiest method of making this determination is to use your power of observation. If the whole system seems to be inoperative, then the possibility of an entire power supply failure is very likely. If, on the other hand, only one or two sections of the system seem to be inoperative, then the problem is most likely to be limited to one or more of the individual output blocks in the system.

Start by taking one of the planning sheets described earlier and filling in some of the required information. Once the details of the type of equipment and its problem are completed, the next step is to state the problem that exists. Be aware that the statement of the problem may change as you complete some of the initial diagnostic steps. This system is identified as a splitting or separating type of processor. Indicate the steps involved in troubleshooting the splitting system and establishing a set of brackets around the area suspected to have failed. The brackets should be placed as seen in Figure 11–5.

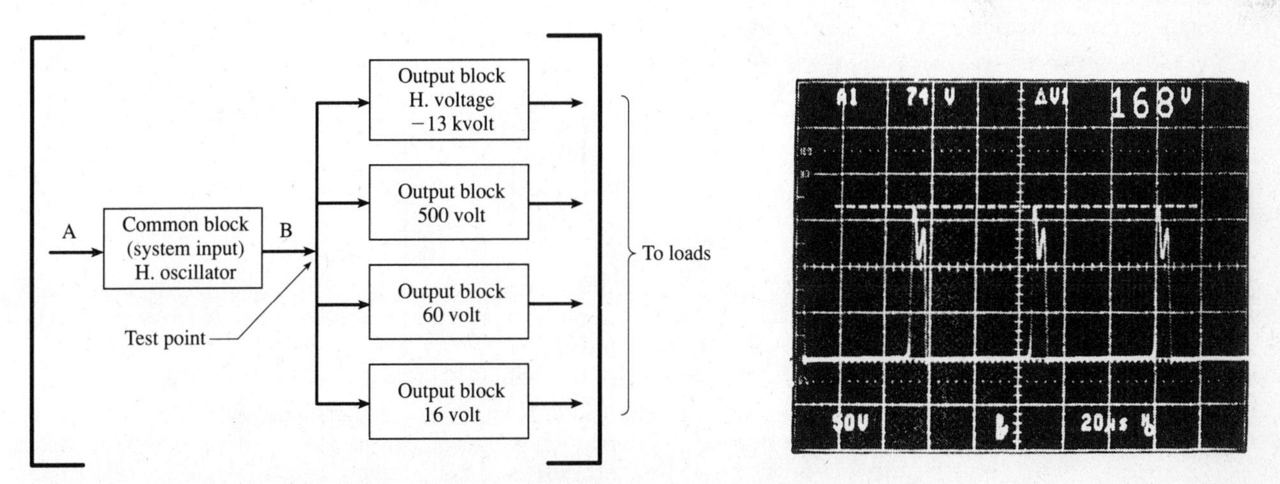

Figure 11–5 **A set of brackets will enclose the entire section of the scan-derived power supply. The initial test point is identified at the center of the block diagram.**

Splitting Signal Path Analysis

1. Determine whether the input point is functional. This requires locating the point to make the test and determining what equipment to use. Also determine what to expect to find if the system is operating properly. In this circuit the input to the system is the horizontal oscillator. The best way to test this oscillator is to evaluate its output, since oscillator blocks actually have no input, but are able to create an output signal. The initial set of brackets enclose the oscillator block and the blocks where the individual levels of operating voltage are created. These brackets are shown on the block diagram of Figure 11–5.
2. Check one of the outputs to this system. This system has only one output—the output of the horizontal oscillator block. The type of test equipment to use is the oscilloscope and the expected waveform is shown in Figure 11–5 at point B. This waveform is what you should expect to see if the input is correct. All of this information can be entered on the planning sheet. At this point the output of the oscillator is either correct or it is incorrect. An incorrect output could be a low amplitude of signal, a wrong frequency of signal, or a total lack of any signal. This test will determine the next step to plan.

   If the signal voltage at point B is correct, then you must consider which, if any, of the output blocks is nonfunctional. If the signal voltage at this point is not correct, then the problem is localized to the horizontal output block. Figure 11–5 has a set of brackets surrounding the entire system. One of these brackets is to be moved as a function of the next step. Let us continue by assuming that none of the output signals is present.

   A lack of correct signal voltage at point B requires moving the right-hand bracket to the point of measurement. This will reduce the area of suspicion to the horizontal oscillator block. The suspected area is identified as being between points A and B on the block diagram. The next step is to review the schematic diagram to diagnose this section of the monitor. There is little need to check other sections of the system if the output of the oscillator is missing or incorrect.

   If the output of the horizontal oscillator is correct, then the left-hand bracket must be moved to point B for analysis of the rest of the system. Point B is the point of separation for this system; this is shown in Figure 11–6. Moving the left-hand bracket to this point will reduce the area of suspicion to the transformer and its output circuits. It is necessary to return to the second step of the procedure to analyze the splitting system for this next step.
3. Check one of the inputs to this system. There are several outputs shown on the diagram in Figure 11–6. Any one of these can be checked at this stage of the diagnosis. If the output of this block is correct, then the problem is most likely to be

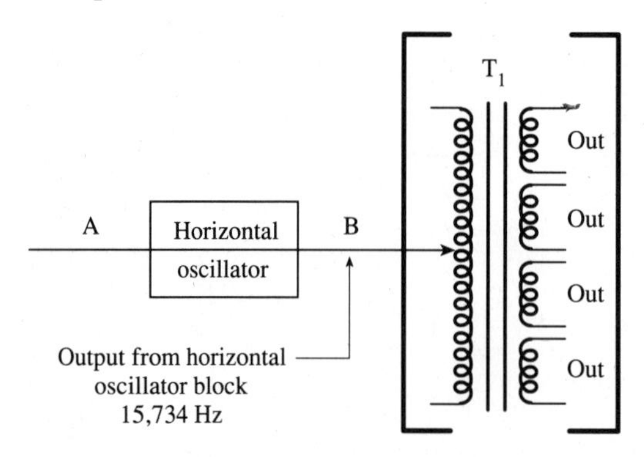

Figure 11–6 **The point of distribution of the voltage is identified at the point marked B in this diagram of the scan-derived power supply.**

found in one of the other output blocks. The planning sheet should show this necessary step.
4. Check the output of the other blocks. This may sound repetitive, but the problem does exist in one or more of the output blocks. This next step requires checking each block's output to determine which of them is inoperative.
5. Move brackets until only one is identified as the one in which the problem exists. Some of these steps may be accomplished by observation. Use of the block diagram and the schematic will help to identify the specific paths for the signals and the creation of the desired operating voltage levels. Information on the schematic may show that one of the output blocks delivers its power to a specific block in the system. If this section of the system is operating, then it is not necessary to test its input or its output. Use of the powers of observation are considered one of the steps for analyzing and locating the specific nonfunctioning section.

The visual diagnosis will answer several questions and aid in locating the initial set of brackets used for troubleshooting. If the entire unit is not functioning, then the premise could be that the complete power source or its own source of power is not performing properly. If, on the other hand, only one section of the system is malfunctioning, then the premise has to be to concentrate on that section of the power source from which it obtains its voltage and current for normal operation.

DIAGNOSING INDIVIDUAL OUTPUT CIRCUITS

Diagnosis of individual output circuits is accomplished in the same manner as diagnosing any other type of power source. Figure 11–7 shows blocks of these four output power sources with their respective circuits. Each of the output blocks has a different level of output voltage. Each of these blocks is essentially a half-wave rectifier system. The method of testing most of these supplies requires an oscilloscope. Any exception to this statement is due to the presence of voltage values beyond the range and capabilities of the oscilloscope's input circuitry. Measurements are made at the output points of each of the blocks. These points are identified as the point where the capacitor connects to the individual rectifier diodes.

The output voltage values are shown. The rectified DC voltage should have a minimum level of ripple voltage. Any one of three conditions will exist when these individual points are evaluated: no volt-

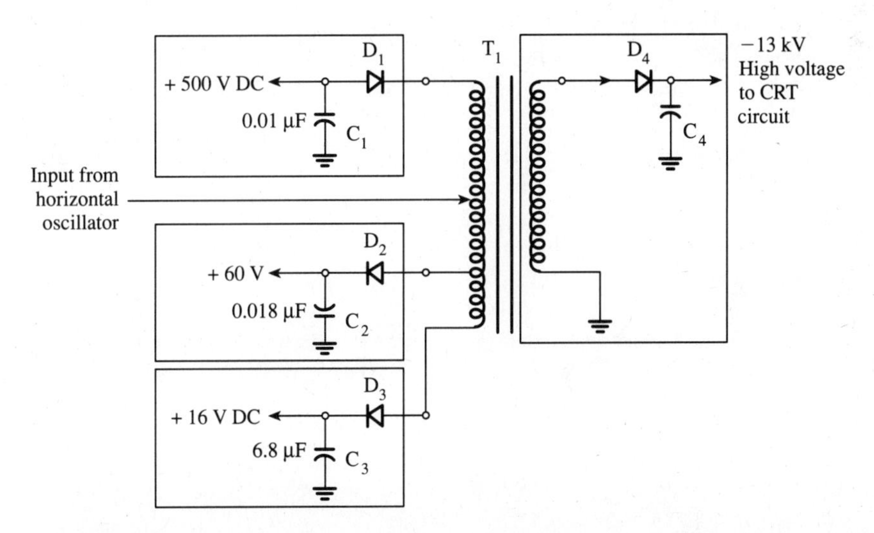

Figure 11–7 **Block diagram of the scan-derived power supply with individual circuit schematic diagrams shown inside each block.**

age, a higher than normal level of ripple, or a complete lack of any filtering action.

Lack of Output Voltage A total lack of output voltage indicates a failure of either the specific winding of the transformer or an open rectifier diode. Since this section of the power supply system is a series circuit, the rule for analysis of the series circuit is used. A test is made at or near the midpoint of the circuit. This test is made with the oscilloscope, since the output of the transformer is a high-frequency signal. If the signal is present, then the diode has failed. If no signal is present, then the winding of the transformer has failed.

Higher than Normal Ripple Voltage Level A higher than normal level of ripple voltage indicates a partial failure of the filter capacitor or an excessive amount of load current in the circuit. Often placing another capacitor in parallel with the suspected capacitor will provide information about the condition of the original capacitor. A reduction of the level of ripple voltage by use of the second capacitor indicates a reduction in the actual value of capacitance in the original capacitor. It must be replaced, which should solve this problem.

Lack of Filtering Action A lack of any filtering action indicates total failure of the filter capacitor. Replacing it should return the system to normal operating condition. An alternate method of testing is to use a meter to measure the levels of DC voltage in the system. If the voltages are incorrect, or if the levels of ripple are high, the service technician would use the oscilloscope. The service technician would use one instrument to make all of the tests instead of using both the meter and the oscilloscope. This is much more efficient and provides the required information about both the output voltage and ripple with one measurement. Remember, efficiency makes you more productive.

Another method of diagnosis and service for the scan-derived power system is substituting another source of signal for the horizontal oscillator. If you suspect that this block is not operating properly, injecting a signal from a second source could easily show that the transformer and its output circuit were operational. All that is required is the use of an alternate signal source capable of reproducing the level, frequency, and form of signal normally obtained from the horizontal oscillator circuit of the system. This substitute signal would be injected at point B on the drawings in Figure 11–7. If the output voltages are present with this alternate signal source, then the area of suspicion would be limited to the horizontal oscillator circuits.

POWER SUPPLY REGULATION

The use of voltage regulation is critical for the highly sophisticated electronic equipment in use today. Many of the components used in digital circuits require a constant voltage level for proper operation. Circuits designed to control a variety of machinery also require a constant, controlled voltage level. Consumer electronic devices often need this same type of critical control of their operating voltages. All of this is accomplished by the use of one additional block in the power supply section of the unit. This block is included in the diagram shown in Figure 11–8. It is outlined with a dashed line in this diagram and consists of two sections. One of these is the voltage regulator; the other is the feedback, or control, section.

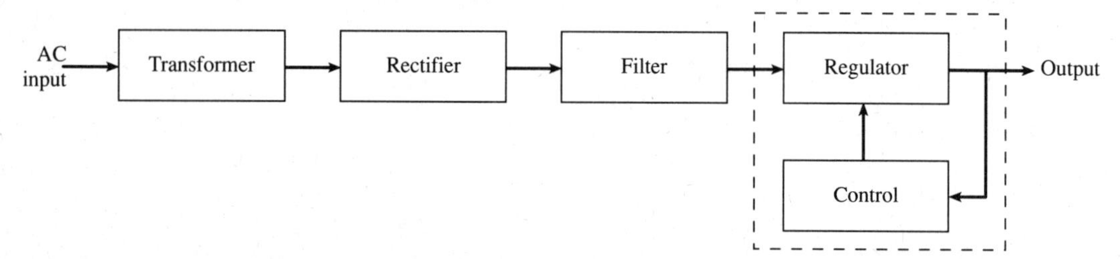

Figure 11–8 The dashed lines encompass the voltage regulation section of this power supply.

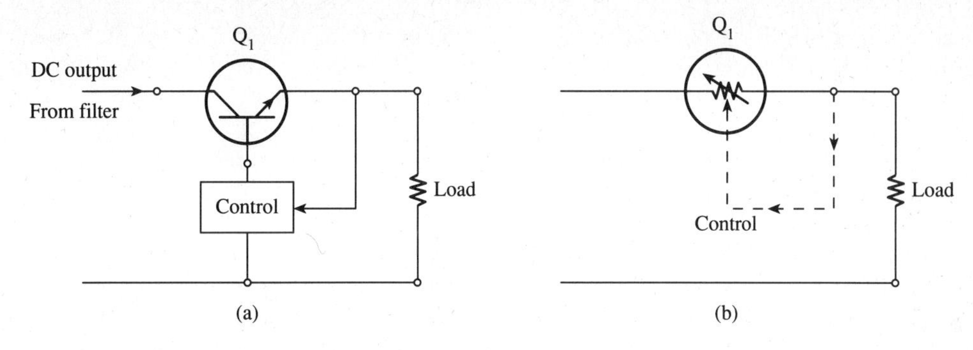

Figure 11–9 Schematic diagram for a voltage regulator circuit using a transistor as the regulator device (a). Part (b) shows the effect of the collector-to-emitter resistance on this circuit.

A very simple voltage regulator is shown in Figure 11–9(a). This circuit consists of a regulator transistor and a controller. The regulator transistor is connected in series between the output of the rectifier and its filter section and the load. The circuit could be considered to be similar to that shown in the (b) portion of the figure. The emitter-to-collector connections of the transistor Q_1 act in a manner that is similar to that of a variable resistor. This concept is explained below.

The transistor's internal resistance can be controlled by the voltage difference between its emitter and base elements. When the emitter element's voltage level is held constant, changes at the base element control its internal resistance and the quantity of current flow through it. If the base voltage is maintained at a fixed level, then voltage changes at the emitter will control the internal resistance of the transistor and its current flow. It does not matter which one of the two elements is kept at a fixed voltage level as long as one is and the other has a varying voltage level applied to it.

Load values seldom remain at a fixed level of resistance. Normally the amount of resistance will vary with the demands for current in the load. The basic rules for current flow bring to mind how this works. If the voltage in a circuit is maintained at a constant level, the current flow depends upon the amount of resistance in the circuit's load. Varying the load resistance will change the amount of current.

In this circuit the voltage drops that develop across the terminals of the transistor and the terminals of the load will add to equal the applied voltage from the DC power source. This is a loop, or series circuit. Keeping this in mind, when the resistance of the transistor changes, the voltage drop across its terminals also changes. These changes are directly proportional to the ratio of the individual resistance values of all of the components in the circuit.

Using Figure 11–10 as a reference, note that the zener diode is used to maintain a constant voltage on the base of the transistor Q_1. Changes in the value of resistance of the load will affect the voltage drop that develops when current flows through the circuit. The reference voltages shown in this figure

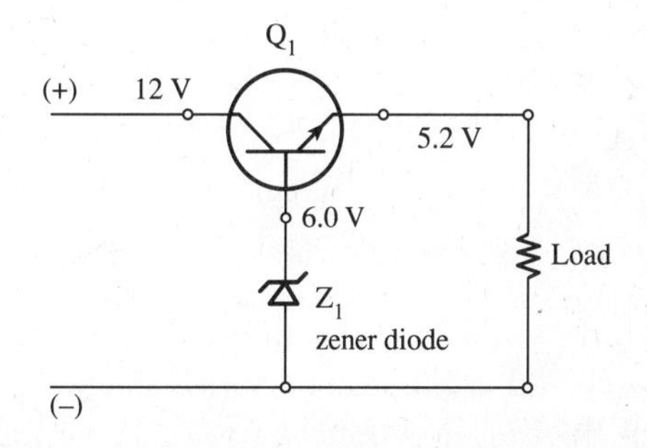

Figure 11–10 A zener voltage regulation diode, Z_1, is used to hold the base of this voltage regulator transistor at a constant voltage level.

have a value of 6.0 V at the base of the transistor and the cathode of the zener diode. The design voltage for the load and the emitter of the transistor is 5.2 V. When the resistance of the load decreases during its operation, the voltage drop across the load will also decrease, according to Kirchhoff's voltage law.

A decrease in the resistance value of the load and its resulting voltage drop will also affect the voltage difference between the emitter and base of the series transistor. The voltage difference will increase. This increase in the level of emitter-to-base voltage will make the transistor's internal resistance decrease by changing the bias on the transistor. A decrease in the original value of internal resistance of the transistor will result in a decrease in the voltage drop that develops across its emitter-to-collector terminals. The voltage drop, or loss, is then picked up by the load resistance. This cycle is a continuous one and results in a fixed value of voltage for the load.

When the opposite conditions exist, and the load resistance value increases, the reverse occurs. An increase in the voltage level of the load will reduce the emitter-to-base voltage of the series transistor and reduce its conduction. This results in an increase in emitter-to-collector voltage for the transistor and a reduction of the voltage across the load.

Some of the more sophisticated voltage regulation circuits will use an integrated circuit in place of the zener diode. The reason for this is that the IC offers a greater range of voltage selection for the regulated output. Many ICs in use as voltage regulator references include a variable resistance. This variable resistance is used to establish the desired level of output voltage for the circuit.

SERVICING THE SERIES VOLTAGE REGULATOR

Service for the series voltage regulator is accomplished in a manner that is similar to any other series circuit. You must first confirm that it is actually a series circuit. Then the rules for voltage drop and current flow in the series circuit can be applied as one component of the analysis. Since the circuit is primarily in a linear form, the half-split rule of analysis can be applied. One of the first places to test is the output of the rectifier and filter system. The correct voltage level must be present if the transistor regulator is to work properly.

If the correct voltages are found at the output of the rectifier and filter, then one of the brackets that was originally placed around the entire rectifier and regulator system has to be relocated. This is the left-hand bracket and it would be moved to the point of the successful test. The balance of the area inside of the brackets at this time includes the regulator transistor and the load. The next test is at the center of these components, or the emitter of the regulator transistor.

The test to be made at this point is measurement of the voltage at the emitter of the transistor, which will provide the information required to isolate the defective component. It will be either the transistor or the load and most likely will be the series voltage regulating transistor.

THE SWITCHING VOLTAGE REGULATOR

The one major drawback to a circuit using a series transistor for a regulator is the amount of heat that is generated during operation of the transistor. This heat is one part of the operating power of the circuit. The heat does not perform any practical amount of work and, as a result, is a waste of energy for the circuit.

One way to limit development of this wasted power is the use of circuits that will reduce the amount of time the regulator is on. When the duty cycle is reduced to 50 percent of the total time, the amount of heat that is generated (and wasted) is reduced by one half. This process is accomplished by using a switching regulator circuit.

The block diagram shown in Figure 11–11 is similar to that used earlier. The major difference is the development of the square wave to control the function of the series regulator transistor. The blocks are the same as those in Figure 11–8. The difference is the inclusion of the square wave output from the control block. One circuit using this system is shown in Figure 11–12. It uses the voltage divider network of resistors R_1 and R_2 to establish a bias, or reference

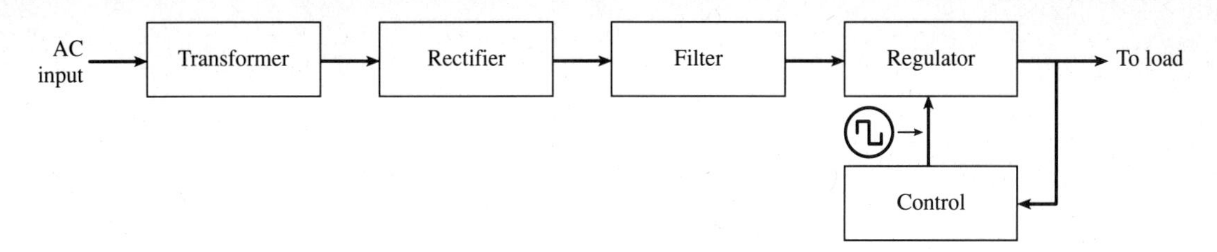

Figure 11–11 **This block diagram uses a square wave output from the control block to maintain constant load voltage and current conditions.**

point, for the switching regulator control block. The output of this block is in the form of a square wave. The width of the individual pulses can be adjusted by this control block. Their widths are used to control the amount of time the series transistor Q_1 is conducting. All of this is controlled by the feedback from the voltage developed across the load, since the load is essentially in parallel with the voltage divider network.

SERVICING THE SWITCHING REGULATOR

Servicing the switching regulator is done in a manner that is similar to the one described in the previous section. A voltmeter will provide information about the level of voltage present across the terminals of the load. An oscilloscope is required to validate the output of the switching-regulator control block. This is measured at the base of the series regulating transistor, Q_1.

Since this is essentially a series circuit with a feedback system, you must look at both. The rule for examining a feedback system requires checking the forward path first. In this circuit, the forward path consists of the transformer, rectifier, filter, and series regulator. Using the premise that the hardest-working components fail more often than any others in the circuit, you should first examine the rectifier and series transistor circuit. The block diagram in Figure 11–13 has a set of brackets around the entire system. These are where the initial bracket set should be placed.

After insuring that the AC input voltage is present, the first test is at the midpoint of this circuit. This is indicated as Test Point #1 on the drawing. The test should be made using an oscilloscope because this is the best equipment to measure the output from the full-wave bridge rectifier. A failure of any one of the diodes in the bridge circuit is best seen using the oscilloscope.

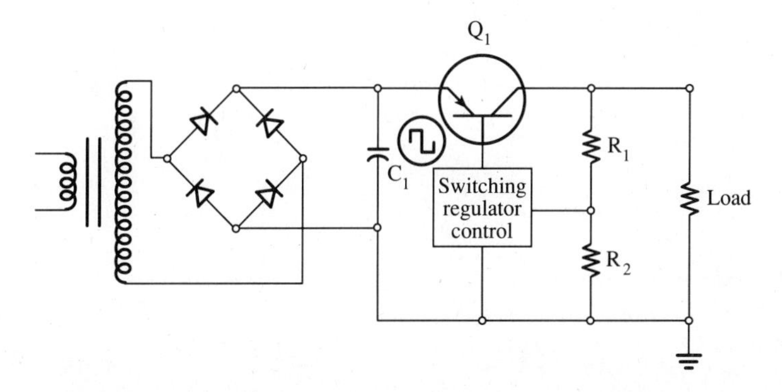

Figure 11–12 **Schematic diagram for a switching power supply regulation system.**

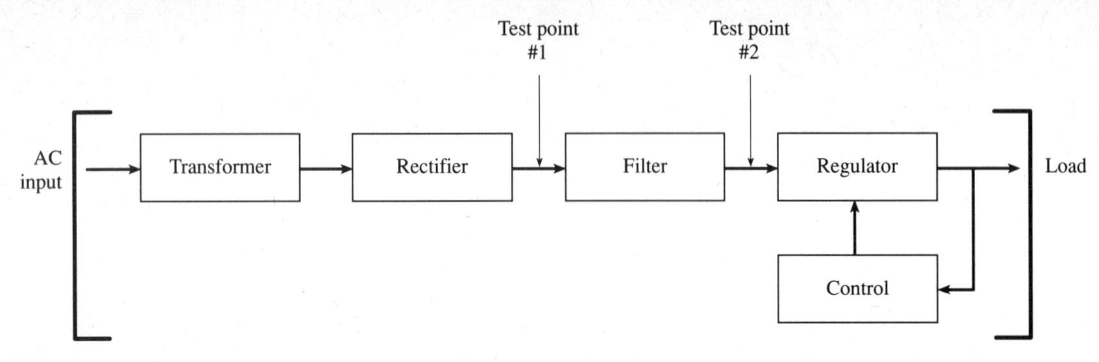

Figure 11–13 **Initial bracketing of the power supply extends from its input to the output of the regulator circuit.**

In a practical sense, the action of the filter circuit is included as a part of this measurement. It is impossible to separate filter action from the output of the rectifiers without actually cutting the leads or circuit board foil between the two blocks. Therefore, the measured waveform at this point should be a DC voltage with a very small amount of ripple voltage present. One test at the midpoint will actually evaluate the whole system except the feedback and regulator portions.

If the observed waveforms are correct at this test point, the left-hand bracket is moved to the test point. The second test would normally be made at the midpoint of the remaining circuit. This is identified as Test Point #2 on the drawing. A test at this point is redundant, since the results of the first test will include filter action. In reality, the left-hand bracket will be moved to this second test point after the first test is made *if* the results of the test indicate that the filter system is operating properly.

The balance of the system still between the brackets is shown in Figure 11–14. This system is identified as a feedback type of system. The rule for a feedback system requires testing the forward path first. If the forward path is operating properly, then the problem is located in the feedback portion. If the forward path is not operating correctly, then it must be examined first.

The process of testing the forward path of any feedback system is to inject a known value of signal or voltage at the output of the control, or feedback, path. In this example, this would be a square wave pulse and it would be injected at the base of the series regulator transistor. If this corrects the problem in the circuit, then the brackets should be moved to include just the feedback system. If this

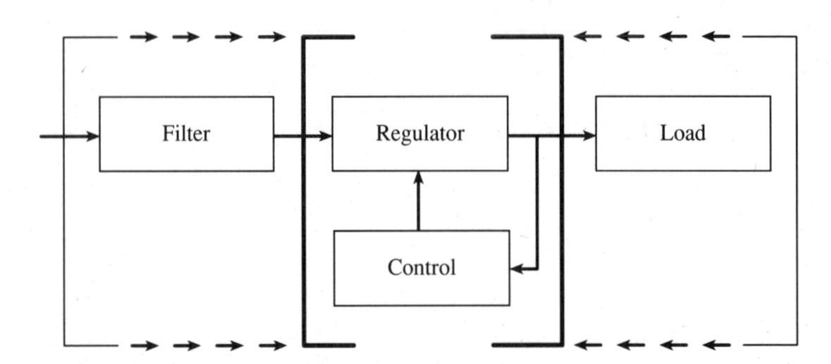

Figure 11–14 **Brackets are continuously moved until the problem is located in the regulator section, as shown here.**

does not correct the problem, then the brackets should be moved to surround the forward path, or the transistor circuit section.

HIGH-FREQUENCY POWER SUPPLIES

Once the concept of using the horizontal scanning frequency of the television receiver and computer monitor was found to be practical, using high frequencies for power supplies was expanded. The designers of electronic circuits explored frequencies much higher than the 15,734 Hz used for these devices. Their efforts produced power supply systems operating at much higher frequencies. Some of these supplies use frequencies ranging between 25 kHz and 100 kHz. The advantages of these higher frequencies are smaller components and lower weight. The combination of these two factors created complete assemblies of power supplies that are much smaller and more efficient than those operating at frequencies of 60 Hz.

The block diagrams for these high-frequency power sources are very similar to those used for scan-derived systems. A system of this type is shown in Figure 11–15. The block titled "High Frequency Switch" makes this supply different from those described previously. The high-frequency switch is actually an oscillator circuit. Its operating frequency ranges from 25 kHz to 100 kHz. The blocks to the left of the switch develop the normal DC operating power for this oscillator. The output of the oscillator is connected to a high-frequency transformer. The output of the transformer is rectified and filtered. This results in the required values of DC voltage and current necessary to operate the load. The control block is one part of the voltage-regulating portion of this system.

This type of system is not utopian. For every advantage, it seems to have some sort of disadvantage. The major disadvantage of high-frequency power sources is the creation of radiated signals. These signals can radiate from the circuits used to develop the high frequencies. Power sources using this technology must be adequately shielded and filtered to eliminate these unwanted signals. If this is not done, the unwanted signals can, and will, create interference for other electronic devices. This interference may cause the other devices to produce unwanted outputs or may even make them inoperable.

SERVICING THE HIGH-FREQUENCY POWER SOURCE

Servicing the high-frequency power source is accomplished in the same manner as for other power sources. The system block diagram is examined. The system is primarily a linear type, with one section of feedback. The primary portion is examined first using the rules for linear system analysis. Once the input voltage is validated, the next test is made at or near the midpoint of the system. In this system the midpoint would be at the filter block. An oscilloscope reading will determine if all of the blocks to the left of this point are operating correctly. A display showing the proper amplitude of voltage and a small quantity of ripple will indicate that all is correct up to this test point.

The next test is made at the output of the high-frequency switching circuit. Again, the oscilloscope is used to provide the necessary waveform information. If this section is working properly, then the feedback system is analyzed using the process described in the previous section.

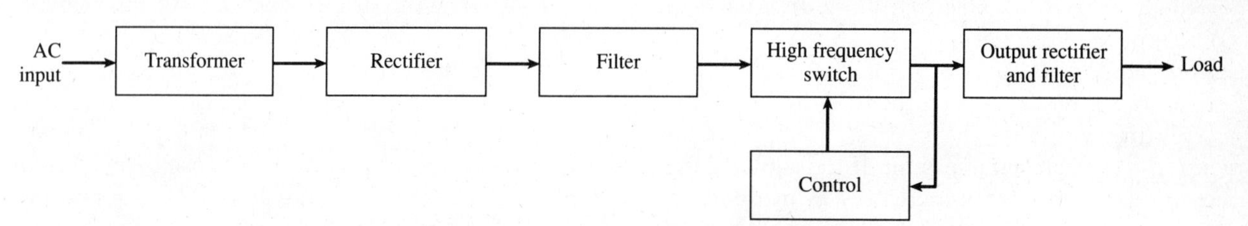

Figure 11–15 Block diagram for a high-frequency power supply.

REVIEW

Use of the high-frequency power supply has reduced the size and weight of this part of the electronic device. The concept of a magnetic field moving past a conductor will create a voltage on that conductor. An increase in the speed of movement of the magnetic field will allow a reduction in the size of the components used for this purpose. The high-frequency power supply used in modern electronic devices operates at frequencies ranging from 20 kHz to 100 kHz.

The concept of using the scanning frequency of the TV receiver and the computer monitor is not new. Manufacturers of both TV sets and monitors have used the 15,734 Hz scanning frequency to develop the high voltages required to operate cathode-ray tubes. It was fairly simple to add some additional windings onto the high-voltage flyback transformer. These extra windings are used to develop the required low voltages for operating additional portions of the TV or monitor.

Operation of the high-frequency power supply requires a low-frequency power source to start the system. Often the system consists of the typical power supply components and an oscillator circuit. Once the oscillator circuit is operational, a switching system turns off the start-up section. The system then uses the high-frequency signal from the oscillator as its power supply source.

Troubleshooting the scanning circuit power source requires the use of an oscilloscope. You must look for the proper amplitude, frequency, and shape of the signals. If the output of the oscillator is correct, the next place to evaluate is the output of the high-frequency transformer. This transformer often has several low-voltage windings. Each of these windings is used as a source for individual low-voltage power supplies.

The method of diagnosis for the scan-derived power source requires using the rules for diverging, or splitting, systems. The first test is made at the point of the separation, or split. If the signal is correct at this point, then the next test is made at one of the outputs. If the signal is not correct at the splitting point, then the problem is between the input and the point of separation. In either case, the rules are clear and easy to follow.

Most of the output power supplies found in scan-derived power supplies are half-wave rectifiers. These are serviced in the same manner as any other half-wave power supply. They are essentially a series circuit, or closed loop. Using the half-split rule for troubleshooting, the method of diagnosis for the series circuit is to make the first measurement at or near the center of the circuit. Often the trouble is in one of the rectifier diodes. The only other component in this half-wave system is the filter capacitor. A measurement with the oscilloscope will provide information about each of these components.

Voltage regulation is an important part of the power supply. Almost all voltage regulators sample the output voltage and, by using a feedback network, adjust the voltage level from the regulator system. One of the basic regulator circuits has a transistor connected in series between the DC power source and the load. The base of the transistor is connected to a reference voltage source. Changes in voltage at the emitter control the internal resistance between the collector and emitter. These changes adjust the output voltage of the transistor and maintain a constant voltage for the load.

More sophisticated voltage regulator circuits use a switching circuit to control the base voltage of the series transistor. The use of the switching circuit offers less power for the regulator circuit since it is on for only a portion of the total time.

High-frequency power sources were developed as they were from the scan-derived circuits. The reasoning is very simple: An increase in the frequency of operation will permit even smaller components and permit some very high power levels to be created. Some high-frequency power supplies deliver well over 50 A of current at voltages close to 5.0 V. One additional block is found in this type of power source—a high-frequency power oscillator. Use of this section is required if the desired levels of output power are to be produced.

REVIEW QUESTIONS

1. What is the basic electrical concept used to generate a voltage on a wire?
2. Identify the various frequencies used in power supplies.
3. Why was the term "scan-derived power supply" used?
4. What are the advantages of using the scan-derived power source?
5. What type of power supply circuit is usually used for the output of the scan-derived power source?
6. What is meant by a start-up section for a power supply?
7. What is the difference in value between capacitors used in 60 Hz power supplies and those used in high-frequency power supplies?
8. What type of signal processing system is used for the scan-derived power supply?
9. Why can you use your powers of observation when localizing the problem area for systems using scan-derived power supplies?
10. Explain how the bracket system is used in a high-frequency power supply.
11. How do you identify total failure of a filter capacitor?
12. Explain why the oscilloscope is preferable for evaluating the power supply.
13. Explain how the series voltage regulator's transistor is controlled and is able to maintain a constant output voltage.
14. What is the advantage of using a switching voltage regulator instead of one where the base voltage of the transistor is locked by a zener diode?
15. What advantage is there to using the high-frequency power supply?
16. What is the disadvantage of using the high-frequency power supply?
17. Explain why a transformer requires an alternating current voltage instead of a direct current voltage.
18. What troubles can typically be found in a power supply operating at a frequency higher than 60 Hz?
19. A high-frequency power supply's output voltage is zero. List the steps you would take to locate its problem area.
20. The planning sheet used for establishing a procedure for analysis includes a section for feedback. How is this section used?

Chapter 11 Challenge One

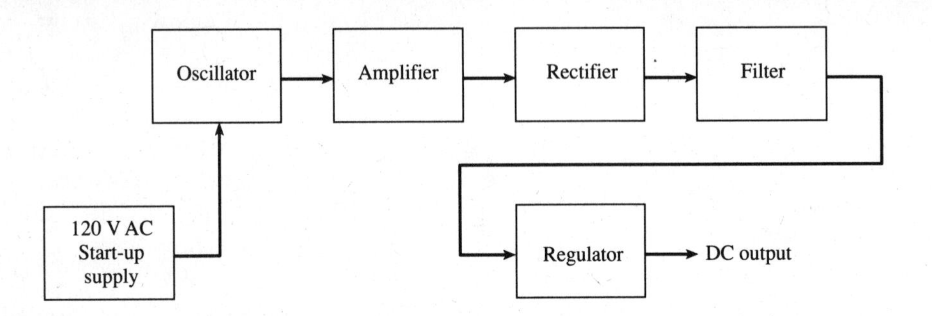

Problem: No output from this high-frequency power supply to the load.

Develop a plan to analyze and locate the specific problem area. Use this process:

Identify the test area or point.
Identify the type of testing equipment to use.
State the expected result if the measurement shows the circuit to be working properly at that point.
Continue to develop the plan, using these statements:
If OK, then ____________________

If not OK, then ____________________

Chapter 11 Challenge Two

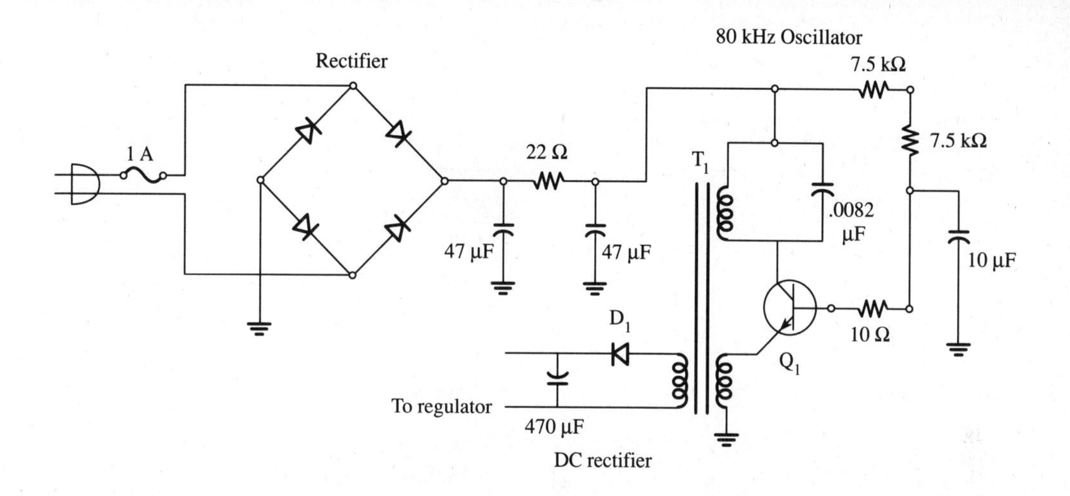

Problem: No output from terminal on this high-frequency power supply to the load.

Develop a plan to analyze and locate the specific problem area. Use this process:

Identify the test area or point.
Identify the type of testing equipment to use.
State the expected result if the measurement shows the circuit to be working properly at that point.
Continue to develop the plan, using these statements:
If OK, then ______________________________

If not OK, then ______________________________

Chapter 11 Challenge Three

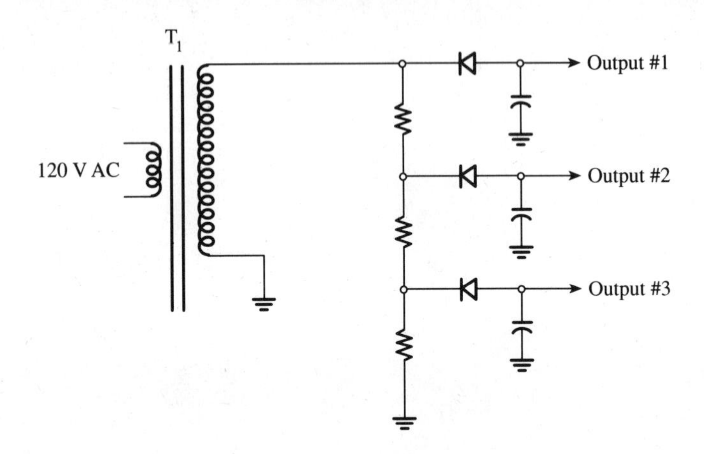

Problem: No DC at any output terminal of this power supply to the loads.

Develop a plan to analyze and locate the specific problem area. Use this process:

Identify the test area or point.
Identify the type of testing equipment to use.
State the expected result if the measurement shows the circuit to be working properly at that point.
Continue to develop the plan, using these statements:
If OK, then ________________

If not OK, then ________________

Chapter 11 Challenge Four

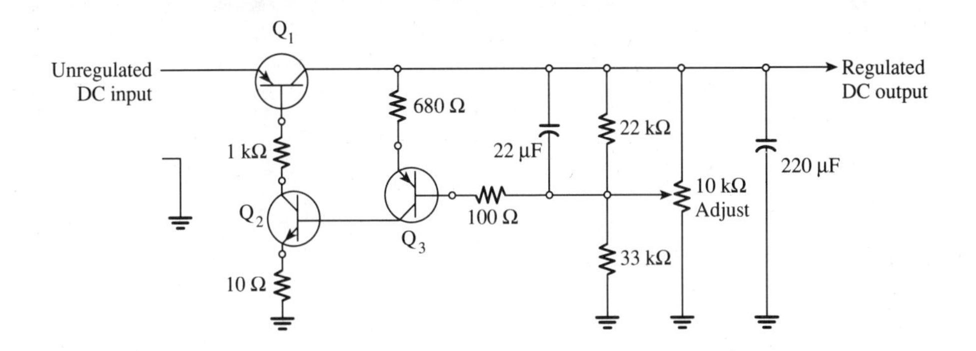

Problem: No output from this power supply regulator circuit to the load. The output at the filter capacitor is correct.

Develop a plan to analyze and locate the specific problem area. Use this process:

Identify the test area or point.
Identify the type of testing equipment to use.
State the expected result if the measurement shows the circuit to be working properly at that point.
Continue to develop the plan, using these statements:

If OK, then ______________________

If not OK, then ______________________

Chapter 11 Challenge Five

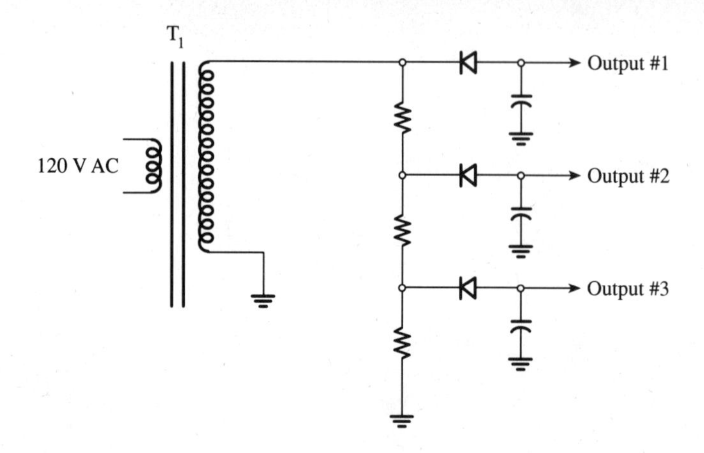

Problem: No DC voltage present at output #2 of this power supply to its load.

Develop a plan to analyze and test the specific problem area. Use this process:

Identify the test area or point.
Identify the type of testing equipment to use.
State the expected result if the measurement shows the circuit to be working properly at that point.
Continue to develop the plan, using these statements:
If OK, then ____________________

If not OK, then ____________________

CHAPTER 12

Repairs and Replacements

INTRODUCTION

Once the specific electronic problem is located, the next question the service technician must answer is, "Where are the replacement parts?" There are several answers to this question. The exact answer depends upon factors such as the size of the repair company, the available stock of replacement components, and the physical distance between the customer and the repair center offices.

Persons responsible for servicing in the field, or away from the repair center, are often identified as field service engineers. Technically, the title "engineer" is not correct, for they do not actually design or engineer the installation. The field service engineer's primary function is to act as an electronic service technician and to repair equipment not housed on the premises of the parent company. This equipment may be owned by the parent company and leased or rented to another company, or it may be owned entirely by the other company. In either situation, the equipment must be repaired, and the person who probably has the best training and knowledge of this equipment is one trained by the parent company.

The material presented in this chapter will assist the student/reader in identifying sources of replacement components. In addition, purchasing concepts will be discussed. Finally, the material will aid the student/reader in developing techniques for the proper removal and installation of replacement components.

OBJECTIVES

Upon completion of this chapter, the student/reader should be able to:

1. identify sources of replacement components;
2. understand the value of quantity purchasing;
3. understand how to remove defective components;
4. recognize the need for effective soldering; and
5. understand how to properly install replacement components in existing equipment.

SELECTING REPLACEMENT COMPONENTS

The electronic service technician's role often includes ordering and specifying replacement components as one part of the repair process. This responsibility may be limited to ordering a replacement circuit board and installing it, which is often available from the manufacturer of the equipment. This process may be the simplest way of obtaining the desired board or module. This, of course, is based on the premise that the manufacturer has the board available and can supply it immediately. It also assumes that the service technician has valid access to the manufacturer's replacement parts department.

The previous sentence may suggest that replacement components and boards are not always available. Unfortunately, this may be true. Several years ago, one of the major manufacturers of television receivers was unable to provide sufficient quantities of replacement boards for its receivers. The result was that, when the board was ordered, this manufacturer sent a kit of resistors, capacitors, and transistors to the service company. The service technician had to remove the old components and install these new parts, since adequate quantities of replacement boards could not be supplied.

In another situation, the input transistors in an oscilloscope owned by a service company were accidentally destroyed. Replacement parts were available only from the manufacturer of the oscilloscope. This oscilloscope was marketed by a major company that it had identified a limited number of parts organizations from which to purchase replacement parts. Other, nonauthorized companies could not buy from these sources. The transistors were unique to the oscilloscope and were not available from any other source. The manufacturer had used a distinct set of part numbers for these transistors, which could not be cross-referenced to any other source. The service company's attempt to locate them elsewhere was unsuccessful. Because the oscilloscope could not be used during this period, the service company did not have it available to process other items for repair.

This incident did have a happy ending. A call was made to the manufacturer explaining the problem. The manufacturer made an exception to its rule of not selling parts to unauthorized accounts and did ship the required parts. This entire process took time away from the repair company's ability to service customer equipment. While test equipment is not destroyed deliberately, it can and does happen. The ability to obtain replacement parts for repair is critical. When these parts are only available from the manufacturer, and there is limited access to its parts department, the time and cost of obtaining them can be expensive to the service organization.

Sources of replacement parts are often limited to

one of three kinds: the original factory or company, independent companies that produce universal or exact replacement components, or independent companies that remanufacture the original boards or modules. These sources are discussed below.

ORIGINAL FACTORY SOURCES

Original factory sources are often the place to obtain the exact replacement components required for servicing. When you are employed by the same company, these parts are usually readily available. Factory service representatives, or field service engineers, have access to the parts departments of their companies. Independent servicers also often have access to these companies. The addresses of these companies are listed in major publications. One such source is the National Electronic Sales and Service Dealers Association and the International Society of Certified Electronic Technicians, of Fort Worth, Texas. These combined organizations publish an annual *Professional Electronics Yearbook and Directory*. This directory contains the names of major companies involved in the manufacture and distribution of consumer products.

Another source of names and addresses of companies offering tools, diagnostic products, repair services, and consulting services as well as other products used in the electronic service industry is *Service News* magazine. *Service News* publishes its *Annual Buyers Guide* each January. Published by United Publications, Inc., of Yarmouth, Maine, the guide contains a wealth of information that is invaluable to the servicer. Additional information is often available in any of the several magazines dedicated to one or more facets of electronic installation and servicing. Several of these sources are listed in the Appendix.

UNIVERSAL REPLACEMENT COMPONENTS

Universal replacement components are available from a large variety of sources. Often a local, independent electronic parts distributor will provide all of the components required for a repair. Many of these independent companies have access to manufacturers who provide exact replacement components. In some cases, these same manufacturers also made the identical components for the manufacturers of the original set or device. Names and addresses of these independent electronic parts distributors are usually found in the local telephone directory.

Other independent sources of components often advertise in magazines devoted to a specific aspect of electronic installation, sales and service. One of the better ways of locating these magazines is to review directories of publications in a local library. It is also possible to write to a specific industry's trade association to obtain information about sources of components. These, too, are often listed in the directories at a local library.

One more source of information may be the annual Consumer Electronics Show held in Chicago, Illinois, each June and in Las Vegas, Nevada, each January. Qualified electronic personnel may obtain badges for entry into the show from its sponsors, the Electronic Industries Association in Washington, D.C. Many other national and regional trade shows are also open to electronic service personnel.

Be aware that not all parts marketed as "universal" are truly universal in their electrical specifications. Experience has shown that some of the solid-state components marketed by a variety of sources as direct replacements do not actually have the exact same specifications as the component being replaced. These sources provide manuals showing what they recommend as substitutions for the original part number. The method of determining that the replacement part will work properly is often done by computer analysis.

A good many of these substitutions will work properly as replacements for the original parts. In some circuits these "exact replacement" parts do not work in exactly the same manner as the original component. Be alert to this situation and use universal replacement components with some caution.

Tolerances of Replacement Components The ability to provide exact value replacement components is not always possible. The electronic service technician needs to know what the acceptable tolerance parameters are for a variety of components. Unfortunately, these vary from a very low tolerance of ±0.5 percent to well over 50 percent. The experienced service technician understands this range and is able to select the correct component tolerance for the specific application.

The service literature of almost all manufacturers includes a list of the component parts used in the device. This list often includes the values of the components as well as the part numbers used by that manufacturer. A sample of one such parts list is shown in Figure 12–1. This parts list uses a reference number for the specific component. While each manufacturer often uses a proprietary set of standards to identify components, the letters R, C, and L have common uses. This parts list uses terms such as "C205" to identify a specific capacitor. The second column provides the manufacturer's own part number assignment. The third column identifies the type of component, its specific value, any tolerance parameters, and a maximum operating voltage level for the original part.

Using this same chart, the tolerance values for capacitors C120, C151, and C216 are interesting to note. These capacitors are specified to have a capacitance value of 0.01 μF and a tolerance range from a value of +100 percent to a value of -0 percent. In other words, the value of the replacement capacitor can fall between an upper level of 0.02 μF and the exact original value of 0.01 μF and still be acceptable for correct circuit operation.

When selecting a replacement capacitor to use in the circuit, any value within the range stipulated by the tolerance rating is acceptable. The service technician must also recognize that, under emergency conditions, the value may be created by connecting two or more components in series or parallel, as required, to make up the specified value. One other item to remember is that, when the tolerance parameters of the replacement components are better than the original, there is little or no reason not to use the better-rated unit. After all, the tolerance factor is used to identify the maximum extremes for the value of the replacement unit. In addition, the capacitors used as the example here have a voltage rating of 50 V. This means that the *minimum* voltage rating for a replacement capacitor is 50 V. When space permits, a capacitor with a higher voltage rating could certainly be used.

These rules are valid for almost all components used in electronic devices. Tolerance values, as previously stated, are maximum deviations permitted for proper circuit operation. Any component rated with a ±1 percent tolerance could be used to replace one rated at ±5 percent *as long as its value and operating voltage ratings are within the limits of the original unit.*

During those times when the vacuum tube was the only major active conductive device used in radios, the two filter capacitors used in the power supply were rated with values of 30 μF and 50 μF and an operating voltage of 150 V. The tolerance ratings for these capacitors were +100 percent to −50 percent! In other words, any capacitor with a rating of between 100 μF and 25 μF could replace the original 50 μF value and still be acceptable. This is true for almost any replacement. The major factors to observe are whether the replacement unit will perform adequately and whether its operating voltage rating is high enough to keep it from failing due to voltage breakdown.

These factors are also valid for resistors and inductors, as well as for semiconductors. One very common semiconductor is the silicon diode used in power supplies. These diodes are provided with a current rating and an operating voltage rating. A typical current rating for single power supply diodes is 1 A. The voltage rating can vary from a low of 50 V to a high of 1000 V. When replacing these diodes, it is possible to use the 1000 V rated unit for any one with an original voltage rating of less than this value. It is often possible to use a diode with a higher current rating, since the actual current flow is limited to the demands of the load.

PURCHASING POWER

The concept of using one value of diode to replace an entire group of diodes whose original ratings were less than that of the replacement unit brings up another interesting concept. This is the idea of purchasing quantities of commonly used components in bulk instead of purchasing them individ-

C101,117, 166	ECCD1H151K	Ceramic 150 pF ±10% 50V
C102	ECEA25V4R7	Electrolytic 4.7μF 25V
C106	ECSZ16EF10Y	Electrolytic 10μF 16V
C107,109, 113,115, 120,122, 126,130, 132,151, 154,167, 213,216, 902	ECKD1H103PF	Ceramic 0.01μF +100% −0% 50V
C108,118	ECCD1H180K	Ceramic 18 pF ±10% 50V
C110	ECCD1H560K	Ceramic 56 pF ±10% 50V
C111	ECKD1H103MD	Ceramic 0.01μF ±20% 50V
C114,121, 123,912	ECCD1H220K	Ceramic 22 pF ±10% 50V
C116	ECV1ZW50P32	Trimmer 2~50 pF 100V
C125, 153	ECEA16V10L	Electrolytic 10μF 16V
C127,202	ECCD1H100D	Ceramic 10 pF ±0.5pF 50V
C129,204, 211	ECCD1H820KP	Ceramic 82 pF ±10% 50V
C131	ECEA25V3R3L	Electrolytic 3.3μF 25V
C155	ECEA25V100L	Electrolytic 100μF 25V
C156	ECEA16V47L	Electrolytic 47μF 16V
C157	ECCD1H181K	Ceramic 180 pF ±10% 50V
C158,910	ECKD1H102MB	Ceramic 0.001μF ±20% 50V
C159	ECAG25ER47F	Electrolytic 0.47μF 25V
C160	ECKD2H101KA	Ceramic 100 pF ±10% 500V
C162	ECQM05103MZ	Polyester 0.01μF ±20% 50V
C163	ECEA10V33L	Electrolytic 33μF 10V
C164	ECCD1H221K	Ceramic 220 pF ±10% 50V
C165	ECCD1H680K	Ceramic 68 pF ±10% 50V
C171	ECEA35V47L	Electrolytic 47μF 35V
C201	ECCD1H020C	Ceramic 2 pF ±0.25pF 50V
C203	ECCD1H050D	Ceramic 5 pF ±0.5pF 50V
C205	ECEA16V220L	Electrolytic 220μF 16V
C207,214	ECEA50V1L	Electrolytic 1μF 50V
C208	ECQM05103MZ	Polyester 0.01μF ±20% 50V
C209	ECQM05473MZ	Polyester 0.047μF ±20% 50V
C212	ECCD1H120J	Cermic 12 pF ± 5% 50V
C215	ECQM05472MZ	Polyester 0.0047μF ±20% 50V
C901	ECCD1H010C	Ceramic 1 pF ±0.25pF 50V

Figure 12–1 Typical list of component parts as provided by the manufacturer will include schematic identification, the manufacturer's part numbers, and a description of the values and ratings of the individual parts.

ually on an "as required" basis. Two examples of this type of purchasing power will explain why this is important to the service technician. Each of these examples shows how the service technician can make a legitimate profit by taking advantage of quantity purchasing power.

Using the previously described diode as an example, a 1 A and 1000 V rated silicon diode may have a legitimate list price of $2.00. When purchased in single-lot quantities, the cost to the service technician will be about $1.20. The profit on the part is $0.80. This represents the typical trade discount of 40 percent of the list price when components are purchased individually.

This same diode can be purchased in larger quantities and in bulk from any one of many parts distribution companies. The price quoted in one catalog for these diodes in quantities of 100 is $0.09 each. The profit margin on the sale of one diode is over 95 percent! This is a legitimate profit. The decision to purchase 100 of these diodes was based on their use. This one item, used in many electronic devices often fails and has to be replaced. The electronic service technician is wise to evaluate the number of similar items that are purchased during a given time period. When the cost of 100 units is equal to or less than the retail sale price of five units, the quantity purchase may be an excellent decision.

The second example of this is related to the purchase of fuses used to protect test equipment. A local educational institution used a fairly high quantity of fuses in its testing equipment. The purpose of the fuse was to protect the electronic circuits inside a multimeter from misuse by beginning electronics students. The cost of this fuse in individual quantities was $.90. Purchasing the same fuse in bulk (quantities of 25 at a time) reduced the cost to $.36, or less than half the individual-unit cost. This is just one example of how quantity purchasing of often-used components can save the purchaser money. Profit was not the motive in the second example. What happened was that the educational institution was able to purchase other items with the money saved by using quantity purchasing power. This is a classic example of being able to stretch a budget without compromising the quality of the materials.

REMOVING THE DEFECTIVE COMPONENT

Before removing any component suspected of being defective, the service technician most consider how to remove it and what to do if the replacement part does not correct the problem. The process of removal also requires some care and attention to installation of the original part. Several things must be considered and recognized before any component is removed from the circuit or system.

MOUNTING THE PART IN THE SYSTEM

How the part is mounted in the system is one of the first questions to be answered. Removal of the suspected part is just one part of the replacement process. You must look at the mounting, the lead dress, and any other significant factors used in the original installation. For example, was the part mounted on a heat sink? If it was, you have to look for any nonconductive devices used as insulators between the case of the part and the mounting surface.

Often the case of a power transistor is also its collector connection. Power transistors used in many electronic devices have their collector connected to a positive voltage source. The mounting surface, or heat sink, is often connected to circuit common. An insulating device, such as a thin piece of plastic or mica, is used as an insulator between the case and the heat sink. If this is not replaced during the repair process, the result is a direct short circuit between the positive source and circuit common. The use of this insulating material is shown in Figure 12–2. This exploded view of the assembly of the power transistor is typical for this type of mounting. This insulator is required when reassembling the replacement power transistor.

Another concern is proper placement of the device's leads. Electrolytic capacitors are polarized. The capacitor's positive (+) lead is normally connected to the more positive terminal of the circuit. If the leads of the capacitor are accidently reversed, the result is buildup of a gas inside the case of the capacitor. Expanding gases exert pressure equally on all surfaces of the device containing them. Ulti-

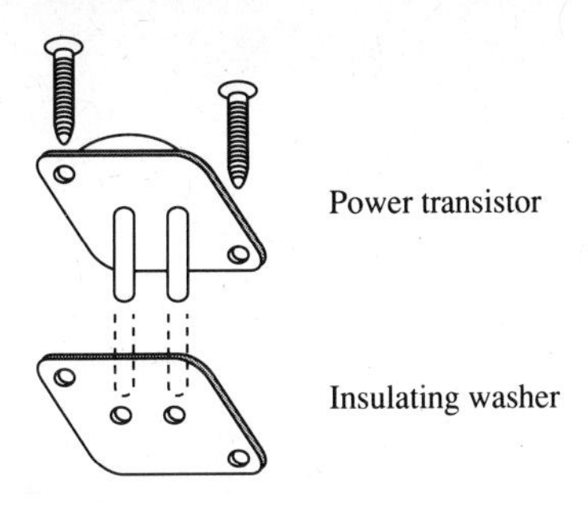

Figure 12–2 Many power transistors use an insulating washer between the case and the mounting surface. This washer must be replaced when installing a replacement transistor.

mately, the gases will escape. This process often results in a small explosion with damage to the surrounding area.

The service technician is required to observe the polarity of the electrolytic capacitor when installing a replacement unit. If this polarity is not observed during the removal process, and a new capacitor is accidently installed with a reverse polarity, the result could be as shown in Figure 12–3. Not only does this capacitor have to be replaced again, but there could be damage to other components *or to any people in the immediate area* when this explosion occurs.

Another major factor in the removal and replacement of parts is placement of the leads of any integrated circuit component. All integrated circuits have some sort of locating arrangement. This may be a small tab sticking out from one side of its metal case, as shown in Figure 12–4(a). Flat pack types of integrated circuits are marked in another way. There may be an indentation on one end of the case. A method of showing this is to place a small dot of paint on one corner of the body of the IC. A second method is the use of a recessed area on one end of the IC's body, as shown in Figure 12–4(b). The service technician needs to identify the original end of

Figure 12–3 The effects of installing an electrolytic capacitor with its polarity reversed will damage the capacitor and can produce an explosive result. (*Photo by J. Goldberg*)

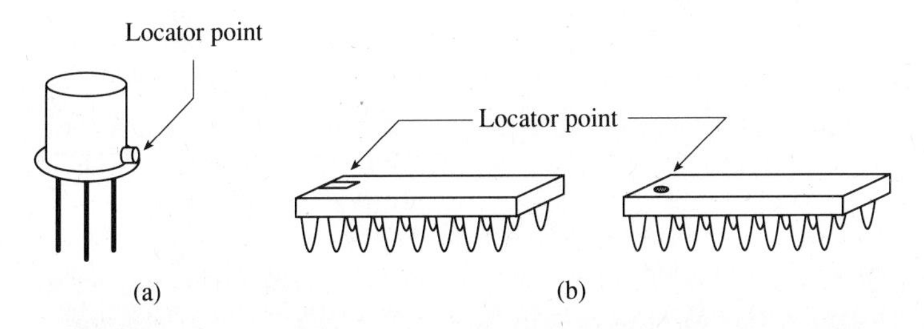

Figure 12–4 Transistors and integrated circuits have a locating mark on their cases. The mark is used to identify the pin connections and placement on the circuit board.

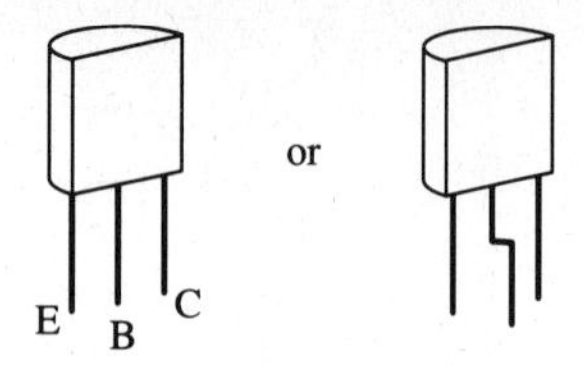

Figure 12–5 Plastic body transistors usually, but not always, have their leads configured as shown on the left. It may be necessary to bend them to fit existing circuit board mountings, as shown on the right.

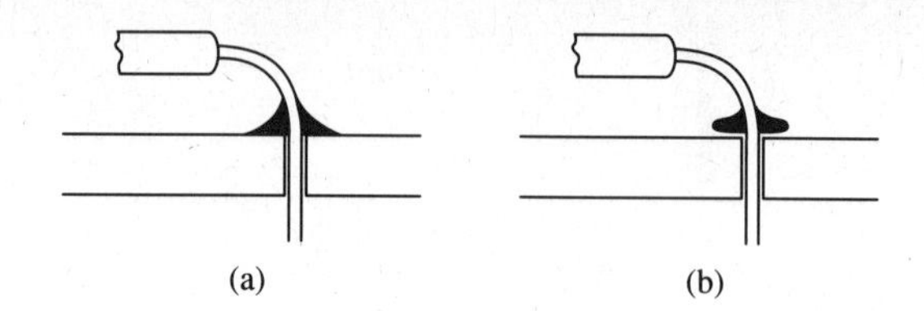

Figure 12–6 Example of a correct solder connection to a circuit board in seen in (a); a cold solder joint and poor connection is illustrated in (b).

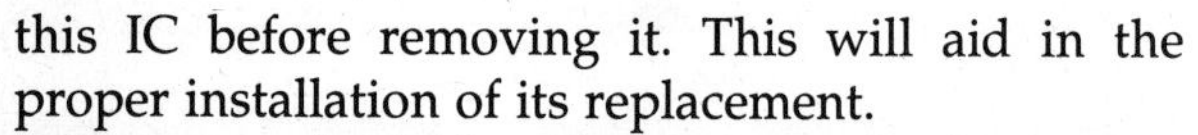
this IC before removing it. This will aid in the proper installation of its replacement.

Transistor lead identification can also be confusing. The more traditional plastic body for transistors is shown in Figure 12–5. The more common lead termination is also shown in this figure. This does not mean that all plastic body transistors use this lead termination. Also, there are times when one of the leads is bent away from immediately under the body of the transistor. These, too, can be installed incorrectly. The service technician needs to recognize the original lead placement and both the placement and the lead termination of the replacement transistor.

In summary, the service technician must use the talent of observation before removing any components from the circuit. It should go without saying that, if you don't remember the placement of the original leads—and you have not written it down—the possibility of your installing the replacement part incorrectly is likely. An incorrect installation could cause further damage to the system. If this did not occur, the system still would not operate correctly and the service technician would be spending additional time identifying the problem and making the repair—a repair now caused by incorrect installation of the replacement part.

SOLDERING

One of the more critical operations performed by electronic service technicians is soldering. This, like so many other operations, is an art. An improper solder connection will often create its own set of problems. Even if these do not occur immediately, they often can happen later and often in the beginning or middle of a very critical operation. Much has been written about the process of proper soldering. It is not the purpose of this book to present detailed material on either soldering or desoldering of electronic components.

There are just a few cautions you should be aware of during the process of desoldering. These include being careful not to damage adjacent components, not lifting the copper conductive path from the circuit board base material, avoiding any solder bridges between board paths, and attempting not to destroy the component being removed in case it is still functional. A final word of caution is to be certain that the solder connection is valid. The quality solder connection appears on the left side of Figure 12–6. The right-hand illustration shows what is commonly known as a cold solder joint. An insufficient amount of heat was used when making the connection, which resulted in failure of a complete solder melt and flow around the joint to be soldered. Cold solder joints have a habit of failing. The technique of properly soldering should be practiced by all involved in electronic servicing.

COMPONENT REPLACEMENT

Some of the precautions described earlier also apply to installation of the replacement component. These include lead dress, component polarity, safety cautions, static electricity problems, and those related to physical size and how to combine individual components to replicate the original component value.

LEAD DRESS

Lead dress describes the placement of the wires, or leads, of the replacement component. One of the major concerns when replacing components is the length of the leads of the replacement part. The operating frequencies of many electronic devices have increased dramatically in recent years. Many of these devices include resonant circuits as one part of their circuitry. Lead length and placement is of major concern as the operating frequency reaches into the gigaHertz range. A lead as short as a quarter-inch can act as a resonant circuit at these very high frequencies. Leads that are improperly routed in the circuit can create some feedback paths and result in internal self-oscillation at undesired frequencies. The service technician must be aware of this and attempt to keep the length and placement of the replacement component's leads as close to those of the part removed as possible.

COMPONENT POLARITY

Component polarity was briefly discussed earlier. The polarity of semiconductors is of prime importance. The polarity of electrolytic capacitors is equally as important. When in doubt about the capacitor's polarity, the method of determining the placement of the positive lead is to connect it nearest to the positive point in the circuit.

Diodes must be installed with the correct polarity or they will not operate properly in the circuit. The diode will still change from a low-resistance device to a high-resistance device with the polarity of the applied voltage. If the diode is installed in a reverse position, the output voltage will be opposite the polarity required for correct circuit operation.

Reversing the emitter and collector leads of a transistor will not permit it to operate properly. Often the collector voltage on an incorrectly installed transistor will be the same as the level of the source voltage. This will also affect adjacent circuits and make the entire system inoperable.

SAFETY

Safety is of utmost importance in any electrical or electronic system. There are several different concerns related to safety.

The primary safety concern for all service technicians deals with the possibility of electrical shock. Certain basic steps have been identified by several manufacturers of electronic equipment. These are all related to electrical safety and are focused on the correctness of the service repair process. These are:

1. Be sure that all components are positioned in a manner that will avoid the possibility of adjacent component short-circuit conditions. The possibility of inducing a short circuit is very high when any unit that is not in its original container, or case, is transported from the customer's office or home to the service center.
2. Be sure to replace all insulators and barriers between sections after working on any electronic device.
3. Check for frayed or broken insulation on all wiring, including the AC line cord.
4. Fuses and certain resistors and capacitors may carry special designations, such as "flameproof." These will be identified on the schematic diagram and on the parts list, and they must be replaced with components equal to the original value for both safety and liability purposes.
5. Once the repair is completed, the service technician should perform an AC leakage test on all exposed metal parts on the cabinet, input and output terminals, etc., to minimize the possibility of electrical shock.

One of the concerns mentioned above is the replacement of any component originally identified as a safety designated part. Resistors, for example, can burst into flame under certain abnormal operating conditions. Manufacturers of consumer products are required to identify those parts that are known to cause fires when they fail. Suppliers of these parts produce them in a way that will inhibit any flames from developing during the failure mode.

Flameproof components are identified on schematic diagrams. They are also identified in the parts lists of the devices. Each manufacturer has a personal method of identifying these parts. Some schematic diagrams and parts list show them as "fail-safe" parts, others identify them by enclosing them in a box on the diagram or by shading the box con-

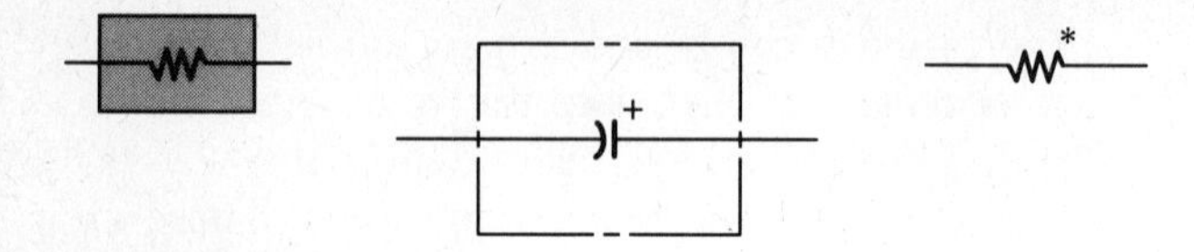

Figure 12–7 Safety-related components are identified on the schematic diagram with markings such as these.

taining the schematic symbol for the part. Several of these symbols are shown in Figure 12–7.

The service technician has an obligation to the customer to replace these with components having the same rating. Failure to do so could result in litigation for damages on the part of the customer. Not only is this costly in money, it also costs in time away from repairs for the service technician. Consider, too, the mental anguish involved should someone be injured or killed due to the failure to install the proper replacement component in the circuit.

The safety check described earlier is conducted by using an AC voltmeter, a 1500 Ω resistor, and a 0.15 μF capacitor. The circuit for this test is shown in Figure 12–8. One lead of the test set is connected to a good earth ground, such as a water pipe or electrical conduit. The capacitor and resistor are wired in parallel and connected across the leads of the AC voltmeter. The other voltmeter lead is then connected to the case of the unit and to the individual input and output connectors. A voltage reading of 0.3 V rms or less is an acceptable value for this reading. This value represents a current flow of 0.2 mA. Any reading higher than this represents a safety hazard. The electronic service technician must locate and correct this undesired and unsafe leakage before returning the unit to the customer.

COMPONENT PRECAUTION

STATIC ELECTRICITY

Static electricity is one of the more recent concerns affecting electronic service technicians. The concept of static electricity and its ability to destroy components is very real. It is very difficult for many of us to realize that the simple process of picking up a

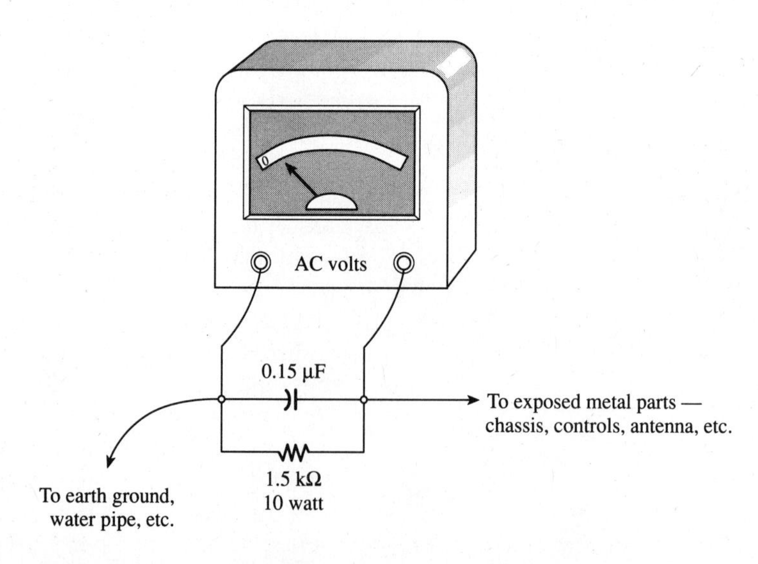

Figure 12–8 Circuit used to test for AC leakage between the power lines and the exposed metal surfaces of the device.

part with our fingers can destroy a semiconductor device. Much of the clothing we wear is made from manmade fibers, such as polyester threads. The interaction of the fibers as they rub against each other can create a static electrical charge. The process of moving your body on the seat of a chair can also create static electrical charges. Many of these charges are too small to be observed, but often their voltaic value can exceed 5000 V. The duration of this charge is very short and you may not even realize that it existed.

The moment of truth comes when the service technician realizes that the new component does not work properly, if at all. Just by picking it up and handling it, the service technician may destroy the new part. This description is not meant to frighten you, but rather to make you aware that this is indeed a serious problem and steps to minimize the effect of static electrical discharge need to be taken.

The solution to the problem of static electric discharge can be very simple. The answer is to neutralize the possibility of the static charge buildup. This is accomplished in several ways. One of these is to simply touch the case or chassis of the device being repaired. If this device is connected to the AC power line with a three-wire power cord, the case is at zero volt potential. The reason for the zero volt potential is that the grounding lead, or green wire, of the AC power line is attached directly to the chassis or case of the unit for safety purposes. Touching the chassis or case will neutralize any voltage charge on the technician's body. The service technician should maintain contact between his or her body and the case at all times.

A second method requires a "grounding strap" connected between the service technician's wrist and the common or ground connection on the work station. These straps are readily available commercially from many electronic parts sources. Shown in Figure 12–9, the device consists of a conductive

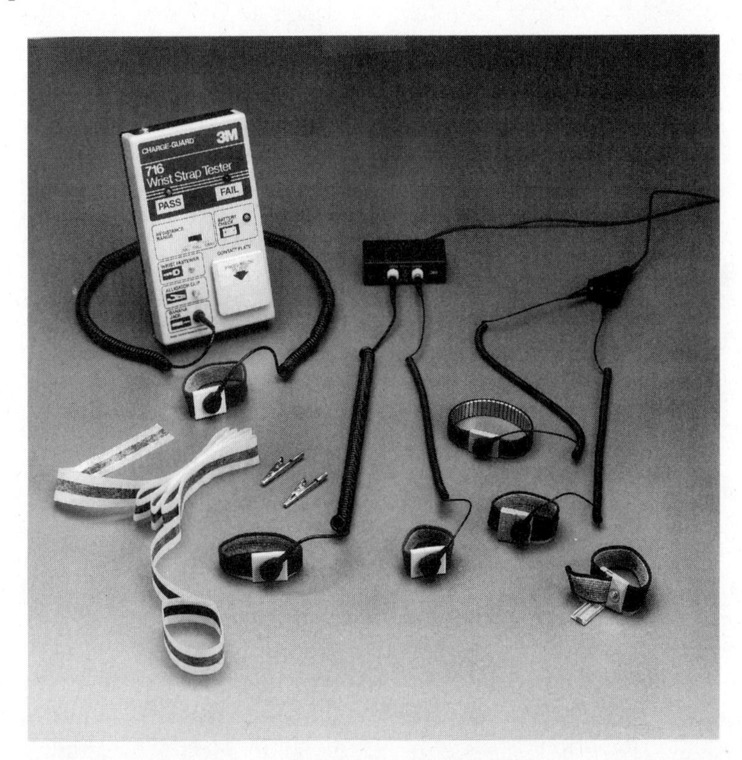

Figure 12–9 **One method of minimizing the effects of static electrical buildup is shown here. (*Courtesy 3M Electrical Specialties Division*)**

wrist strap and a chain connecting the strap to the circuit common connection. When working with any type of semiconductor device that is sensitive to static electrical discharge, the service technician must insure that the buildup of the static charge cannot damage the equipment.

LEAD PROBLEMS

Lead problems and the need to properly route leads of replacement electronic components has always been a concern for the electronic technician. Lead dress relates to both the physical problems that may arise as well as the electronic problems that can be created.

The physical problems of concern to the service technician as they relate to the placement of leads include the possibility of pinching or chafing wires and the lack of shielding or insulation. The service technician has to be concerned about the probability of a lead being caught in the cover of the device when it is closed. There is also some concern that leads may be pinched when the front or rear panel of the device is reattached to the rest of the unit as it is being "closed up" at the end of the repair.

Usually leads carry either voltages or signals. If one or more of these leads are pinched by the metal case, the probability of the insulation around the lead's conductor being broken and exposed to the metal case is high. If this conductor should touch the metal case, either the signal will be short circuited or the operating voltage will be short circuited to circuit common. In either of these situations, the device would be inoperative.

Another major factor related to lead placement is concern about adequate shielding of signal-carrying conductors. Electronic signals radiate from any conductive surface. Wires are conductive surfaces and signals will radiate from the wiring in a unit. If the original conductor was shielded, then the replacement wire must also be shielded. Connections to circuit common from the shielding must also be made if they were on the original installation.

A third factor to be considered when reinserting a component is to keep leads away from any moving parts in the unit. Moving parts, such as the moveable head on a printer, seem to be able to reach out and grab a wire that is out of place. Before completing the installation, the service technician should cycle the device to be sure that all wiring is properly routed.

If the device is installed in an environment where it is in motion, the technician also has to be aware of the effects of vibration on the components. All parts must be firmly fastened so that vibration is not a factor for future failure. Any wiring added to the unit must also be able to withstand motion. This means that only stranded wiring should be used, since solid wire has a greater tendency to break when it vibrates or moves.

SIZE OF REPLACEMENT PARTS

Physical size of replacement components always seems to be somewhat of a problem for the electronic service technician. At times, the replacement component may be physically larger than the size of the original component. If the device has adequate space, then the issue of size is not critical. When the space available for the replacement component is limited, the service technician may be required to do some creative installation techniques. One of these, and perhaps one of the more simple ones, is to replace a horizontally mounted component with one that is vertically mounted. Figure 12–10 illustrates this concept. The original resistor was mounted parallel to the circuit board. The available replacement component was significantly longer than the part that was removed so the only readily available solution was to mount the resistor in a vertical plane.

Some words of caution are required for this type of modification. First of all, the lead that is exposed, or uppermost, on the board should be the one with the lowest level of voltage for the component. This lead should be covered with an insulating tubing for safety reasons. Second, when making this modification, be sure that the cover, case, and any other circuit boards or modules will still fit after the change has been made. If both of these condi-

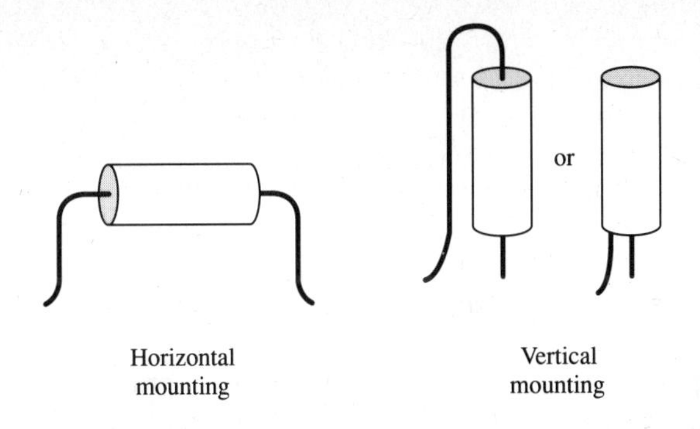

Figure 12–10 Horizontally mounted components can be mounted in a vertical position, as illustrated here, when size and space do not permit the installation of the replacement component in the same manner.

tions are acceptable, then the modification will be acceptable.

PARTS SUBSTITUTION

The concept of parts substitution was partially discussed earlier. This section will present information about the advisability of combining the values of two or more components. This may be necessary when you are away from the home parts depot and you need to return the equipment to operating condition immediately. The basic theories related to individual components and to series and parallel combinations of these components still apply. There is little reason why they cannot be used when servicing in the field.

COMBINING RESISTOR VALUES

Resistor values can be combined in series, parallel, or combinations of both series and parallel values. While this practice is not encouraged, it can be done under emergency conditions. The first things to consider when required to do this are what is available in resistor values and what final resistance value is desired. A value of 8200 Ω will be used for the example of the desired resistance required to return the unit to normal operation. There are several methods available to the service technician attempting to construct this value from an assortment of other resistance values.

Using the series circuit as a starting point, several different values can be connected to achieve the desired 8200 Ω value. One method of accomplishing this is to connect three 2700 Ω resistances in series. This will provide a value of 8100 Ω, well within the limits for a 5 percent or 10 percent tolerance rating. A second method of creating the desired 8200 Ω value would be to connect a 4700 Ω in series with a 3600 Ω resistor. This, too, would not be the exact value of the original resistor, but it certainly would be within acceptable tolerance levels. Actually, any combination of two or more resistor values whose total resistance falls within the tolerance parameters is acceptable as a replacement for the original value.

In an alternative process, two resistors can be connected in parallel in order to provide this same value of 8200 Ω. One method of accomplishing this is to place a 15 kΩ resistance in parallel with an 18 kΩ resistor. Using the general rule that the total resistance of resistor combinations in parallel is always less than the single lowest value, the equivalent resistance for these two values will be less than 15 kΩ. Since the two resistors' values are close to each other, the equivalent resistance for this combination will be close to, but higher than, half of the value of the 15 kΩ resistance. The equivalent resistance for these two resistance values in a parallel combination is 8181 Ω. This value falls within the acceptable tolerance parameters for the original resistance.

The process of combining two or more resistances to obtain a desired value is easily accomplished by applying the rules for series and parallel resistances. One additional factor that must be resolved when doing this is the wattage rating of the original resistor and the rating of those used to replace it. The power formula for resistances either in parallel or in series combinations states that the individual power values add to equal the total power in the circuit. This indicates that the power ratings of each individual resistor can be less than that required for the one original resistance value. The major factor here is that the total power dissipated by all of the resistors used in either the parallel or the series combination will be equal to, or greater than, that of the original resistor.

COMBINING CAPACITORS

Capacitors may also be combined in both series and parallel combinations when the desired value of capacitance is not available. The fundamental rule to remember is that the total value of capacitors in series combination will be less than the value of either of the individual capacitors. The basic rule for capacitors connected in series is easily remembered when you realize that only the outer plates of the capacitor are charged by the external power source. Placing two capacitors in a series circuit will reduce the total capacitance since the outer plates are now farther apart. When you think about the relationship of the outer plates of the capacitor and the effect of moving them apart as it relates to the value of capacitance for the capacitor, you will be able to understand this concept.

The second method of connecting capacitors is to place them in a parallel combination. This effectively increases the total plate area of the capacitor. An increase in total plate area will also increase the amount of charge that develops on the plates of the capacitor. This, then, also increases the value of the amount of capacitance.

All of the possibilities of combining components in series or in parallel must be related to the basic rules for series and parallel circuits. One of the rules for the series circuit is that the current is constant throughout the entire circuit. This rule must be considered when any two components are series connected. Typically, the current capacity of the lowest-rated unit determines the amount of current that will flow in any series circuit.

When connecting semiconductor diodes in series, their current rating is also limited by the value of the lowest-rated diode in the string. Placing semiconductor diodes in series will increase the total voltage rating for the string. If, for example, each diode has a reverse voltage rating of 1000 V, then four diodes could be connected to a 4000 V source. In practice, some amount of underrating should be considered. In a practical substitution, the suggestion would be to use five diodes and provide a 20 percent safety factor. Any component operated at its maximum rating will have a tendency to fail more rapidly than one operated under its maximum ratings.

Semiconductor diodes may also be connected in parallel when additional current capacity is required. Here, the rule for the parallel circuit applies. In the parallel circuit, the total voltage must be the same, and the current can and does vary through each branch of the circuit. Semiconductor diodes connected in parallel must have the same voltage rating. The individual current ratings can differ without any ill effects on the operation of the circuit.

Finally, the process of selecting components to use as replacements in original equipment is not always an easy job. The service technician has to consider physical mounting, component specifications, soldering techniques, and lead dress as parts of the replacement process. This is not to say that this process is difficult; it really is not very hard to do. The important thing is to remember to check everything before, during, and after the installation. This procedure will often reduce the total time of the repair since it will eliminate costly repeat work due to a failure to do the job properly the first time.

REVIEW

The location and procurement of replacement components is one of the major functions of the repair process. There are several ways to procure these components. These include obtaining them from the manufacturer of the device, from independent replacement parts distributors, and from remanufacturers of original equipment parts. Which source to use depends upon the priority of obtaining the parts, local access to parts, and whether the parts are available. The successful electronic service technician will use all of these sources as needed.

Selection of the specific component will depend upon the factors of quality, tolerance, availability, and price. Many original equipment sources are listed in publications available from the National Electronic Sales and Service Dealers Association, in the advertisements of many major magazines dedicated to one or more specific aspects of electronic service, and also in *Service News* magazine's *Annual Buyers Guide*.

Universal replacement components are just as the name implies: parts that will normally replace several other parts. The service technician must be aware that not all universal replacements will function properly in every circuit. Care must be exercised in selecting these parts.

The tolerance of the value of both the original and the replacement part has to be considered. As long as the replacement part measures within the tolerance limits of the original part, the substitution should be acceptable. It may be necessary to combine two or more individual parts in a series, parallel, or combination series and parallel construction under emergency conditions. This is acceptable if the replacement combination will physically fit in place of the original and if the value of the combined components is within the original part's tolerance range.

The voltage rating and the component's electrical rating have to be considered when an original part has both values indicated on the parts list for the unit. Series, parallel, and series-parallel combinations may be used for voltage rating values as well as for component value ratings.

The use of purchasing power to reduce the cost of individual units is an acceptable practice for those in the service industry. When you are able to save money on the cost of one item by purchasing it in quantities, this savings is often reflected in a legitimately higher profit for the company. The practice must be evaluated against the number of units sold or used in a given time period in order to be practical.

Mounting replacement components can be an interesting process. This is particularly true when the replacement part is larger than the one it replaces. Creative planning often will permit the part to be safely inserted in a different place or vertically instead of horizontally. This practice is acceptable if the system will be restored to its original working condition and if the original component is not readily available.

Before removing any defective component the service technician should identify how it is placed in the circuit; any unusual mounting problems; and how the wires, or leads, of the unit are placed. The replacement component should be installed in the same manner.

Another important factor in the replacement parts process is the use of successful soldering techniques. A poor solder connection may prematurely fail or may not work at all. Solder bridges between two adjacent circuit board runs can introduce additional problems into the unit being repaired. All of these will take time away from the service technician's ability to repair equipment quickly. Recalls and reevaluation of the reason the system is not working are also costly in time and money to the service organization and technician.

The importance of safety cannot be understated. The service technician must be aware of any components with a safety rating; these have to be replaced with other safety-rated components. In addition, leakage from the unit to the power source and AC power lines should be tested for safety reasons. This test will minimize the possibility of either the technician or the customer being exposed to electrical shock conditions and the possibility of electrocution.

Another area of concern is static electricity. A small static electrical charge can destroy semiconductor materials. The use of static-eliminating equipment is important for those servicing electronic equipment.

The electronic service technician's training and thinking have to be broad enough to consider all of the issues raised in this chapter. These must be considered in addition to the application of the rules for circuit analysis and the basic rules presented by Ohm, Kirchhoff, and Watt. When all of these are remembered and applied, the service technician will be successful.

REVIEW QUESTIONS

1. Why may it be necessary to obtain universal replacement parts?
2. Name two or more sources for listings of places to obtain replacement parts for repair purposes.
3. What are the limitations of using universal replacement parts?
4. Under what conditions can a 2 percent tolerance rated part replace one having a tolerance factor of 5 percent?
5. What effect does the voltage rating have on the ability to use a universal replacement part?
6. Explain the concept of bulk purchasing and why it should be considered.
7. What markings are used on an integrated circuit and why should they be observed before the IC is removed from the circuit?
8. What can occur when the leads of an electrolytic capacitor are accidently reversed during installation?
9. What is a solder bridge and what effect does it have on a circuit or system?
10. What is meant by the term "cold solder joint"?
11. What is the purpose of a leakage test?
12. How is a leakage test performed?
13. What are flameproof components and why are they used in equipment?
14. What effect can static electrical discharge have on semiconductors?
15. State two ways in which static electrical discharge is controlled.
16. A 15 kΩ resistor with a 2 watt power rating has to be replaced with a combination of 1 watt resistors. Using common 5 percent tolerance rated resistors, describe the combinations possible for this replacement.
17. A 100 μF capacitor having a voltage rating of 100 volts and a tolerance rating of +50 percent to -25 percent is to be replaced. What is the acceptable range of values for the replacement?
18. Using the values of the capacitor in question 17, if the available capacitors have capacitance ratings of 200 μF, how can they be connected to be used as a replacement?
19. A semiconductor diode with a rating of 1500 volts and 5 A is to be replaced with diodes rated at 1000 volts and 1 A. What combination will accomplish this?
20. What precautions should you take when you replace a horizontally mounted resistor with a vertically mounted resistor?

CHAPTER 13

After the Repair

INTRODUCTION

The testing and evaluation performed after the physical repair is completed is as important as the actual repair. The electronic service technician has to be able to evaluate the success of the repair. This evaluation includes a burn-in time and a complete retesting to ensure the correctness of the actual repair. It is very embarrassing to return a repaired unit to a customer only to discover that the problem was not fixed. Several steps need to be performed prior to returning the unit to the customer. These steps will aid in determining that the repair was done correctly and that the unit functions properly. This is not to say that something else could not fail during delivery or reinstallation of the equipment; all this does is minimize the probability of failure at this level.

OBJECTIVES

Upon completion of this chapter, the student/reader should:

understand the need for testing after completion of the repair;

recognize the term burn-in and its application during the repair process;

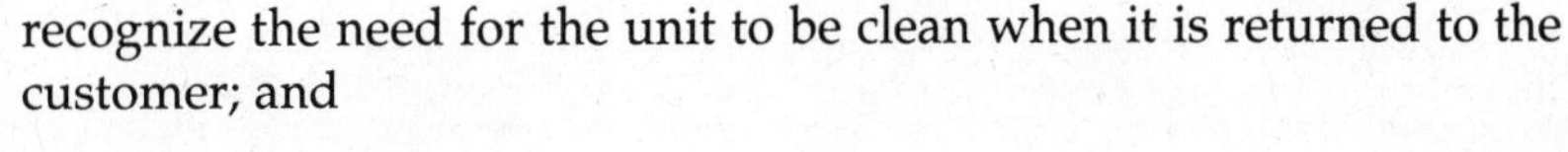
recognize the need for the unit to be clean when it is returned to the customer; and

recognize the need to analyze the repair procedure as one component of the learning process.

RETESTING THE UNIT

The funnel approach to diagnosis and repair of electronic equipment is not completed when a replacement component is installed in the unit. The chart shown in Figure 13–1 is a repeat of the one illustrated in Chapter 1. At the bottom of the funnel a box containing the words "Final checkout" is shown. This box is as important in the repair process as those shown above it for the diagnosis process. No repair is complete until it is tested under field operating conditions. This is an essential component of the overall repair process.

One specific example of this is when an automobile's cruise control system failed. The correct diagnosis called for realigning the position of a shutoff switch connected to the brake pedal. The service technician adjusted the switch, but did not road test the repair. The customer drove the car out of the service facility and, when he tried to use the cruise control system, found that it still was not working properly. He had to return to the service facility and lose additional time until the repair was performed correctly. In addition, the customer decided not to return to that repair facility in the future since it failed to complete the repair properly the first time.

This illustrates the need to test the repair after it is completed. This repair facility lost a customer because it did not properly repair the problem initially. The customer's return cost the service technician time away from other income-producing work. Needless to say, this is not desirable. Too many recalls can result in replacing the technician with one who is more competent.

This same type of retesting is required for electronic equipment repair. Both automobiles and elec-

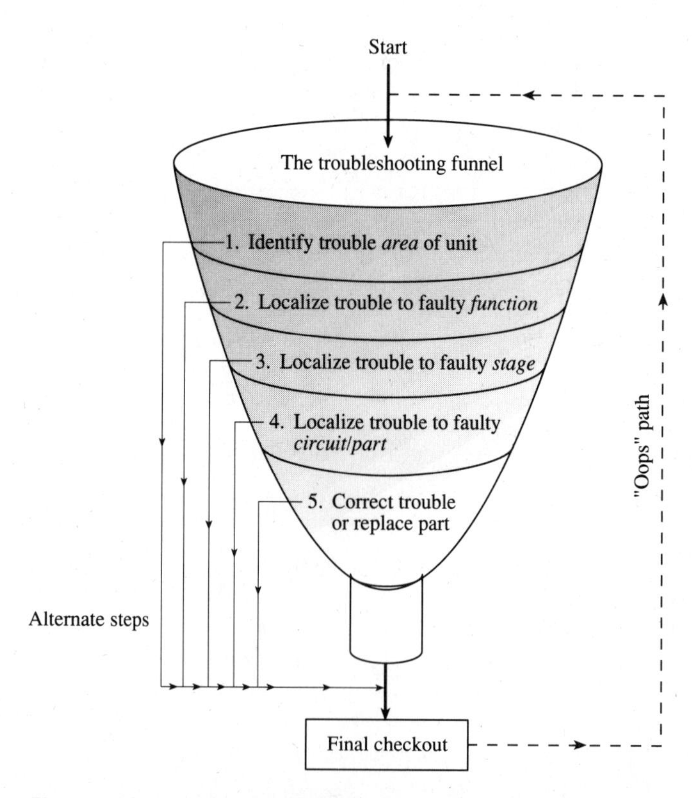

Figure 13–1 **The funnel process, as described in Chapter 1, is an effective method of testing and evaluating the repair.**

tronic devices are complex and require testing after they are repaired. This additional time will create customer goodwill and minimize recalls of the product.

BURN-IN TIME

This term identifies the time period required to let the new parts settle in and perform as they are designed to work with each other. Burn-in time is essential for any units that are assembled from discrete parts. One example of this is a computer. Several different parts make up the assembled computer; these often are purchased from different sources and assembled at one facility. After assembly the individual parts need to be tested in the unit. This ensures their correct operation in the assembled unit.

The specific amount of burn-in time depends upon the type of equipment and the number of discrete components installed in it. Typically an assembled computer is tested for a minimum of 48 hours. Individual electronic devices will each have their own time period for burn-in.

CUSTOMER RELATIONS

This concept cannot be overemphasized. Too many newspaper and magazine articles have been written about how service organizations take advantage of unsuspecting customers. Customers sometimes feel that they may be overcharged or unnecessary parts may be installed in order to inflate the service bill. This is a very real situation and each service organization must be aware of it and work to minimize this feeling of distrust. This actually occurs in only a small percentage of the total times equipment is serviced. However, the fact remains that these few times are the ones people remember.

I know one very successful service organization that has a procedure for minimizing recalls and maximizing customer relations. This procedure, which has several steps, seems to be very successful. The steps are described below.

MEETING THE CUSTOMER

The customer is met initially by a person who looks neat and clean. If the repair requires the customer to bring the unit to the facility, the physical area of the repair facility also is clean and neat. The person making the initial contact with the customer also presents a clean and neat appearance and is pleasant to talk with.

The "customer greeter" looks over the equipment to see if anything is damaged or there are any parts obviously missing, etc. Any missing or damaged parts are listed on the service order. This person is also trained to be a good listener. Often the customer will describe exactly what the problem is while talking to the technician. The careful listener will note this description on the work order. This will make the service technician's diagnosis and repair much simpler.

INITIAL SERVICE

Initial service is performed by an individual whose sole purpose is to disassemble the unit and clean all of the loose parts. If the device is a consumer product, the case and the control knobs are carefully cleaned. The purpose of this is to return the unit to the customer in better condition than when it was brought in.

If the unit has a removable back, it is taken off to save time for the service technician who actually makes the final diagnosis and repair. Removing the back of the unit can be done by someone who is learning the repair business and not by the highly experienced and relatively highly paid technician. There may be times when only the experienced technician has the knowledge and expertise to do this; under these conditions the person who performs the initial steps is bypassed.

The final step in this initial process is to place all loose components, such as knobs, screws, feet, etc., in a plastic bag or container clearly identified as belonging with the specific unit. By doing this, the service facility ensures that all parts are available when the repair is completed and none is lost or misplaced.

ACTUAL SERVICE

Actual servicing of the nonoperational unit is the next step. This is performed in a dedicated service area by the experienced technician. Each service area, or booth, has the necessary testing and alignment equipment required for the type of repair being performed. At times, there may be a backup at one of the booths due to the specific type of specialized equipment needing servicing.

The knowledge and expertise described in this book is applied by the service technician at this time. Each of the steps described in earlier chapters is followed. In this manner the technician is able to efficiently diagnose and repair the device.

TESTING THE REPAIR

Testing the repair is the next step in this process. The service technician makes all required tests on the repaired equipment. These may include an alignment of both mechanical systems and electronic circuits. Once the service technician is satisfied that the repair is successful, the unit is sent to another area for further testing.

This area is away from the service technician's booth. It consists of shelving and the appropriate connections for auxiliary test equipment. One example of this would be servicing television receivers. An antenna distribution system is connected to the receiver. The receiver is turned on and left on for a predetermined time period, ranging from four to 24 hours in length. If the receiver operates properly during the entire time period, it is ready to be reassembled and returned to the customer.

Note that the phrase "reassembled" is used here. The receiver's back or case may not be reattached until the end of this test. The decision about this is left to the individual service company. One company may feel that reassembly of the unit is critical to its final checkout. Another company may prefer to do this after the unit has operated successfully for the entire post-testing period. If the second method is used, the unit is rechecked after is has been completely reassembled.

COMPLETING THE REPAIR

Successful completion of this post-testing period is when all loose parts are reinstalled in the device. At this time a person with limited electronic service experience is assigned the job of making certain that all of the parts are successfully reinstalled in the unit. A final cleaning is given to the case and control panel.

The final step in this process is packaging the repair and notifying the customer that the work has been completed. If there is no specific time period for the customer to pick up the repaired unit, it may become dusty in the storage area. This company has a policy of wrapping all of the repaired units in plastic bags and sealing them.

A customer receiving this wrapped, cleaned, and repaired unit often feels that the process was done professionally. Concern about the quality of the repair is minimized. This type of approach goes a long way toward maintaining customer confidence.

FIELD SERVICE CALLS

Field service calls should be performed in a similar manner. Because only one person is generally sent on a field service call, the service company should attempt to instill a feeling of confidence in the customer. The customer's initial impression is extremely important. When the service technician is clean, neat, and knowledgeable about the product, the customer feels confident in both the service technician and the company providing the service. When the opposite is true, the customer may be wondering why he or she is using this particular company.

The need to listen to the customer was stressed in an earlier chapter. The importance of this part of the service process cannot be overemphasized. Again, the customer will often identify the specific problem. When the service technician listens to what is said and applies this information as one part of the diagnosis of the service problem, the result often is a rapid and correct repair.

Here, too, the final step of the repair process is

checkout of the equipment. There may not be a long period of time available for the burn-in or testing period. The service technician has to adjust for available time and conditions. This becomes a choice at the time of the repair.

POSTANALYSIS

Once the repair is successfully completed, the service technician should take a few moments to analyze what was done and how it was accomplished. The planning sheet presented earlier has one column identified as "Feedback." The purpose of this column is the same as what is being described here. You need to analyze what was accomplished during the repair process. Often, when you stop to do this, you can identify a better procedure for future repairs. This is part of the learning process.

The world's greatest electronic service technician (remember this person?) will constantly review what was done and how to do it better the next time. Learning really never stops. When you recognize that this process is a continual one and you are willing to adjust your personal behavior, then growth and success will continue.

Looking at the repair process, the analysis should return to the beginning diagnosis of the problem. The steps involved are directly related to the funnel shown earlier in this chapter. Always start with the biggest section of the unit. Consider the type of signal or current flow process that is involved. Establish a set of brackets around the entire area of the problem. Use the service literature available from the device's manufacturer as a guide in this process.

Once the brackets are established, the next step is to determine what type of flow is involved. Then, and only then, can the proper diagnosis process be recognized. Use the process that best describes the type of flow involved, whether it is signal or current flow.

Make the first test at the recommended point in the circuit. Use the results of this test to determine where the next test is to be done. Locate the test point and establish the proper test signal. Select the correct type of testing equipment for the test. Set the controls on the testing equipment for the expected measured values. Know what to find if the test is successful at that point. This knowledge should be obtained from the service literature prior to actually performing the test.

Determine the order of testing after the first one is completed. In other words, if the test results support the service literature, where will the next point of testing be? Also determine where the next testing point will be if the tests indicate failure. This is a repetitive process and is done until test results indicate either a component or a module operating in an incorrect manner.

One of the final steps of the test process is either to replace or repair. This decision is determined by what is available and how long the equipment can be out of service. There is no single correct answer to this question. The final choice is left to the service technician at the time of the repair.

Another component in the service process is writing up the service invoice. This document becomes a part of the permanent record for the equipment. It is extremely important that the service technician write a concise and clear description of the problem. It goes without saying that handwriting is an important component in the production of this report.

The final step, other than presenting the bill, is testing the unit for successful operation after the repair has been completed. Often several operational cycles are required to ensure correctness of the repair. This is another step in the process that cannot be rushed or overlooked. The service technician does not want to return to the same problem as soon as the repair is completed. Taking the time to correctly diagnose and repair the system or unit is of utmost importance in the repair cycle.

REVIEW

The material in this chapter describes the final components of the repair process. These include the process of testing the correctness of the repair after it is completed. Testing is accomplished by both initial operation of the unit and burning in the repaired device. Failures in electronic devices often occur within the first few hours of its operation. This is true for both new equipment and equipment that has been repaired. Keeping the unit on for a time period of one or two days will often provide information about its continued operational success.

Recalls to repair something that was previously fixed are both expensive and embarrassing to the service organization. Efforts to minimize this are a very important part of the overall service process. In addition, companies whose reputations include frequent recalls are those which often fail.

Another very important thing to consider for those in the field of servicing is customer relations. Both personal contact and telephone relations are critical to the success of the service organization and individual. The first impression is the one that lasts; this is true for a telephone call and for the individual making the actual service call. The importance of this cannot be overemphasized.

The ability to return to the customer a unit that is clean and has all of its original parts goes a long way toward maintaining customer confidence. The process described in this chapter is just one example of how the successful service organization operates.

Finally, analysis of the repair should aid the service technician's learning process. When you take the time to analyze and review the repair process, learning is successful.

REVIEW QUESTIONS

1. Why is it important for the service technician to test the repair once it is completed?
2. Explain the term "burn-in time."
3. What is the typical length of time for the burn-in process?
4. Describe the steps discussed in the chapter for meeting, greeting, and writing up the service order.
5. What is the difference between a field service call and one conducted in the service shop?
6. Why is the write-up after completion of service on the product important?
7. Explain the term "postanalysis."
8. Why is it important for the field service technician to present a clean personal appearance?
9. Do you agree with the statement that a customer, if given enough time, will exactly explain the problem to the service technician?
10. As a review of this book, do you feel that knowledge of the laws and rules of Ohm, Kirchhoff, and Watt will be applied by the successful service technician?
11. Why should the service technician refer back to the original planning sheet after completing the service repair?
12. Explain the importance of returning the repaired unit looking as good as, or better than, when it was turned in for repair.
13. Explain the term "feedback" as used in this book.
14. How is the funnel approach to electronic servicing used?
15. What is the first step in the diagnosis process described in the funnel procedure?
16. When is the decision made to either repair or replace sections, blocks, or modules in the defective unit?
17. Explain the importance of using service literature in the repair process.

18. Explain the importance of identifying one of the signal paths or current paths during the diagnosis process.
19. Explain the importance of using a piece of testing equipment with a high input resistance or impedance rating.
20. Explain the importance of observing how the original component was installed prior to its removal and possible replacement.

Appendix

1. Electronic Industries Color Coding System
2. Electronic Industries Resistor Color Coding System
3. Electronic Industries Transformer Lead Coding System
4. Electronic Industries Standard Value Set
5. Sources for Parts Supplier Information
6. Recommended Schematic Symbols

ELECTRONIC INDUSTRIES COLOR CODING SYSTEM

The set of colors and their representative numbers are commonly used by the electronics industry. These number values are found on a variety of components and wires.

Color	Representative number
Black	zero
Brown	one
Red	two
Orange	three
Yellow	four
Green	five
Blue	six
Violet	seven
Gray	eight
White	nine

ELECTRONIC INDUSTRIES RESISTOR COLOR CODING SYSTEM

Color	Significant Figure	Decimal Multiplier		Tolerance
Black	0	1	1×10^1	
Brown	1	10	1×10^2	
Red	2	100	1×10^3	
Orange	3	1000	1×10^4	
Yellow	4	10,000	1×10^5	
Green	5	100,000	1×10^6	
Blue	6	1,000,000	1×10^7	
Violet	7	10,000,000	1×10^8	
Gray	8	100,000,000	1×10^9	
White	9			
Gold		0.1	1×10^{-1}	±5%
Silver		0.01	1×10^{-2}	±10%
No Color				±20%

ELECTRONIC INDUSTRIES TRANSFORMER LEAD CODING SYSTEM

1. **Primary:** Black. If tap is used, black is common to both input windings.
2. **Tap:** Black with yellow tracer, or striping
3. **High-voltage secondary:** Red with red and yellow tracer for center tap, if used
4. **Rectifier filament** (if used): Yellow with yellow and blue tracer
5. **Filament winding #1:** Green with green and yellow tracer
6. **Filament winding #2:** Brown with brown and yellow tracer
7. **Filament winding #3:** Slate with slate and yellow tracer

This system will aid in identifying leads on power transformers. Normal color coding uses the same color for both the start and the end of the winding. When a tap is used, the same color is used for the wire with the addition of a tracer of an alternate color. EIA standards suggest the use of the color yellow for the tracer, or striping, on the tap.

Power transformers for solid-state devices do not follow this color coding system. The best way to identify solid-state power transformer leads is to locate two leads with the same color and then check them for continuity with an ohmmeter.

ELECTRONIC INDUSTRIES STANDARD VALUE SET

Almost all components using the Electronic Industries Standard Value Set will start with this set of numbers:

1.0*	1.5*	2.2*	3.3*	4.7*	6.8*
1.1	1.6	2.4	3.6	5.1	7.5
1.2*	1.8*	2.7*	3.9*	5.6*	8.2*
1.3	2.0	3.0	4.3	6.2	9.1

Each of the above values is available with components rated with a ±5 percent tolerance factor. Those numbers with an asterisk (*) are only found on components rated at 10 percent tolerance.

SOURCES FOR PARTS SUPPLIER INFORMATION

Professional Electronics Yearbook & Directory
NESDA/ISCET
2708 West Berry Street
Fort Worth, TX 76109

Computer Technology Review
924 Westwood Boulevard, Suite 650
Los Angeles, CA 90024

Buyers Guide
Service News Magazine
P.O. Box 995
Yarmouth, ME 04096

Electronic Industries Telephone Directory
Harris Publishing Company
2057-2 Aurora Road
Twinsburg, OH 44087

RECOMMENDED SCHEMATIC SYMBOLS

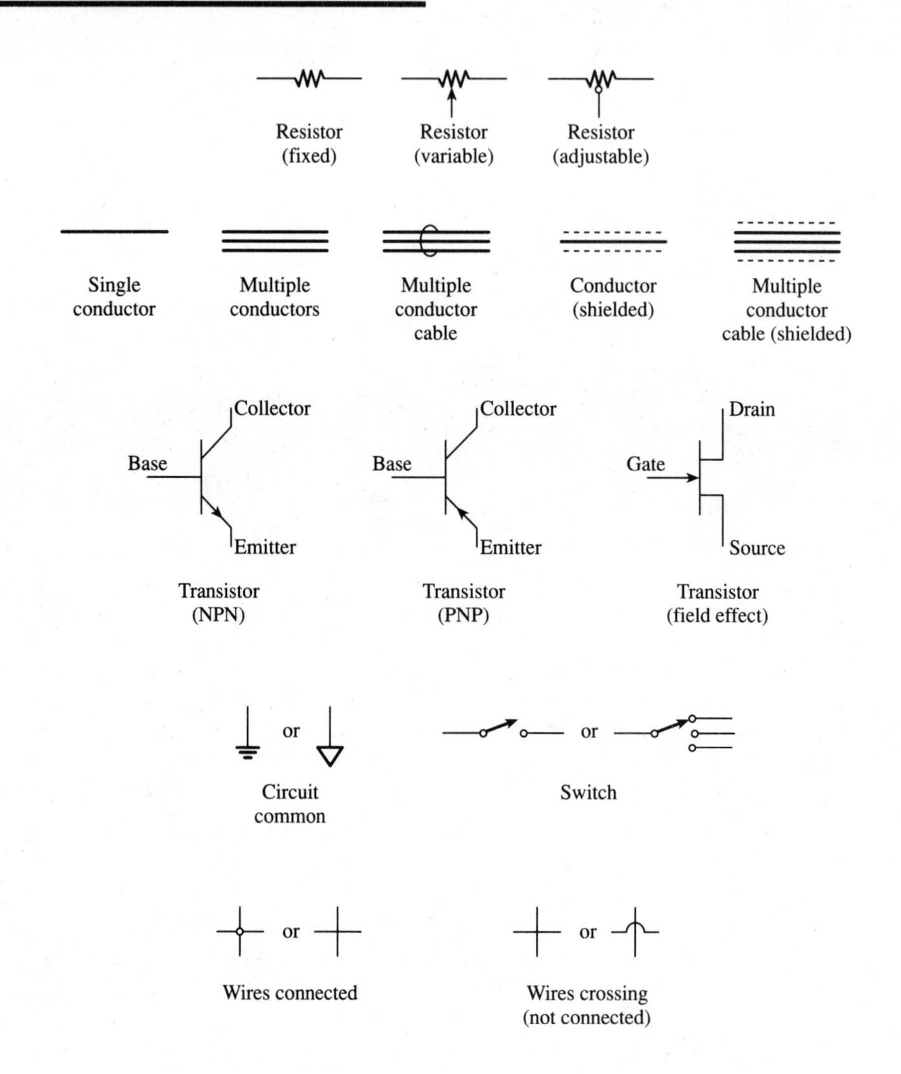

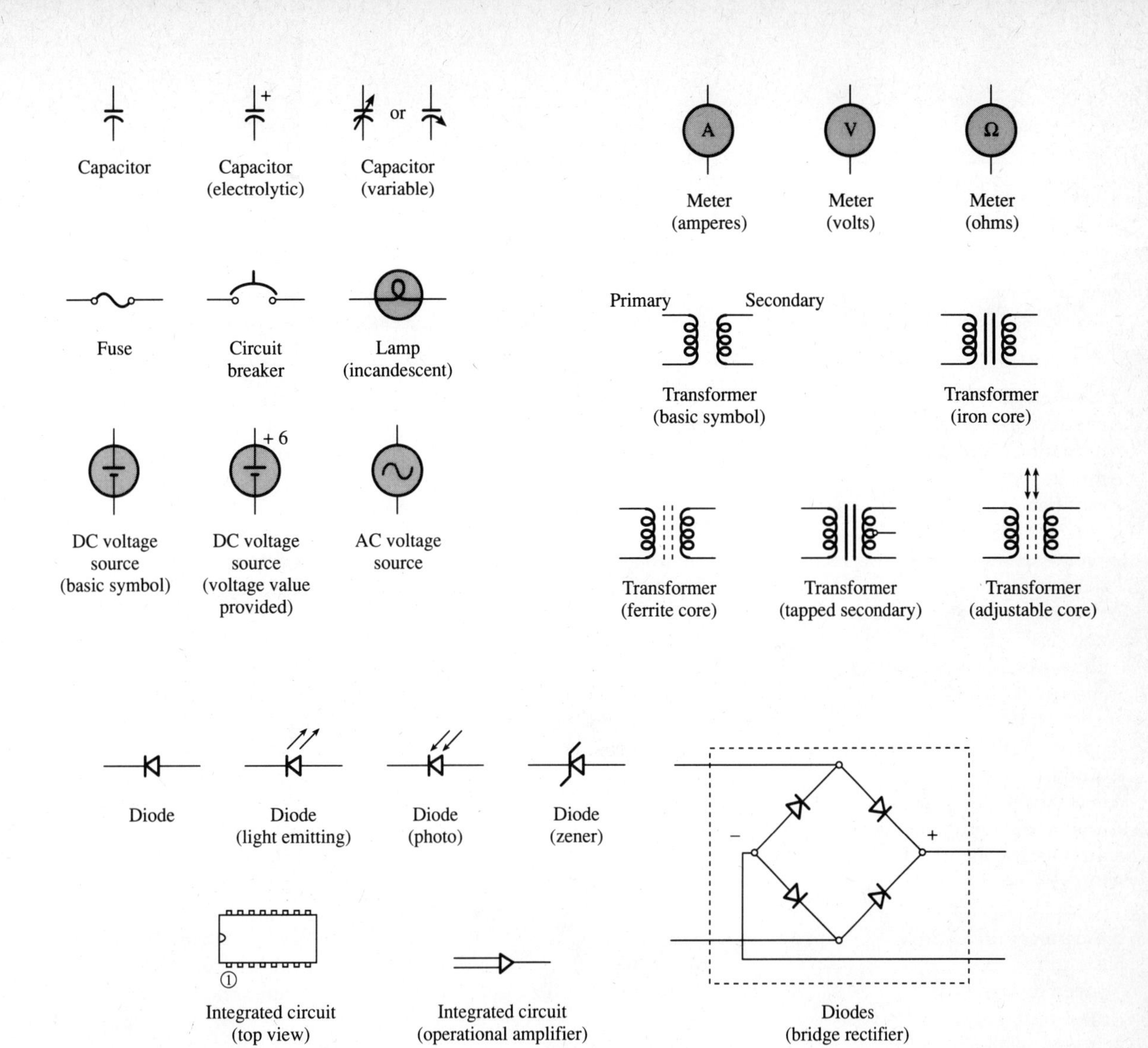
Capacitor
Capacitor (electrolytic)
+
or
Capacitor (variable)
A
V
Ω
Meter (amperes)
Meter (volts)
Meter (ohms)
Fuse
Circuit breaker
Lamp (incandescent)
Primary
Secondary
Transformer (basic symbol)
Transformer (iron core)
+ 6
DC voltage source (basic symbol)
DC voltage source (voltage value provided)
AC voltage source
Transformer (ferrite core)
Transformer (tapped secondary)
Transformer (adjustable core)
Diode
Diode (light emitting)
Diode (photo)
Diode (zener)
−
+
①
Integrated circuit (top view)
Integrated circuit (operational amplifier)
Diodes (bridge rectifier)

Index